The Golden Number

Pythagorean Rites and Rhythms in the Development of Western Civilization

Matila C. Ghyka

Translated by Jon E. Graham

Inner Traditions
Rochester, Vermont • Toronto, Canada

Inner Traditions
One Park Street
Rochester, Vermont 05767
www.InnerTraditions.com

Originally published in French under the title *Le Nombre d'Or: Rites et rythmes pythagoriciens dans le développement de la civilisation occidentale* by Éditions Gallimard, Paris
First U.S. edition published in 2016 by Inner Traditions

Library of Congress Cataloging-in-Publication Data
Names: Ghyka, Matila C. (Matila Costiescu), 1881–1965, author.
Title: The golden number : Pythagorean rites and rhythms in the development of western civilization / Matila C. Ghyka ; translated by Jon E. Graham.
Other titles: Nombre d'or. English
Description: First U.S. edition. | Rochester, Vermont : Inner Traditions, 2016. | Includes index.
Identifiers: LCCN 2016022411 | ISBN 9781594771002 (hardcover)
Subjects: LCSH: Symbolism of numbers. | Pythagoras and Pythagorean school. | Golden section. | Rhythm.
Classification: LCC BF1623.P9 G413 2016 | DDC 133.3/35—dc23
LC record available at https://lccn.loc.gov/2016022411

Printed and bound in India by Replika Press Pvt. Ltd.

10 9 8 7 6 5 4 3 2 1

Text design and layout by Debbie Glogover
This book was typeset in Garamond Premier Pro with Avenir Lt Std, Gill Sans MT Pro, and Cambria used as display typefaces.

Contents

Illustrations

Publisher's Preface

First published in 1931, *The Golden Number* is Matila Ghyka's seminal work on the golden ratio and the hidden history of the Pythagorean tradition in the West. Whether known as the golden number, golden ratio, golden section, golden mean, golden cut, phi (Φ), division into mean and extreme ratio, or as ratios derived from the numbers of the Fibonacci series, the divine proportion—so called by Luca Pacioli in his book of that name—is ubiquitous in nature, art, architecture, music, and literature.

As Ghyka demonstrates, the golden number is the calling card of life, the irrational, governing number of living forms that grow and reproduce. It is no wonder that the Pythagoreans chose the pentagram—every segment of which is in a phi proportion to the other segments—as their secret symbol and "rallying sign."

In part 1, "Rhythms," Ghyka focuses on number and proportion as they were understood by the ancients, for whom number had both a simply numerical as well as a metaphysical or philosophical aspect; and number seen in this philosophical way as immaterial idea and eternal essence of form was the ordering principle by which the world was created. The relationship between two numbers produced a ratio, and at least three terms were needed to make a proportion, a relationship between two ratios. The simplest and most elegant proportion, being the only one that divides a line so that the whole is to the larger segment as the larger is to the smaller is the one produced by the golden section. Proportion allowed "Plato and his Pythagorizing disciples" to see "the same phrase as a proposition of music, geometry, general aesthetics, cosmogony, or metaphysics. They were able to move effortlessly from scientific numbers to pure numbers, from harmony to pure intelligence and, reciprocally, around 'invariants' that here and there marked off the paths of knowledge." It was this understanding of number and how "the ancients were able to manipulate proportions that were irrational or 'potentially commensurable'" that Ghyka and others (Hambidge, Lund, Mössel) discovered in the canons of architecture of Mediterranean cultures whose buildings exemplified harmony, symmetry, and eurhythmy. This ancient stream of understanding number and proportion surfaced again in the Middle Ages among the designers and builders of the Gothic cathedrals. It became more public in the Renaissance in the work of the mathematician Luca Pacioli, and in the art of Leonardo da Vinci, Piero della Francesca, and Albrecht Dürer, all of whom were aware of the divine proportion and employed it in their works.

Shifting gears, and following the Greek practice of seeing analogies between architecture and music, Ghyka moves from outer, visual volumes in space to inner, audible periods of time to assess the numerical basis of prosody and music, the very rhythms of incantation and the "magic" they produce. The final chapter of part 1 discusses the incantatory power of spoken words and their relation to love, both human and divine.

Part 2, "Rites," is an in-depth look at the historical legacy of Pythagoras and his followers. In particular, Ghyka shows how the rites and practices of the Pythagorean brotherhood—the original secret society, and one based on an understanding of number and geometry, fraternal kindness and shared property, personal discipline and right living, musical therapy, transmigration of the soul, and love for all living beings—infused early Christianity, contributed in no small measure to the flowering of Alexandrian Gnosticism and Hebrew Kabbalah, and made up the template for the Knights Templar, the Society of Jesus, and, more importantly, the masons' guilds that built and left their marks on the Gothic cathedrals of medieval Europe. The lineage was kept alive when the masons' guilds—secret and hereditary organizations of practical tradecraft with levels of initiation, secret signs, and special rites—became the brotherhoods and lodges of "speculative" Freemasonry in the eighteenth century, which are still with us today, having taken the blazing five-pointed star as their emblem of identification. The Pythagorean tradition also survives, though in a diminished form, in the pentacles of the Tarot deck and in the words of power and number mysticism found in modern ceremonial and practical magic.

In compelling chapters at the conclusion of part 2, Ghyka surveys esotericism and politics as well as modern physics, biology, and philosophy. In quantum mechanics, which sees the world reduced to number and probability, the author finds the fulfillment of the Master of Samos's dictum that "all is arranged according to number." Surveying living forms and their group behaviors, from single-celled marine organisms to human society—and invoking Vernadsky's term *biosphere,* prefiguring the Gaia hypothesis—Ghyka asserts that the world is one being, united in a single consciousness, as evidence for Plato's Panpsyche, or world soul. In a discussion of Bergson's philosophy of the élan vital, he offers a thoroughgoing critique of the notion that consciousness is merely epiphenomenal to life, and contends that consciousness transcends our ability to reduce or understand it fully.

The Golden Number is a foundational text in what has become known as the field of sacred geometry. It has influenced numerous painters, poets (Paul Valéry), architects (Le Corbusier), photographers (Cartier-Bresson), theater directors (Peter Brook), and philosophers (Deleuze). Even if one takes issue with some of the historical details of the presentation or interprets them differently—understandable given the eighty-plus years since it first appeared—its reach and breadth still have the power to inspire. *The Golden Number* remains a tour de force, a grand synthesis of the Pythagorean tradition and its vitalizing role in the development of Western civilization.

Ehud C. Sperling
July 22, 2016

Foreword

Dear Sir,

Your manuscript has accompanied me in all my little excursions around Paris. I have given it what remains of my presence of mind after a heavy and laborious year. I am sure that it will prove even more savory to me after I have had time to meditate more deeply on its full and valuable substance.

This book was needed: now it exists. It concentrates all that is precise in aesthetics. The range of your knowledge is a marvel to me. First and foremost I admire the personal stamp you have given to such a vast and complex subject. The eternal desire to link physical and biological morphology to the science of forms created through human effort and sensibility, the need to compare and combine the architecture of nature's structures with the constructions of the artist, the mathematics that emerge in the former and the seemingly random formulas used in the arts—these are your subject, one whose full extent you have successfully explored with the organization of its parts and the enunciation of its problems.

Your analysis of phi is a poem!

Your song is that of prodigious and protean *expression*. You celebrate its grandeur, whose ubiquity and proliferation bring to mind some considerable "invariant" of our sensory system, with a scientific rigor and a kind of enthusiasm that I find totally delightful. This is because I believe—and I have made this a precept of my personal aesthetics—that there are powers of passion and "sentiment" in the order of the mind that are as strong as those that can be found—albeit more rarely—in the order of the "heart."

However, I am unable to refrain from observing that the number phi, with its wonderful properties, may seduce artists into using it without considering the materials, size, and location of their works. But in all constructions, whether they are machines, buildings, or works of art, the major question of the likeness between the project, or model, and the work itself arises. What is possible or suitable on one scale will be impossible or inappropriate on another. Even in the mechanical domain, this question is only imperfectly resolved. With regard to aesthetics, I do not know if this question has ever been posed in all its generality.

The mind has a tendency to conceive forms, connections, and the dependency of their parts without regard to either matter or size. Pure geometry lives on this ignorance. It has no concern for units of measurement and declares itself "true" at any scale.

But what distinguishes the practitioner and the artist is, on the contrary, their ability to maintain their personal *temperament* throughout the creative process, or the most intimate level of exchanges, between what they desire and seek, and what is offered them by the knowledge they hold of their subject and the final, real state of their work. These observations make it easy to imagine a conflict between the *defining features* that go into producing works of art, each of which is a specific solution to a problem that will never be repeated in exactly the same way, and the generality of the aesthetic principle that the number phi represents and clarifies. Therefore this number should not be used blindly and callously. It should not be seen as an instrument that allows the artist to work without relying on his own skill and intelligence. Quite the contrary! It should inspire the artist to develop these qualities, and this is where the exceedingly remarkable properties of your *golden number* come into play.

Reading your work, I cannot help thinking to some extent of literature. Unfortunately, this art takes a backseat to the others with respect to the search for intrinsic relations and the observation of proportions and formal requirements. It has no *golden number.* I have always dreamed of writing a book that would be secretly armed with reasoned conventions founded on the precise observation of the relationship between the mind and language. I have always backed down given its excessive difficulty: the immense task of recreating a notion of literature that would be clear enough to make it possible to reason about it. What is more, age is asking me: "What would be the point?"

I think and predict that your remarkable synthesis will earn what it deserves. Not only will it be given welcome, it will become influential. The balance between knowledge, feeling, and power in the arts has been broken. Instinct can only supply pieces. But great art must correspond to the whole man. The *divine proportion* is the general standard of measurement.

A kind of mysticism or esotericism (that may well have been necessary) formerly reserved these difficult and delicate truths to the proponents of these secret doctrines. Did these restrictions harm the advance of these studies or did they instead succeed in keeping alive into our day the results of experience become traditional principles that would have perished over the course of the ages without this occult transmission of powers? I do not know. Jealousy has its virtues and depths. Secrecy is seductive and life bestowing. But our time wants everything to be produced for everyone. It also seeks to define everything. Perhaps by subjecting the magic problems of art to its examination, our age may find it merely involves, in the end, to uncover those methods in the domain of shrewd sensibility that hold the same power as those that have already proven to be so fruitful in our analysis of the Expanded Universe.

Paul Valéry

Paul Valéry (1871–1945), French poet, essayist, critic, and philosopher, was a pivotal figure between symbolism and modernism. Most famous for his poetry, including "The Young Fate" and "The Cemetery by the Sea," he also studied law, mathematics, architecture, and language. He was elected to the Académie Française in 1925, founded the Collège International de Cannes in 1931, and was named the first chair of poetics at the Collège de France in Paris in 1937.

Introduction

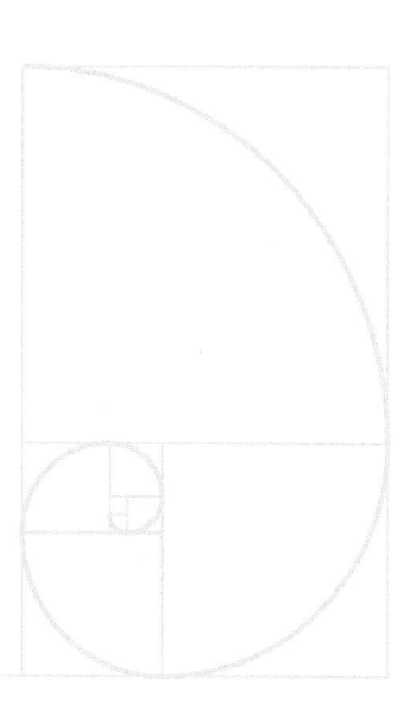

This book is both sequel and amplification of my previous book, *Esthétique des proportions dans la nature et dans les arts* (The Aesthetics of Proportion in Nature and the Arts),[1] in which I attempted to outline the following:

1. A mathematical theory of form. I started with an inventory of possible geometric forms and concluded with a comparative study of symmetries and natural forms characterized by morphological types—those of crystalline systems and living organisms in particular.
2. A presentation of the evolution of the ideas on proportion, harmony, and the geometrical canons that were broadly used during the great ages of Mediterranean art in the composition of architectural designs.

As the written source texts are quite understated in this regard and as no graphic evidence has survived (at least from antiquity), not even

1. Published in the collection *La Pensée contemporaine* (Paris: Éditions de la N. R. F., 1927).

the drawings that would once have illustrated the treatise by Vitruvius, the study of these canons, or rather procedures for composition and establishing proportion, is limited to the study of a number of hypotheses that have been recently put forward on this subject. What they share in common is that the graphic procedures used to demonstrate them end up producing outlines in which the theme of the whole is reflected and reproduced, in accordance with a more or less obscured rhythm, in the separate parts. The laws of analogy (as stated by August Thiersch), of the repetition of the fundamental form, of identity in variety, of the same and the similar—all these different names for the same principle and observation derive effortlessly, in fact, from the very concepts of symmetry and analogy as defined by the ancients. The ἀναλογία (analogy) of Plato and the Pythagorean arithmeticians is nothing other than proportion (equality, the equivalence or agreement of two or more ratios), especially geometric proportion; and συμμετρία (symmetry), according to these same philosophers and to Vitruvius, meant the commensurability between the whole and its parts, a connection determined by a common measure between the different parts of the whole, and between these parts and the whole (this is Vitruvius's actual definition, and the word *symmetry* held this meaning—which is quite different from its current meaning—until the end of the seventeenth century).[2]

The most interesting of these hypotheses also serendipitously agree with the Pythagorean theory of the harmony of the spheres, and the philosophical and cosmological ideas expounded by Plato in his most Pythagorean dialogue (*Timaeus*), and the speculations it inspired on analogy, the correspondence between the macrocosm (the universe created by the Great Ordering One) and man, the microcosm. At the end of the passage mentioned earlier, in the only architectural treatise handed down from antiquity, Vitruvius cites the human body as an example of the eurhythmy produced by an ideal "symmetry."

2. As late as 1650, Fréart de Chambray wrote: "Symmetry . . . overall union and interdependency of all the parts of a building."

A new hypothesis has been made known to me since the publication of my previous book. It involves a new system or canon establishing graphical proportion that synthesizes and reconciles to some extent the two principal theories presented and discussed in detail in my *Esthétique des proportions* (those of Fredrik Lund and Jay Hambidge). This new hypothesis, presented by Ernst Mössel,[3] concerns the partition or polar segmentation of the fundamental circle, or circle of orientation. This theory offers a very rational, practical point of departure; hence my initial notion of adding a "booklet" to my book.

The abridged account of the evolution of Mediterranean architecture and related aesthetic doctrines that accompanied presentations of the theories of Cook, Lund, and Hambidge in my first book included some Pythagorean philosophy (or Platonic philosophy, which often amounts to the same thing). I often encountered, as a symbol or dominant form of the design, a geometric figure (the pentagram) specifically associated with Pythagorean traditions (and one I have also found as a dominant presence in the study of living forms). I often frequently, if not to say most frequently, found proportions that arose from themes based on a particular ratio—the golden section. This ratio, this algebraic invariant, can also be abstractly and directly engendered using a very simple logical operation, in fact the simplest that can be carried out if one remains true to the Platonic concept of proportion. I had also discovered this ratio in biology, in numerical patterns, as an abbreviated symbol of the living form (while at the same time in opposition to the patterns of crystalline balance of nonliving forms), and as an impulse of growth. This "golden number" summarizes arithmetically and algebraically the properties of the dominant geometric form (the pentagram) mentioned above.

In the study of the transmission of these symbols or diagrams by generally anonymous intermediaries who had rarely been initiated into their true meaning or properties, I stumbled upon new sources that

3. Ernst Mössel, *Die Proportion in Antike und Mittelalter* (Munich: C. H. Beck, 1931).

provided the clues necessary to mark out this mysterious route with greater precision. Despite my wishes, these additions to my booklet became a new book—in two parts.

While ideally it is better to have read or at least skimmed through my *Esthétique des proportions* (or to at least have it on hand as a ready reference) before reading this new work, I have endeavored to make this book stand on its own, while abbreviating what was presented earlier and reducing the mathematical explanations to a strict minimum.

I would like to offer my apologies here, retrospectively, for the accumulation of arithmetic, algebra, and geometry that may have been responsible for making my first book so unappealing to the vast majority of readers. My first excuse is, as I said earlier, that I was seeking to make a reference work that would free the reader of the need for a cumbersome library of all the reference books needed for the study of the "science of space" and all the mathematical conceptions presented in the book.

Then, while accepting that the geometric perspective may not be the only or even the most important one in an aesthetic theory of form, I would like to recall the argument made by Edmond Picard in *Le Paradoxe de l'avocat* (The Lawyer's Paradox): every case, even the worst and most desperate case, should be pleaded fully without holding anything back. It is through the treatment and polishing of all its facets that the edges and form of the crystal emerge, and sometimes, for a judge, the cutting edge, the "thread" of truth. This is why it seemed helpful to me to dig so deeply, as deep as was compatible with the context and size of the work, in the mathematical study of the notions of ratio, proportion, and harmony.

Of course the purely mathematical speculations from which these three concepts arise can be pushed much further. Plato, among others, did this with his riddle of the number of the world soul in the *Timaeus.* And Plato's mathematical meditations are not eccentric musings that can be expelled from all his doctrines but the very marrow, the living fountain of his full conception of harmony, love, and the cosmos. They are also the foundation of the aesthetic concepts of analogy, of similarity in diversity, and variety in the similar, for which Vitruvius has provided us a very clear echo, and which, as long as they were still living (either as the philosophical contemplation of the macrocosm-microcosm paradigm during the Middle Ages, or as the creative trigger of rhythm at the beginning of the Renaissance), they played an important formative role in the evolution of all Mediterranean art.

But as I promised earlier to restrict the mathematical content of this new book to a strict minimum, and as it is high time I introduced it properly, here is its basic outline.

After redefining the idea of proportion and its related notions at the beginning of the book, I will describe the formal and rhythmic similarities and divergences between life and inorganic matter. Here I will give some details on the rhythm of the proportions in the human body (its morphology and that of plants and marine organisms were extensively discussed in my book on the aesthetics of proportion).

I will then present Mössel's new theory on Egyptian, Greek, and Gothic designs, and develop more thoroughly some questions touched on in my *Esthétique des proportions:* the composition of architectonic volumes, optical corrections, poetic rhythm, and relations between rhythms, rites, and magic, which will bring us to the end of part 1.

In the second part, stimulated by Salomon Reinach's courteous skepticism concerning the uninterrupted transmission of Platonic-Pythagorean mathematical esotericism,[4] I will attempt to complete the pathway of this transmission and show that Pythagoreanism, and the geometry it never cast aside, was passed down through a golden chain not only in art (Plato, Vitruvius, da Vinci, Pacioli, and various master builders) and in mathematics (Plato, Nicomachus of Gerasa, Pacioli, Kepler, Descartes, Russell, and Einstein), but in other domains as well, the most interesting of which, if not to say sensational, was indicated to me by Reinach himself.

4. *Revue Archéologique,* October/December 1927.

Powerful branches have sprouted from the magnificent rootstock left by the Master of Samos [Pythagoras] forming the noble fan of the "traveler's palm" [*Ravenala madagascariensis*]. And from this gigantic, more than two-thousand-year-old tree, the Tree of Knowledge and the Tree of Life, flows the life sap that has been what I call (after Flaubert), the "law of number." It was the guiding principle and vivifying basis, as long as it was understood, not only in Mediterranean art but also during the entire intellectual "adventure" (in Bergson's sense of the word) of Western civilization.

PART 1

Rhythms

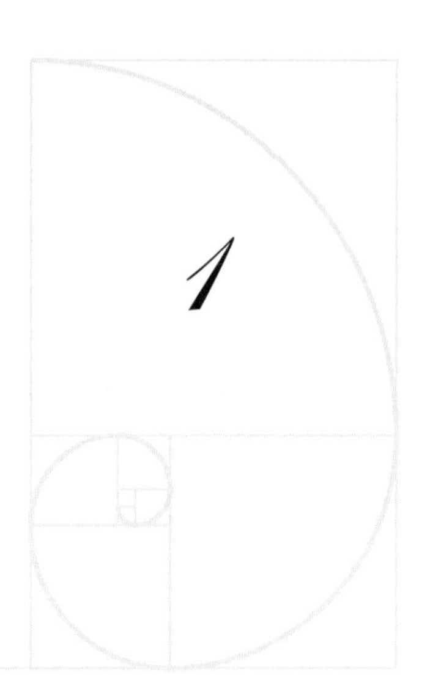

1

From Number to Harmony

Daughters of the golden numbers
Strengthened by the laws of heaven,
Upon us falls and slumbers
A honey-colored God.

Paul Valéry, "The Canticle of the Columns"

When we seek to redefine and rethink the so-called elementary notion of number, ratio, and proportion, one possible method is to return to the Greek sources dealing with this subject. Despite its lack of originality, this procedure does not lead to a mental trajectory that can be characterized as requiring the least amount of effort. However, as we find in Plato and in the sole complete treatise on numbers left us by antiquity—the one by Nicomachus of Gerasa[1]—starting points and

1. Nicomachus of Gerasa (a Greek colony in Palestine founded by veterans, γέροντες, of Alexander), known as "the Pythagorean," lived during the first century CE. He most likely pursued his studies in Alexandria. Two of his books have survived into the present in their entirety, *Manual of Harmonics* and his *Introduction to Arithmetic;* a good portion of his *Theologumena arithmetikes* (*The Theology of Numbers*) has also been saved by a compilation due to Iamblichus (Rome, fourth century), author of the famous *Life of Pythagoras,* who also wrote an important commentary on Nicomachus's *Introduction to Arithmetic.* The most famous translation (Latin) of this book is the one by Boethius (Rome, fifth century), which had an enormous influence throughout the entire Middle Ages.

I relied mainly on the English translation in the University of Michigan Studies series (New York: Macmillan, 1926), which is preceded by a valuable study on Nicomachus and Greek mathematics by Frank Eggleston Robbins and Louis Charles Karpinski.

often fairly clear conclusions, I will be content to provide only their definitions and sometimes their commentaries.

We know that Plato's conception of number and the importance he attributed to it ("Numbers," he said in the *Epinomis,* "are the highest degree of knowledge," and also "Number is knowledge itself") are derivatives of the most orthodox form of Pythagoreanism.[2] Nicomachus was a Pythagorean or, rather, an acknowledged neo-Pythagorean, and his mathematical work is simply a judiciously organized and clearly written compilation of elements borrowed from the works of the Alexandrian school, the titles of which alone have come down to us.

The basis of the following ideas and definitions can be found in the assertion, "Everything is arranged according to number,"[3] which goes back to the fourth century BCE among those who had known the last survivors of the first school founded by the Master in Sicily[4] as

---

2. It is tempting to say that Plato (429–347 BCE) was an initiate who had never sworn an oath of secrecy.

3. Ἀριθμῶ δὲ τε πάντ᾽ἐπέοικεν, probably a fragment of the *Ieros Logos* (*Ιερὸς Λόγος*) or *Sacred Discourse* of Pythagoras, cited by Iamblichus.

Aristoxenus of Tarentum (who in the second half of the fourth century BCE was friends with initiates of the original Pythagorean school) says virtually the same thing in prose: "Πυθαγόρας . . . *πάννα τά πράγματα ἀπεικάξων τοῖς ἀριθμοῖς.*" Cited by Johannes Stobaeus.

4. It is presumably helpful to present here a few short chronological notes on the subject of the life of Pythagoras and the sect or school that he founded. Pythagoras was born in Samos between 580 and 570 BCE and returned to that island after several long journeys (a sojourn in Egypt is quite probable, those in Phoenicia and Chaldea were imagined for symmetrical purposes when the legend of Pythagoras was crafted in the fourth century). He again left Samos because of the tyranny of Polycrates and reached Sicily around 529. His teaching was so successful that his disciples seized political power in Croton first (Crotoniate League), then extended their rule over a series of towns and a large portion of Magna Graecia (Sybaris fell in 510). The Master died about 500 BCE, but Pythagorean supremacy lasted in Sicily until around the year 450. On this date, popular uprisings successively freed the vassal cities, and the leading members of the sect, besieged by the mob in Metapontum, lost their lives in a gigantic fire. Cited as among the very few survivors are Philolaus of Croton (who was accused of selling books holding the brotherhood's secrets to the tyrant Dionysius of Syracuse, books that according to Diogenes Laertius, Plato would have subsequently obtained) and Lysis, who settled in Thebes with the family of Epaminondas.

the most important of his philosophical revelations. Their commentaries and definitions would even appear first and foremost to the reader unfamiliar with Greek mathematics to be wrapped in a disconcerting a priori metaphysics given the honest rationalism that presides over our own initiation into mathematics.

But the reader will gradually notice that the uncustomary tone of this starting point does not prevent reason's most imperious domination of the development and linking of ideas, and that this little moment of Hellenic mental gymnastics will later permit him to effortlessly follow into the present the evolution and vicissitudes of an extraordinarily robust conceptual system that remains as vital as ever despite its crystalline core of pure thought. Perhaps the reader may also discover that the suspect shimmering of the metaphysical-theological cloud was not mere flashiness but clarity, that today the theory of numbers begins to oddly resemble that of Plato and Nicomachus, while awaiting a time for our physics and cosmogony to catch up with that of the *Ιερὸς Λόγος*[5] itself.

We should first note that the word *logos* in Greek means "reason," "the ability to reason," and "relation" (judgment, the essential faculty

---

(cont. from p. 9) Archytas of Tarentum, a student of Philolaus and one of the greatest mathematicians of antiquity, succeeded in rebuilding a Pythagorean state. He was regent of Tarentum and elected as general for seven years in a row. Plato became friends with him on his first journey to Magna Graecia (388–87), and we can accept Diogenes Laertius's claim that Plato owed his initiation into the Pythagorean doctrine to Archytas. The school's ideas and rites continued to be passed on in a more or less esoteric fashion by isolated small groups, with the flame getting its greatest support from the Pythagorean-influenced Platonists. Plato's nephew Speusippus (died 339) wrote a treatise on Pythagorean numbers, a small fragment of which has come down to us. Aristotle's student Aristoxenus of Tarentum, who wrote a life of Pythagoras around the end of the fourth century, claimed he knew the last Pythagoreans, but the mockery of comic poets reveals that the sect survived into the second century. But it was in Alexandria toward the end of the second century BCE that the doctrine that came to be known as neo-Pythagoreanism began to flower openly and took on an important role in the very capital of the Roman Empire as well as in Egypt and Syria.

5. The *Ιερὸς Λόγος* or *Sacred Discourse* attributed to Pythagoras, but most likely written after his death during the Crotoniate period (beginning of the fifth century BCE), and of which Delatte in his *Études sur la littérature pythagoricienne* (Paris: Champion, 1919) was able to restore a certain number of verses.

of reasoning intelligence, is, moreover, the accurate perception of the relationships between ideas or things). This same term, the essential *word* (like the Word in the Fourth Gospel), also means the divine creative intelligence (Nicomachus would casually call the Creator God ὁ τεχνικὸς λόγος, ὁ τεχνίτης[6] θεὸς). Like Plato, Nichomachus made a distinction between two kinds of numbers: the divine number or number-idea, and the scientific number. The first is naturally the ideal model of the second, of that which we generally see as number, but because forms (dependent on quantities, qualities, and arrangements) are the only permanent things in the material world, because the structure of things (copy of the model or paradigm perceived by the *logos* as resulting from idea and number) is their sole reality, the divine number would also be, more generally, the directing archetype of the entire created universe.

In his *Theologumena arithmetikes* Nicomachus discusses this number-idea or pure number; in his *Introduction to Arithmetic* he deals with scientific number. The theory of numbers was in this way divided into two disciplines, the first, arithmology (the mysticism of number) with metaphysical tendencies, was concerned with the pure number. The second, arithmetic in the strict sense of the word, dealt with scientific abstract numbers in accordance with a rigorous syllogistic method like that of Euclid. But this theory of scientific numbers was also addressed to philosophers and not novices. Lastly, there was a third science, or rather technique (the discipline we call arithmetic today), that was accorded very low standing and consisted of calculation with concrete numbers. This was the arithmetic for businessmen or λογσιτιχή. A scholium on Plato's *Charmides* has this to say on the subject:

> Logistics (calculation) is the theory that is concerned with countable objects and not (true) numbers. In fact it does not look at the number in the proper sense of the word but assumes that 1 is the

6. τεχνίτης: someone who works with art; the creation can simply be a rearrangement of chaos, choice.

> unit, and that everything that can be counted is number (in this way, instead of the triad they take 3, in place of the decad, 10) and they apply the theorems of arithmetic to them.

This distinction appears much more clearly if we recall that the Greeks did not use exclusive symbols, numerals, to depict even concrete numbers, but used letters of the alphabet and several additional signs (the Pythagorean mathematicians of Sicily employed groups of dots, which led them directly to the stereometric properties of numbers and to the "figurate numbers" that we shall discuss later). Arab numerals and the decimal system have made calculation (what we call arithmetic) so easy that we have forgotten the distinction between the philosophy of number, the theory of numbers, and calculation, as well as the distinction between cardinal and ordinal numbers. We had to wait for Cantor and Russell to establish their theory of sets to rediscover that the numeral 2, the number two, the dyad or couple, and the idea of duality were very different things. So let's first try to forget numerals and rethink in terms of pure numbers. It will then appear fairly reasonable that our two guides from antiquity accepted the premise that the cosmos was ordered and rhythmical, with number being, according to Nicomachus's expression, the eternal essence of reality.[7]

The principles (ἀρχαί, origins) of number and moreover all things are, he said, borrowing here, too, the terminology of the *Timaeus,* the "same" and the "other"[8] (or the quality of "being the same thing" or of "being another thing").

Modern science has just reached an analogical mental attitude by again eliminating the barriers between mathematics and logic; the

7. "And it was then that all these elements . . . were fashioned by God through the action of Ideas and Numbers." *Timaeus.*

"Original chaos, lacking order and form, and everything that differentiates in accordance with the categories of quality, quantity, and so forth, was given organization and order through number." Nicomachus of Gerasa, *Theologumena arithmetikes.*

8. Moderatus of Gades (a distinguished Pythagorean philosopher and mathematician during the time of Nero, and author of a lost treatise titled "Lectures on

theory of sets, classes, relations of Cantor, Russell, and Whitehead, and Hilbert's axioms are chapters of a single science, the new "logistics" whose elements, symbolic tokens, indiscriminately represent logical fictions, numbers, or geometrical configurations.

As this has cleared the terrain to some extent, I now turn the floor over to Nicomachus of Gerasa as his discussion of the matter in his *Introduction to Arithmetic* will now appear sufficiently clear.

> The ancients, who under the spiritual guidance of Pythagoras first made science systematic, defined philosophy as the love of wisdom. . . . The bodiless things—such as qualities, quantities, configurations . . . equality, relations, dispositions, places, times . . . are in themselves immovable and unchangeable, but can accidentally participate in the vicissitudes of the bodies to which they are attached.
>
> If wisdom accidentally concerns itself with bodies, material supports for immaterial things, it is nevertheless to these things that it especially attaches itself, because these immaterial eternal things constitute the true reality. But that which is subject to formation and destruction . . . (matter, bodies) is not actually by its nature real. . . .
>
> All that nature has systematically arranged in the universe seems in its parts as in the whole to have been determined and ordered in accordance with number, by the forethought and the mind of him that created all things; for the pattern was fixed, like a preliminary sketch, by the domination of number preexistent in the mind of the world-creating God, number-idea purely immaterial in every way, but at the same time the true and the eternal essence, so that with reference to it, as to an artistic plan, would be created all these things, and time, motion, the heavens, the stars, and all the cycles of everything.

We should note that Nicomachus of Gerasa was not a distinguished

Pythagoreanism") said (quoted by Porphyry): "The Pythagoreans call 'One' the idea of identity, unity, equality, concord, and sympathy in the world, and 'Two,' the idea of the 'other,' discrimination, inequality."

philosopher or mathematician like Eudoxus or Diophantus. He was a professor and Pythagorean essayist who worked for the elite. His great ambition appears to have been the composition of a treatise on harmonics for a noblewoman to whom he dedicated a short manual on the same subject (the larger treatise and the name of the woman have unfortunately not come down to us, but the preparatory manual lets us see that she must have had an education in mathematics and music that was far from common). But a certain grandeur emerges in his prefatory remarks from the liturgical seriousness with which he spells out, through his "mathematical initiation," the fundamental article of the school's credo.[9]

After this introduction comes the definition of scientific number (ἐπιστεμονικός ἀριθμός), which is specifically discussed in this treatise (we shall come across the "divine" number again later).

Number is, he says, either a limited multitude[10] (today we would

---

9. The original Pythagorean brotherhood was a combination of religious sect, philosophical school, and political club. After his death (probably at the beginning of the fifth century BCE) Pythagoras's religious and philosophical doctrine was condensed into the Doric verses of the *Ieros Logos*. After the Metapontum massacre and the dispersal of the Sicilian brotherhood, small coteries reformed in Greece and Calabria. These groups consisted of two kinds of adepts: mathematicians and acousmatics. These latter confined themselves to handing down a ritual formalism they claimed literally followed all their master's precepts. The fanatical and puritanical appearance of their practices, which appear to have degenerated into puerile superstitions in the uncivilized regions where this democratic avatar concentrated its recruitment efforts, excited hostility and ridicule. These acousmatics are in fact the "Pythagoreans" that the comic poets mocked. The acousmatics rewrote a kind of catechism for their purposes that consisted of questions and answers, the *acousmata*. Iamblichus has passed on several fragments based on Aristotle's (lost) compilation on Pythagoreanism.

10. Basic concepts:

μέγεθος	magnitude (continuous);
πλῆθος	multitude of distinct elements (today we say "set");
πηλίκον	quantity;
ποσον	quantum;
ποσότης and ἀριθμός	number;
χῦμα	wave, a series that flows.

call it a finite countable set; strictly speaking, this is the ποστότες), or a combination of monads, which is to say, units. I would like to cite here a passage from a contemporary of Nicomachus, Theon of Smyrna,[11] who paraphrased this second aspect of number: "The Pythagoreans considered all the terms of the natural series of numbers as principles in such a way that, for example, 'three' (the triad) is the principle of three among tangible objects, and 'four' (the tetrad), the principle of all the fours, and so forth." This is almost literally Bertrand Russell's definition of numbers as "classes of classes."

These units, these monads, can be points and then give birth to the simultaneously geometric and algebraic world of "figurate" numbers, whether planes (triangular, square, pentagonal, and so on) or solids (pyramidal, cubic, parallelepipedic, and so on), which the first μαθεματικοι of the Pythagorean sect were already studying in Sicily.[12]

Or, third, they could be a wave, a flow of monads, ποσότητος χύμα έχ μονάδων συγκείμενον.

It is this third definition of number, like a moving series springing

---

11. Theon of Smyrna (fl. ca. 125 CE) wrote a mathematics treatise intended to facilitate the study of Plato (*Expositio rerum mathematicorum ad legendum Platonem utilium*) that has survived into the present. Many of the subjects discussed by Nicomachus can be found in it (figurate numbers, for example). He explicitly states that his arithmetic was taken from the Pythagorean tradition.

12. In my *Esthétique des proportions* I stated how this study of figurate numbers (polygonal, polyhedral, and so forth) or the differential geometry of the ancients used by Plato in his *Timaeus, Theaetetus,* and so on, which, thanks to Nicomachus and his popularizer Boethius, was still highly regarded during the Middle Ages and the Renaissance, and which was the subject of serious meditation by Descartes during the years preceding his "illumination" of November 10, 1619, can be of value even today (study of regular solids in spaces greater than three dimensions, theory of multiple combinations, and so forth). I used it for the study of homothetic growth. It was Nicomachus's preferred area of study, and he formulated Hypsicles's theorem on gnomons or geometrical additions of increasing polygonal numbers.

Plato's nephew Speusippus relied on the works of Philolaus for writing his book on Pythagorean numbers, of which one passage (preserved) speaks of pyramidal numbers. The treatises on the figurate numbers of Philippus of Opus (a student of Plato) and Hypsicles (second century BCE) were lost; fortunately the one by Diophantus of Alexandria has come down to us.

forth from the monad (an abstract translation of "punctual* expansion," or geometric increase of the figurate numbers mentioned above), for which variants can be found among the neo-Pythagoreans Theon of Smyrna and Moderatus of Gades, who is the most original and the most purely Pythagorean. It has also reincarnated in the present with Dedekind's cut numbers and segments.

We can summarize these notions regarding "divine" and "scientific" numbers by positing that in both the perceptible world, in which only the structure, form, and rhythm have the appearance of reality, and in the domain of the pure idea, number is the essence of the Form or the preeminent Form, and we can now move on to the examination of the connections or relationships between numbers.

The general concept of the relationship between two objects or two magnitudes is called σχήσις by Nicomachus. The relationship that he calls qualitative, such as double, triple, and so on, ποὶα σχήσις, or strictly speaking, *ratio* (in the modern sense of ratio-measurement in algebra and arithmetic, where, for example, it involves two lengths, and if a and b are the numbers that measure these lengths in relation to the same unit, the ratio $\frac{a}{b}$ is the "measure" of the magnitude a if one uses b as the unit of comparison) is, as we have seen, called *logos*.

We can find a very rigorous definition as early as Euclid, whose theory of ratios and proportions is based on the work of Plato's student Eudoxus: "A ratio is a sort of relation in respect of size between two magnitudes of the same kind. Proportion (ἀναλογία) is the equivalence of the ratios." Translated into algebra this gives the general equation of the geometric proportion between four magnitudes $\frac{a}{b}=\frac{c}{d}$ (discontinuous proportion).

This ratio $\frac{a}{b}$, the comparison of two magnitudes or the concrete numbers that measure them,[13] is the projection on a mathematical plane

[*Here and throughout, the author uses this term in relation to figurate numbers in its older sense: "of, pertaining to, or having a point or points."—*Trans.*]

13. When a and b are concrete numbers, the symbol $\frac{a}{b}$ can be replaced by the result, the quotient of the arithmetical operation "a divided by b," which is to say, that the ratio $\frac{4}{1}$, for example, is equivalent to 4, the ratio $\frac{8}{5}$ to 1.6. The numbers 4 and 1.6 in this case would

of an elementary operation of judgment: the exact perception of the relations between things or ideas (it is a measure, an ideal "weighing"). The comparison between two or more ratios and the perception of their equivalence, their agreement, and their analogy—an already more analytical operation of intelligence matching and connecting several elementary perceptions or judgments—is the same as a schematic projection on the plane of numbers as the equation of proportion given above, $\frac{a}{b}=\frac{c}{d}$.

When the two intermediary magnitudes b and c are equal, we obtain the continuous proportion $\frac{a}{b}=\frac{b}{c}$ (Iamblichus[14] reserves the term *ἀναλογία* for this continuous proportion, and calls "discontinuous" the geometric proportion that has four different terms *το ἀνάλογον*).

Hence Nicomachus's observation that if ratio was a relationship between two terms and proportion a combination or correlation of at least two ratios, then at least three terms were required to establish a proportion.

We should immediately note here that by pushing the "principle of economy" (*entia non sunt multiplicanda*) further, we can obtain a continuous proportion by starting with only two magnitudes a and b. Their sum $a + b$ provides us the third necessary magnitude, and the

---

be "measurement numbers," ratios of a quantity measured by a unit of measurement. In this sense, all numbers—whole numbers, rational fractions, and even incommensurable numbers (algebraic like $\sqrt{2}$ or transcendental like π)—can represent ratios, that is, can be conceived as measurement numbers. The series of all the "real" numbers (rational, algebraic, transcendental) between 0 and a given number, a, can moreover be placed in "univocal and reciprocal correspondence" with the points of a line segment if we take a as the measure of this segment. There is an absolute, corresponding correlation between the continuity of measurement numbers and geometrical continuity.

The Greeks preferred to reserve the name of numbers for the counting numbers (whole) and gave the name and form of ratios or relations to measurement numbers.

14. Iamblichus lived in Rome among the Neoplatonic and neo-Pythagorean circles of the fourth century CE. He wrote a life of Pythagoras, which has survived, whose value comes from the fact that he used—directly or indirectly—the lost works of Heraclides Ponticus, Aristoxenus of Tarentum, and Timaeus of Tauromenium, all three of whom lived during the fourth century BCE. Timaeus of Tauromenium wrote his book on the Pythagorean society after going through the archives of the cities in Sicily and Magna Graecia.

simplest equation of proportion, $\frac{a+b}{a}=\frac{a}{b}$, immediately gives us the most characteristic continuous proportion (the one the Germans call *die stetige Proportion,* or the preeminent continuous proportion) based on the ratio of the "golden section." Translated into words, this equation would read as: "The ratio between the sum of two given quantities is such that the ratio between one of them (the largest) is equal to the ratio between that one and the other (the smallest)."

Applied to lengths by dividing a line segment AC into two segments AB and BC, and choosing point B so that, $\frac{AC}{AB}=\frac{AB}{BC}$ corresponds to what Euclid called "division in extreme and mean ratio." It is both algebraically and geometrically the most "logical" asymmetrical division, and the most important one because of its mathematical, aesthetic, and other properties.[15]

This ratio was baptized the "divine proportion" by the Bolognese monk Fra Luca Pacioli di Borgo, who wrote a magnificent treatise on this subject, *De Divina Proportione* (Venice, 1509), which was illustrated by his friend Leonardo da Vinci. We will revisit the subject of the divine proportion later on.

---

15. Cf. my *Esthétique des proportions,* chapter 2, for a detailed study of this invariant ratio, the most interesting of the algebraic numbers. I will only remind the reader here that if in the equation for the proportion $\frac{a+b}{a}=\frac{a}{b}$, all the terms are divided by b, one obtains (by replacing $\frac{a}{b}$ by x) $x^2 = x + 1$, an equation whose roots are $\frac{1+\sqrt{5}}{2} = 1.618...$ and $\frac{1-\sqrt{5}}{2} = -0.618...$ The numerical value of the ratio or "measurement number" $\frac{\sqrt{5}+1}{2} = 1.618...$ is the arithmetical expression of the golden section or the golden number. Following the suggestion of Sir Theodore Cook and Mark Burr in *The Curves of Life* (London: Constable, 1914), I have designated it by the symbol Φ. We cannot only draw from the above equation $\Phi^2 = \Phi + 1$ and $\Phi = 1 + \frac{1}{\Phi}$, ($\Phi^2 = 2.618...$, $\Phi = 1.618...$, $\frac{1}{\Phi} = 0.618...$), but more generally: $\Phi^n = \Phi^{n-1} + \Phi^{n-2}$, which is to say that in the "Φ series" or geometric progression (or developed continuous proportion) $1, \Phi, \Phi^2, \Phi^3, \dots, \Phi^n$, every term is equal to the sum of the two preceding terms. This series is the *only* geometric progression that is additive in two ways: it shares the characteristics of both geometric and arithmetic progressions.

It follows from this that the proportional lengths with the terms of this series would possess remarkable graphical and geometrical properties (two lengths whose ratio is Φ make it possible to construct an entire geometrical series through simple additions or subtractions by rule or compass).

The Greeks, Nicomachus among others, generally wrote a proportion (not only a continuous geometric proportion but also the arithmetic or harmonic proportions we shall examine later) in the form of a progression or series, which is to say, they wrote "the proportion 1, 2, 4," or "the proportion 1, 3, 9, 27, 81."

This concept of proportion played a dominant role in their philosophical and scientific speculations. The Pythagorean theory of musical harmony, which is purely mathematical (and which it was generally believed by the ancients should be attributed to Pythagoras personally), is founded on the theory of proportion (its elements were the lengths of the segments of sounding strings [on instruments] being inversely proportional to the number of vibrations). This is similar to how with our new Russellian and Einsteinian terminology the same phrase can indiscriminately represent a theorem of tensor geometry, an equation of absolute differential calculus, a law of physics, or a logical "functional proportion," and to how among Plato and his Pythagorizing disciples, the same phrase could be read as a proposition of music, geometry, general aesthetics, cosmogony, or metaphysics. They were able to move effortlessly from scientific numbers to pure numbers, from harmony to pure intelligence and, reciprocally, around "invariants" that here and there marked off the paths of knowledge.

But the invariants that were familiar to them and still alive and active in the thought of Leonardo da Vinci, Shakespeare,[16] Kepler, and Descartes are no longer ours. Studying them as if they were beautiful inanimate bodies stretched out over marble, I am striving to resuscitate them the better to understand what will follow. Please forgive the repetition and monotony of this procedure.

---

16. There's not the smallest orb which thou behold'st
But in his motion like an angel sings . . .
Such harmony is in immortal souls . . .

The Merchant of Venice, 5.1.58–59, 61

His legs bestrid the ocean, his rear'd arm
Crested the world: his voice was propertied
As all the tunèd spheres, and that to friends; . . .

Antony and Cleopatra, 5.2.100–102

Plato is probably the thinker who has meditated most deeply on proportion and harmony. His riddle on number, or rather the rhythm of the world soul, whose mathematical and musical scheme was not rigorously reconstituted until the nineteenth century (by August Böckh, whose lengthy studies on this subject began in 1807), shows the way in which the Pythagorean tetractys could come into play in questions of "general harmony."

Plato also gave special attention (again in his *Timaeus*) to the proportions between solids, and we are in his debt for a theorem on this subject. We will revisit his ideas on this subject in chapter 4.

The definitions I gave earlier are suitable for proportions in the strict sense, or "analogies" of the "geometric" type, which are based on the equality of two or more ratios,[17] and whose elements are between them like the terms of a geometric progression, $1, k, k^2, k^3, \dots, k^n$, and so on.

In a continuous geometric proportion or progression $a, b, c,$ the middle term between the two others is called the "geometric mean," which is to say, if $\frac{a}{b} = \frac{b}{c}$ (thus $b^2 = ac$), $b = \sqrt{ac}$ is the geometric mean between a and c; for example, the geometric progression 3, 6, 12, in which $6^2 = 3 \times 12 = 36$. In a geometric proportion with four terms (of the discontinuous or discrete type) $\frac{a}{b} = \frac{c}{d}$, we have two means or "medieties," b and c, between the two end terms.

It is this type of (geometric) proportion, specifically the continuous type (which can be prolonged by an indefinite number of terms, and thus reproduces the pulse of a geometric progression with a constant ratio),

17. The ratio $\frac{a}{b}$, or measurement number of a quantity in correlation with another (*logos*), is only a particular case of the concept of relation (σχήσις) between two quantities of the same kind. There are others, for example, the relationship the Greeks called "quantitative" (ποσή σχήσις): differing (as plus or minus) from the other amount by a certain quantity (this is the relation that exists between the terms of an arithmetic series, $a, a + k, a + 2k, \dots, a + nk$). Nicomachus mentions ten types of functional relations of this kind between two quantities, the relations of multiples and submultiples (combined with the notion of simple ratio) included. But in our considerations of geometric aesthetics in the course of this book, we shall only concern ourselves with the base ratio or measurement number $\frac{a}{b}$.

that gives rise to the similarity (homothetics) of the figures in geometry, and the analogous planes or volumes in architecture, which I examined in *Esthétique des proportions* and which I shall refer to occasionally in this book. But in the same way that the Greek notion of ratio or numerical relation is more general than that of ratio or unit of measurement as such, the notion of proportion was also generalized, and "analogy" as such or geometric proportion is only one of ten types of "correlations between relations" demonstrated so generously by Nicomachus and Theon of Smyrna.

The general definition of proportion, Nicomachus says, is the combination of two or more ratios. It does not necessarily involve the equality of the two initial ratios, but can also envision between them a difference or another kind of correlation or comparison.

He adds that for all the types of combinations of ratios (or of relations), that is to say, for every type of proportion, three is the smallest number of terms that can be used.

Earlier we looked at the example of the $\frac{a}{b} = \frac{b}{c}$ type of continuous geometric proportion (the ratio between the first term and the mean term is equal to the ratio between that one and the extreme term: $b = \sqrt{ac}$; for example: 2, 4, 8).

The other two usual types of proportion are the arithmetic proportion in which the mean term exceeds the first by an amount equal to that which it is itself exceeded by the last so that it (this mean term, medeity, or arithmetic mean) is equal to the half sum of the two extremes, $c - b = b - a$, or $b = \frac{a+c}{2}$ (for example, 2, 4, 6); and the harmonic proportion (in which the mean term exceeds the first by a fraction of it equal to the fraction by which the last term exceeds it), $b - a = a\frac{(c-b)}{c}$, or $b = \frac{2ac}{a+c}$ (for example, 6, 8, 12).

These three principal types of proportion, established by the Pythagorean philosophers of Sicily, were probably passed on to Plato during his first sojourn in Magna Graecia by Archytas of Tarentum.[18]

18. Ca. 430–365 BCE (cf. footnote 4 in this chapter). The friendship between Archytas and Plato is confirmed by the latter's *Seventh Letter.*

Archytas was the first to officially deal with the famous problem of duplicating the

Eudoxus and Plato's immediate disciples would raise this number to six by adding three "subcontrary" types. The neo-Pythagoreans Myonides and Euphanor would later invent four others (see Iamblichus, *Life of Pythagoras*) toward the first century BCE, bringing the total number to ten (we have also seen this number dear to the Pythagoreans in the number of the types of relation).

Nicomachus and Theon of Smyrna employed a very elegant "logistical" method for establishing, when given three quantities, these ten kinds of proportion with the help of the "principle of the same and the other" and the "principle of economy"[19] (I applied these same principles

(cont. from p. 21) cube (Diogenes Laertius), which could well be only a particular case of the general problem of inserting two geometric means between two "solid" numbers or volumes. It so happens that Plato found a (mechanical) solution to the problem in question, and he is also recognized as the author of a general theorem on the proportions between volumes, which we shall revisit in a later chapter.

19. Both cite Eratosthenes (276–194 BCE) as inspiration for the method. He was the chief librarian of the famous library of Alexandria and author of a *Platonicos* or mathematical commentary on the *Timaeus*. Nicomachus said in this regard: "The principles governing these questions (on proportions) . . . can demonstrate that all the complex kinds of inequalities and the varieties of these kinds can be drawn from one lone equality, as if from one mother and one root."

And Theon: "In this way we take three quantities and the proportions that dwell within them, and interchange the terms, and demonstrate that all mathematics is constituted by the proportions among quantities, and that its source and its elements can be summarized as the essence of proportion (τῆς ἀναλογίας φύσις)."

Here are the ten types of proportions written in the form of equalities, with corresponding numerical examples:

$$\frac{c-b}{b-a} = \frac{c}{c}\ (1, 2, 3) \qquad \frac{b-a}{c-b} = \frac{c}{b}\ (1, 4, 6)$$

$$\frac{c-b}{b-a} = \frac{c}{b}\ (1, 2, 4) \qquad \frac{c-a}{b-a} = \frac{c}{a}\ (6, 8, 9)$$

$$\frac{c-b}{b-a} = \frac{c}{a}\ (2, 3, 6) \qquad \frac{c-a}{c-b} = \frac{c}{a}\ (6, 7, 9)$$

$$\frac{b-a}{c-b} = \frac{c}{a}\ (3, 5, 6) \qquad \frac{c-a}{b-a} = \frac{b}{a}\ (4, 6, 7)$$

$$\frac{b-a}{c-b} = \frac{b}{a}\ (2, 4, 5) \qquad \frac{c-a}{c-b} = \frac{b}{a}\ (3, 5, 8)$$

This is, moreover, the same combinatory method founded on a certain set of transformations of the initial equality on which Nicomachus first established the ten types of "functional relations" (ratios in the broad sense) between two quantities.

earlier to logically establish the "golden section" starting only from two quantities). In these various kinds of proportions, as in the continuous geometric proportion, the middle term is called the mean or mediety.

Fitting or filling in the interval between two given terms consists of finding the mediety that creates proportion. Plato applies these expressions indiscriminately to proportions in the domains of mathematics, music, and cosmogony. The overall "harmonic question" (in the *Republic*) consists of putting the intervals into proportion by means of the initial terms given in definite ratios, in order to obtain the consonance (συμφωνία) or harmony of the intervals.[20] Inserting the mean term into a syllogism, assembling a chain of syllogisms into a "sorites" and thereby passing a bridge between two islets of consciousness, connecting by the flash of the right metaphor two images immersed in the flow of prosodic rhythm, to join together by the eurhythmy based on the analogy of the forms, surfaces, and architectonic volumes, as said this same Plato in his *Theaetetus* and his *Timaeus,* and as Vitruvius details quite clearly—all these operations are parallel, "analogues" to the creation of musical harmony that was the Pythagoreans' first choice as model or example. In the previously mentioned chapter in which the *Timaeus* discusses the rhythm of the world soul, it is the musical double tetractys of the Pythagoreans: $(1 + 3 + 5 + 7) + (2 + 4 + 6 + 8) = 36$. Plato uses this sum of the first four even numbers and first four odd numbers as a framework for establishing the sevenfold celestial scale[21] whose tones make it possible to orchestrate the harmony of the spheres.

20. In the Greek theory of musical harmony, the interval is the set formed by two tones and the relation or ratio that unites them. Filling the interval is achieved here by inserting between two tones other tones that are joined to the first two by simple ratios so that the new interval between two consecutive tones creates the harmony or consonance (*symphonia*) of the intervals. The tones themselves are symbolized by the two numerical terms framing the interval. These numerical terms are not proportional to the numbers of vibrations as in modern harmonic theory, but to the lengths of the vibrating strings (inversely proportional to the number of vibrations), which in both cases gives for the intervals of the diatonic scale very simple ratios: $\frac{2}{1}$, $\frac{3}{2}$, and so forth.

21. He first combines in a complex progression (1, 2, 3, 4, 9, 8, 27) two geometric progressions (1, 2, 4, 8 and 1, 3, 9, 27) then on two instances "fills" all the intervals with

We do not know if Nicomachus and the other neo-Pythagoreans possessed the key to Plato's riddle. Its entire solution is not found in any of the commentaries on the *Timaeus* that have survived, and, as noted earlier, it was not until fairly recently that modern scholars deciphered it. But the influence this passage had on European thought, and on those who, respectively, presented the correlation between the rhythm of the world soul and that of the human soul, and the morphological importance of the five regular polyhedrons, was equally profound[22] in philosophy (the theory of the macrocosm and the microcosm) ethics, and aesthetics. Pacioli and Leonardo da Vinci drew their canon of divine proportion from the former, and Kepler owes to the latter the discoveries of the astronomical laws that immortalized his name. I mentioned earlier the double tetractys (36) with regard to the Pythagorean underpinnings of the *Timaeus.* The tetractys as such, whose discovery by Pythagoras was considered so important that it is invoked in the sacred oath sworn by the Pythagoreans,[23] was the series of the first four numbers, 1, 2, 3, 4, considered as a series and a set; therefore this series (1 + 2 + 3 + 4 = 10) was truly the "decad" because it was a result of

(cont. from p. 23) both arithmetic means and harmonic means that culminate in a musical scale of 36 terms and 35 tones and *leimmas* instead of the 5 tones and 2 *leimmas* of the classical scale. Those who are interested in the complete details of this "harmonization" or placing it in proportion to the rhythm of the world soul will find them in the scholarly introduction to the edition of the *Timaeus* published under the patronage of the Association Guillaume Budé (Paris: Collection des Univesitaires de France, Société d'Édition "Les Belles Lettres," n.d.).

22. The commentaries on the *Timaeus* by Plutarch (*De animae procreatione in Timaeo*) and by Calcidius and Proclus have come down to us, but among the authors of those that have been lost, I can mention Xenocrates, Eratosthenes, Crantor, Eudorus, Clearchus, Theodorus, Panaetius, Adrastus, and Posidonius of Apamea. Theon and Nicomachus explicitly stated that their treatises were especially useful for the study of the mathematics necessary for understanding Plato.

23. "I swear by him who the tetractys found, and to our race reveal'd; the cause and root, and fount of ever-flowing Nature!" (ὅυ μά τον ἀμετέρα ψυχᾷ παραδοντα τετρακτὺν παγάν ἀενάου ψὺσως ῥίξωμά τ'ἔχουσαν). Distich cited by Iamblichus but dating in this form from the fifth or fourth century BCE. It is the oath taken not to divulge any of the brotherhood's secrets.

the "quaternary" formation (as the fourth "triangular" number).[24] The

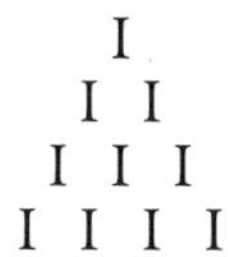

tetractys therefore simultaneously possessed the transcendent qualities of the decad (we shall later see that this number, archetype of the number ten, was the symbolic number of the universe) and the dynamic qualities of triangular increase, which is itself the basis for the generation of all the plane or solid figurate numbers[25] (therefore of the regular solids corresponding to some of them). Lastly, they pertain to the harmonic qualities of the progression 1, 2, 3, 4. In fact, the ratio of 4 to 2 or 2 to 1 represents the octave (the διὰ πασῶν, diapason), that of 3 to 2, the fifth (διὰ πέντε), and that of 4 to 3, the fourth. We can therefore echo Delatte when he says:

24. I have already mentioned the importance of figurate numbers in Pythagorean mathematics. The triangular numbers:

$$1,\ 1+2,\ 1+2+3,\ 1+2+3+4,\ 1+2+3+4+5,\ \ldots\ \frac{n(n+1)}{2},$$

which is to say, 1, 3, 6, 10, 15, 21,..., have tetrahedral or pyramidal numbers as correlations in three-dimensional space,

$$1,\ 1+3,\ 1+3+6,\ 1+3+6+10,\ \ldots\ \frac{n(n+1)(n+2)}{1.2.3.},$$

which is to say, 1, 4, 10, 20, 35,... .

25. The regular plane figurate numbers or polygonal numbers (triangular, square, pentagonal, and so on) can all in fact be drawn from triangular numbers using a process analogous to the one that is the basis of Pascal's arithmetic triangle.

	1	1	1	1	1	1	1
Natural numbers	1	2	3	4	5	6	n
Triangular numbers	1	3	6	10	15	21	$\frac{n(n+1)}{2}$
Square numbers	1	4	9	16	25	36	n^2
Pentagonal numbers	1	5	12	22	35	51	$\frac{n(3n-1)}{2}$
Hexagonal numbers, etc.	1	6	15	28	45	66	$n(2n-1)$

The successive differences or gnomons of the fifth horizontal row, for example, (pentagonal numbers) form an arithmetic series 1, 4, 7, 10, 13, ... , $(3n - 2)$, with the constant interval of $5 - 2 = 3$. In return, one can start from this series to arithmetically or geometrically construct the numbers or figures of the pentagonal series by "gnomonic" growth or fluxion (the general law is given in Hypsicles's theorem $\Delta k = k - 2$). See plate 1.

"The tetractys is the set of four numbers whose ratios represent the essential musical harmonies." The discovery of these acoustic laws, which was considered a brilliant invention in antiquity, was attributed to Pythagoras not only by his disciples but also by specialists from outside his school.[26]

The tetractys came to be identified with harmony itself in the Pythagorean "catechism," a verse of which was preserved by Iamblichus (quoting Aristotle): "Tetractys, pure harmony, that of the Sirens (τετρακτύς ὅπερ ἐστὶν ἡ ἁρμονια ἐν ἧ αἰ Σειρῆνες)."

We can hear these same sirens in Plato's *Republic,* the planetary spokeswomen for the harmony of the spheres.

But it is especially in its form of pure or divine number, that is to say, the decad, that the tetractys becomes the symbol of the universe. Philolaus, one of the Pythagoreans to escape the massacre in Croton or Metapontum in which almost all the members of the original "society" perished, said: "Harmony is the unification of the diverse and the setting of the discordant into accord."[27]

Nicomachus, having spoken of the number "paradigm" that preexisted in the thought of the Creator God, wrote:

> But since the whole was an unlimited multitude . . . it needed an Order. . . . Now, it is in the decad that a natural balance between the whole and its elements preexisted. . . . It is why through his reason, the Ordering God (literally, the "God who arranges with art") used the decad as a canon for everything . . . and it is why the harmonic relations of the whole and the parts of all things in heaven and on earth are based on this canon and are ordered according to it.

26. Delatte, *Études sur la littérature pythagoricienne.*

27. Only brief fragments remain of the *Περὶ φύσεψς* and other treatises by Philolaus that Iamblichus (*Life of Pythagoras*) described as the first books (published circa 440 BCE) presenting Pythagorean ideas to the public; his work also discussed the decad. Stobaeus.

A distich fom the *Ieros Logos* cited by Syrianus also mentions the decad as "key" to the universe. Speusippus discussed the decad among other things in his work on Pythagorean numbers. Archytas of Tarentum devoted a treatise to the decad that has not survived.

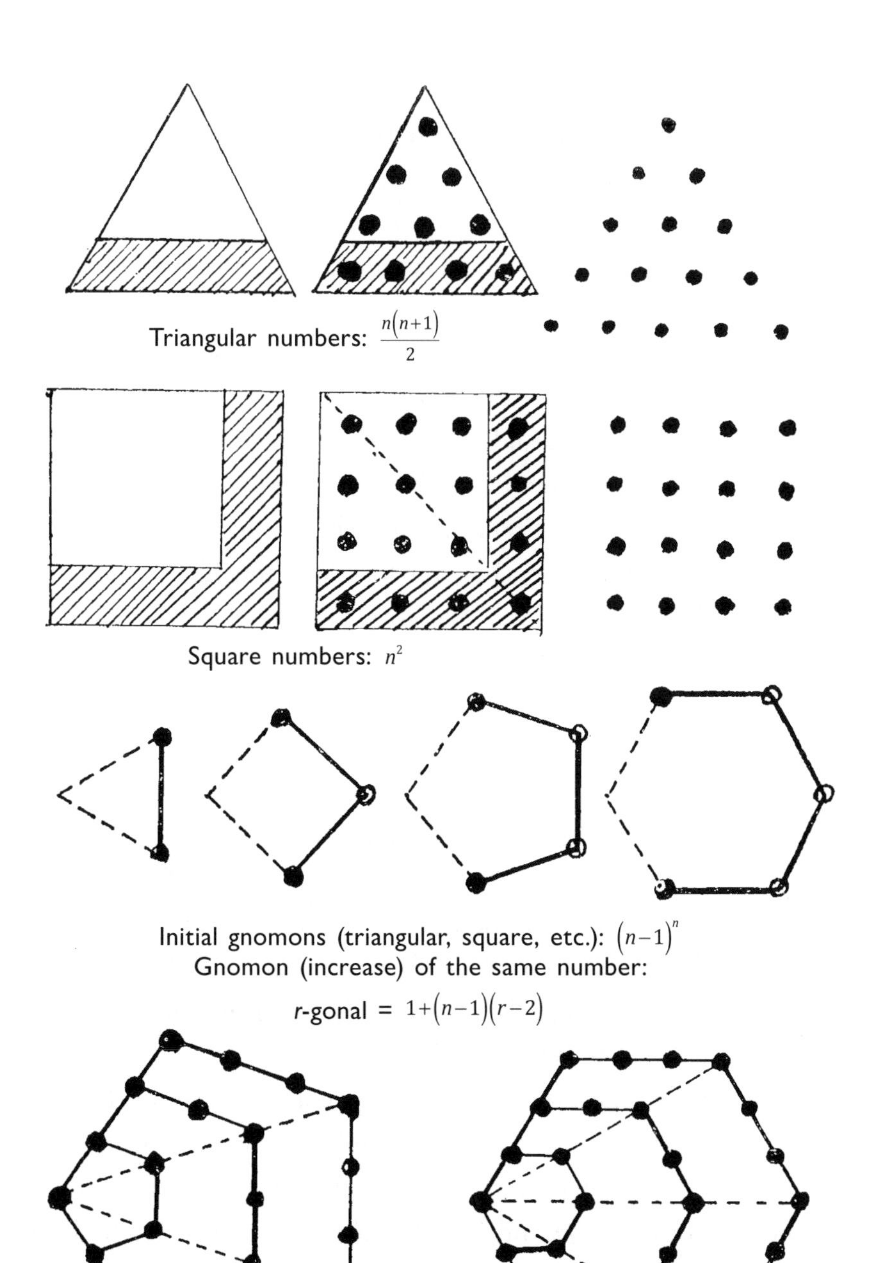

Pentagonal numbers: $\frac{n(3n-1)}{2}$ Hexagonal numbers: $n(2n-1)$

r-gonal numbers: $=n+\frac{n(n-1)}{2}(r-2)$

Plate I. Polygonal figurate numbers and gnomons

We shall later see the same expressions used by Vitruvius in his discussion of the eurythmy of an architectural composition.

And later, in his *Introduction to Arithmetic* (bk. 2, chap. 22) Nicomachus writes: "It is the number ten that, according to Pythagorean doctrine, is the most perfect of possible numbers. . . . It is in accordance with this idea that we noted the ten types of relations and categories, and ten appears to have even established the divisions and forms of the extremities of our hands and feet, and countless other things that we shall note when opportunity arises."

And in his *Theologumena,* Nicomachus calls the decad the All (Πᾶν), "for it has served as a measure for everything like a square and a string in the hand of the Great Ordering One."

Moving now from the decad to its half—we meet one of the most brilliant personalities of the "society of numbers," the pentad and the characteristics of the number five. In arithmology or number mysticism the number five partially pertains to the essence and importance of the decad as being its half and its condensed image, but it is also the ΓΑΜΟΣ (*gamos*), Aphrodite's number as goddess of the fruitful union, the "generatrix" of love, and the abstract archetype of reproduction. Five is, in fact, the combination of the first even number, which is female, matrix, scissiparity (two, dyad) and the first whole odd number, which is male, asymmetrical, (three, triad).

The pentad is also the number of harmony, realized in the health and beauty of the human body. Its graphic image, the pentalpha or pentagram (pentagonal star), will therefore be simultaneously the symbol of creative love and living beauty, of balance in the health (beauty, harmony, and health are interchangeable or connected qualities for Plato) of the human body, of the projection of the soul on the material plane, and a reflection of the great rhythm of the "world soul" or universal life.

This synchronism between the rhythms of the "well-tuned" individual soul and that of the universal soul is clearly specified in the *Timaeus:* "Then, in this body in which a wave (life) was perpetually flowing in and out, they (the gods) introduced the periodic movements[28] of the immortal soul."

In this regard, Plato also mentions a bit later the purifying and regulating role of music.

> For harmony, whose movements are akin to the regular revolutions of our souls, is not regarded by a man of intelligence as a gift from the Muses for enjoying an irrational pleasure, as appears to be the case today. To the contrary, the Muses gave it as an ally to our soul when it seeks to restore order and unison to its periodic movements when they are in chaos. Similarly, rhythm, too, was given us by these same Muses for the same purpose: to correct the irregular and graceless ways that prevail among mankind generally.[29]

We know that this idea of analogy, of the correspondence between the structure (number) and the rhythm of the cosmos and those of the human, between the macrocosm and microcosm as they would later come to be known, would inspire and enrich more than two thousand years of both profane and religious philosophy. Most particularly, Alexandrian Gnosticism and its tenacious offshoots in the Middle Ages and Renaissance imagined and wove together multiple systems out of this premise whose basic frameworks we shall encounter later in the form of either darkened spider webs or still solid, sparkling ideological tapestries.

The pentagram became and remained the emblem of the microcosm, but we know from a passage by Lucian that this geometric symbol of the number five, that is to say, love and living eurythmy,

---

28. "The rhythm is perceived periodicity. . . . Every periodic phenomenon perceptible to our senses stands out from the whole of irregular phenomena . . . to act alone on our senses . . . and gradually our breath, our heartbeats, our thoughts, and our sorrows are all dancing on the subdued but persistent rhythm that we do not think we can hear." Pius Servien [S. Coculescu], *Essai sur les rythmes toniques du français* (Paris: Presses Universitaires de France, 1925).

29. Translation by Les Belles-Lettres (Association Guillaume Budé).

was already the secret rallying sign of the Pythagorean society.[30]

In the course of our investigation we shall follow its unfaltering career as a sometimes suspect symbol of various avatars. Starting with the next chapter, we shall see how as a mathematical emblem of "harmonic correspondences" it was well chosen, for it distinguishes itself from all the other star-shaped polygons with its unbroken and endlessly recurrent rhythm based on the preeminent form of continuous proportion, the one that, in accordance with Plato's pledge, "gives to itself and to the terms with which it joins the most complete unity."[31] This is the proportion that is more characteristic of the morphology and the laws of growth of living organisms.

This attempt to rethink the notions of number, ratio, and proportion as they were conceived of during "antiquity" naturally brings us fully to the notions of rhythm (periodicity or recurrence in time or space resulting from a series of chords or proportions) and the harmony that organically flows from them, and we have incidentally seen (if it can be called an incident) that for our Greek instructors of "mental gymnastics" the perception of ratios and proportions was identified as one with the elementary operation of judgment and creative choice in

---

30. "τό πεντάγραμμον, ῷ συμβόλῳ πρός τους ὁμοδόξους ἐχρῶντο ὑγίεα πρός αὐτῶν ὀνομαξετο." ("The pentagram that the initiates adopted as a symbol was an emblem of health for them.") This very important detail for the study of the transmission of geometric symbols in general is confirmed by a scholiast of Aristophanes. The Basel Museum owns an alabaster disk from the Alexandrian era (probably a mold for ritual cakes) bearing a pentagram carved into it with the letters ΥΓΕΙΑ ["health"] precisely at the point of each of its five branches. Cf. Paul Perdrizet, *Negotium perambulans in tenebris: Études de démonologie Gréco-Orientale* (Strasbourg: Publication de la Faculté des Lettres de l'Université de Strasbourg, 1922).

31. *Timaeus.* Here is the complete passage in which Plato introduces geometric proportion: "But two terms cannot form a beautiful composition without a third; for they must have a bond that brings them both together. Of all the bonds, the most beautiful is that which makes the most complete union of itself and the terms it combines; and it is proportion (ἀναλογία) that is naturally suited to achieve the most beautiful of such a union."

Vitruvius says much the same thing: "The proportion the Greeks called *analogia,* consonance between the whole and its parts."

general, and that intelligence, in its function as knowing or creative synthesis, led to harmony, or was harmony, itself.[32] In brief, the Beautiful, the True, and the Good are ONE in this harmonic conception of knowledge and life.

It is understandable that this "aesthetic" attitude of general philosophy would be a fortiori fully reflected in its corresponding art, especially in the major art of architecture, the harmonization of space. It is more than likely that the architects who were contemporaries of the thinkers who established the harmonies and the rigorous mathematical correspondences for abstractions like the world soul were at least as rigorous in finding the proportions of the temples they built for the worship of such "geometry-minded" deities. Moreover, these temples speak, or rather, to borrow Paul Valéry's expression, they sing.[33]

We have not yet decoded their melodic themes, but we now have a glimpse of the key. The recent hypotheses of Hambidge and Lund that I examined in detail in my previous book, and that of Mössel, which I will present in this one, all endeavored to approach the question using the illumination provided by various Pythagorizing passages of Plato's *Theaetetus, Timaeus,* and *Philebus.* If their geometric armature is a bit more subtle than the empirical triangulations of Viollet-le-Duc or Dehio, this is not, at least for the Greek temples, an argument against them.

The texts and definitions that form the contents of the present chapter may perhaps give us the means to estimate the likelihood of

---

32. In a domain of intelligence that at first glance may seem entirely foreign to harmony, the art of war, we should take note of the most "symphonic" battle in history, the battle of France in 1918. Marshal Ferdinand Foch experienced it as rhythm. "But no, I was merely the conductor of the orchestra . . . an enormous orchestra obviously. . . . But you could say I kept the beat quite well!" His disdain targeted those he called "tambourine players." *Revue Universelle,* April 15, 1929.

33. Goethe also says in the Second Faust:

> *Der Saülenschaft, auch die Triglypey klingt;*
> *Ich glaube gar, der ganze Tempel singt!*
>
> (*Translation:* The pillared shafts, even the triglyph, ringing;
> I think that the whole temple is singing!)

this, but they do make it immediately possible to detect their inspiration in the sole work of this genre from antiquity to come down to the present in which a man of the trade (Vitruvius) briefly speaks to us about architectural composition.

> Symmetry consists in the harmony of measure among the various components of the work, among these elements taken separately and as a whole. As in the human body . . . it arises from proportion—what the Greeks called *analogia*—consonance between each part and the whole. . . . This symmetry is governed by the module, the standard of common measure (for the work in question), that which the Greeks called the πόσοτες (the "number"). . . . When every important part of a building is thus properly set in proportion by the right correlation between height and width, between width and depth, and when all these parts have their place in the total symmetry of the building, we obtain eurhythmy.

Vitruvius placed great emphasis on this perfect "symphony" of the play of proportions in the human body, and on the play of analogous correspondences, sometimes even numerically identical, that the architect should establish in the eurythmic plan of sacred buildings. His entire third book is devoted to this parallelism; comparisons and similarities borrowed from music alternate in it with purely geometrical precepts. The fact that he refers to Philolaus and Archtyas of Tarentum at the beginning of his treatise makes it less surprising to observe that the architectural adviser to the first emperor uses the same expressions as the Pythagorean Nicomachus of Gerasa, who, two centuries later, sought to initiate an unknown noblewoman of the Flavian era into the knowledge of the harmony of the spheres. Nor is it surprising to find in one of his glosses, the last one, in which he repeats his emphatic praise of the "symmetry" of the human body, that this reference to the decad-tetractys naturally emerges: "The ancients held the number Ten as sacred . . . sprung from the monad."

This "symmetry" of Plato (*Theaetetus*) and Vitruvius (which, let me repeat, has nothing in common with what we currently call by this name) thus results from the bond, from the *commodulatio* that connects by means of the standard of common measure (the module) all the elements to each other and with the whole, this bond that can moreover be simple linear commensurability (all the important linear dimensions being multiples of the standard measure), or be formed by more complex functional relations (let me remind you of the ten kinds of proportion listed by Nicomachus).

But it is when this symmetry, this metrical correspondence, is obtained by the continuous chain of proportions, by recurring analogy, and in addition when this analogy manifests favorably in both the forms of the major parts and the relationships between these parts and the architectonic whole, that eurythmy appears.

Like the Great Ordering Being of the *Timaeus,* the architect has cut out, unrolled, and set up his pageant of forms; he has "harmonized," connected his concordances, by filling the intervals by means of the requisite "medieties," and if, in a lightning flash of creative passion, his personal rhythm beating as one with a higher rhythm, he has obtained the grand consonance of the symphony that holds his work of stone or marble and makes it vibrate like an invisible bow, this work then lives, and like the temples of Greece and Sicily, like the Gothic cathedrals, like the sirens in Er's vision, it sings.

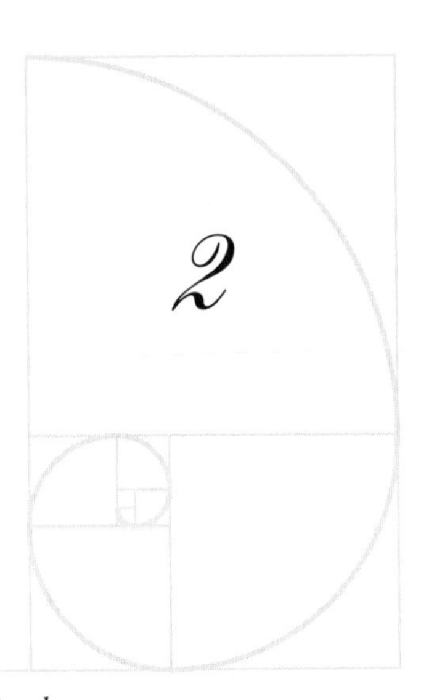

2

The Divine Proportion

Ah, Beauty! Syren, fair enchanting Good . . .
Still Harmony, whose diapason lies,
Within a Brow; . . .

Samuel Daniel, 1592

We have seen in the preceding chapter that the simplest unequal division (asymmetrical) of a magnitude obtained by applying the "principle of economy" (of concepts and operations) was the one that established between the initial magnitude and its two parts the proportion known as "mean and extreme ratio" or "golden section"; if a and b are these two parts (line segments when it involves a length) we have $\frac{a+b}{a} = \frac{a}{b}$, from which we get $\frac{a}{b} = \frac{\sqrt{5}+1}{2} = 1.618....$[1]

Let me remind you that this ratio $\frac{\sqrt{5}+1}{2}$, which I have named Φ in echo of Sir Theodore Cook in order to simplify the writing and calculations, can be found in the geometric figures derived from the regular pentagon (especially in pentagrams or star pentagons) and the regular convex or star decagon[2] (see plate 2).

1. By forming the equation $\left(\frac{a}{b}\right)^2 - \left(\frac{a}{b}\right) - 1 = 0$, and by taking $\frac{a}{b}$ for the unknown from which we get $\left(\frac{a}{b}\right) - \frac{1+\sqrt{5}}{2}$. The negative root $\frac{1-\sqrt{5}}{2} = -0.618...$ is the inverse of the positive root $\frac{1+\sqrt{5}}{2} = 1.618...$, and we will always have (because of the fundamental equation $\frac{a^2}{b} = \frac{a}{b} + 1$), if we call Φ this ratio $\frac{a}{b} = \frac{1+\sqrt{5}}{2}$ of the golden section, then $\Phi n = \Phi n^{-1} + \Phi n^{-2}$, whatever the value of n may be.

2. Cf. the *Esthétique des proportions*. For example, if in a circle with the radius R we

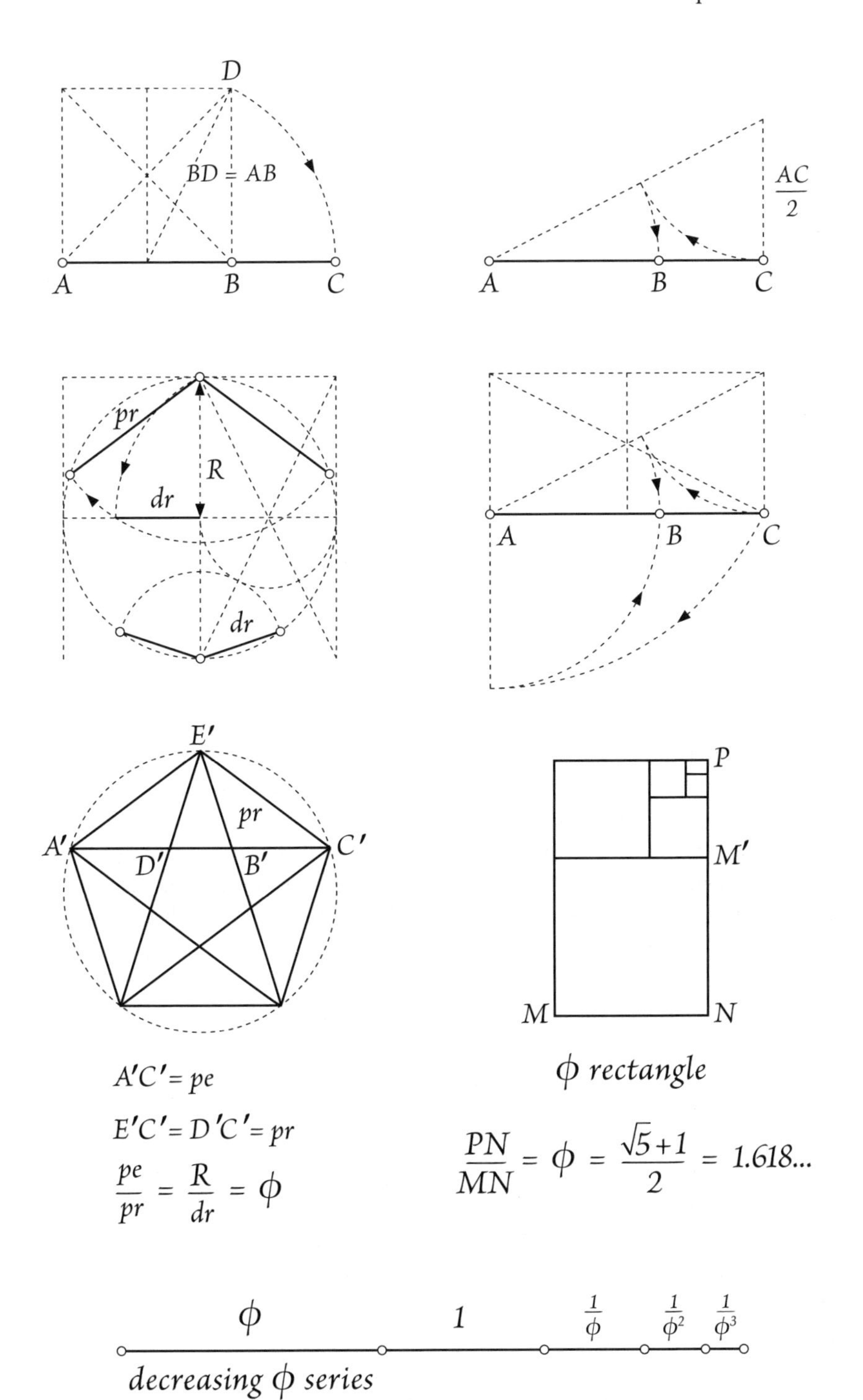

Plate 2. Golden section, pentagon, pentagram, phi rectangle

It is through the explicit intermediary of the golden section, whose rigorous construction has been public knowledge since Euclid, that Claudius Ptolemy in his *Almagest* was able to solve the graphical geometry problems of how to find the sides of the regular pentagon and decagon inscribed within a given circle.

As the regular dodecahedron and icosahedron (regular polyhedrons that, respectively, have 12 pentagonal faces, 30 edges, 20 vertices; and 20 triangular faces, 30 edges, and 12 vertices) are the spatial expansion of the regular pentagon,[3] it should come as no surprise to find the

---

(cont. from p. 35) inscribe the regular convex and star pentagons (with sides *pr* and *pe*) and the regular convex and star decagons (with sides *dr* and *de*), we get: $\frac{pe}{pr} = \frac{de}{R} = \frac{R}{dr} = \Phi$.

This Φ ratio dominates all the proportions of the figures obtained this way, and by drawing, from the starting point of an initial pentagon, an increasing or decreasing series of concentric pentagrams, we get a graphic representation of an endless linear Φ series.

We should recall here that the side of the regular hexagon inscribed within the circle is equal to radius R (this is the main reason for the preeminence of hexagonal symmetry in all examples of the isotropic, crystalline equipartition of space).

3. The 20 vertices of the dodecahedron are the vertices of 4 equal regular pentagons placed two by two in parallel planes, the ratio between the length of the sides of the large and small pentagons being equal to Φ, just like the ratios between the respective distances of the 4 planes. The 12 vertices of the icosahedron coincide with those of 3 rectangles of module Φ perpendicular to them. The dodecahedron and the icosahedron are "reciprocal" [or "dualing"] bodies (like the octahedron and the cube), because by connecting the centers of the faces of one of them we obtain the other, and so forth (see *Esthétique des proportions*). Here a very important fact from the point of view of Greek theories on "symmetry" and proportion comes into play: while in the plane the triangle, the square, and the pentagon were "irreducible" to each other morphologically, we can move in space from the dodecahedron (or icosahedron) to the cube, and from the cube to the tetrahedron. For example, the 12 vertices of the icosahedron (and 6 of its edges) are on the surface of a cube, the 8 vertices of this cube coincide with 8 of the vertices of a dodecahedron whose edges are equal to that of the icosahedron. The other 12 vertices of the dodecahedron and 6 of its edges are located on the surface of another concentric cube, which fully envelops it in such a way that the length of its edge and that of the first cube's edge should be in the Φ ratio. In similar fashion, the 6 edges of any tetrahedron can be set as diagonals on the 6 faces of a cube, and the 4 vertices of the tetrahedron coincide with 4 vertices of the cube (the 4 remaining vertices of the cube and the 6 other diagonals producing another tetrahedron).

This will allow us to appreciate the importance of a passage that will be cited several times later in this book, written by Campanus of Novara, on the role of the golden section as harmonic bond between the five Platonic solids.

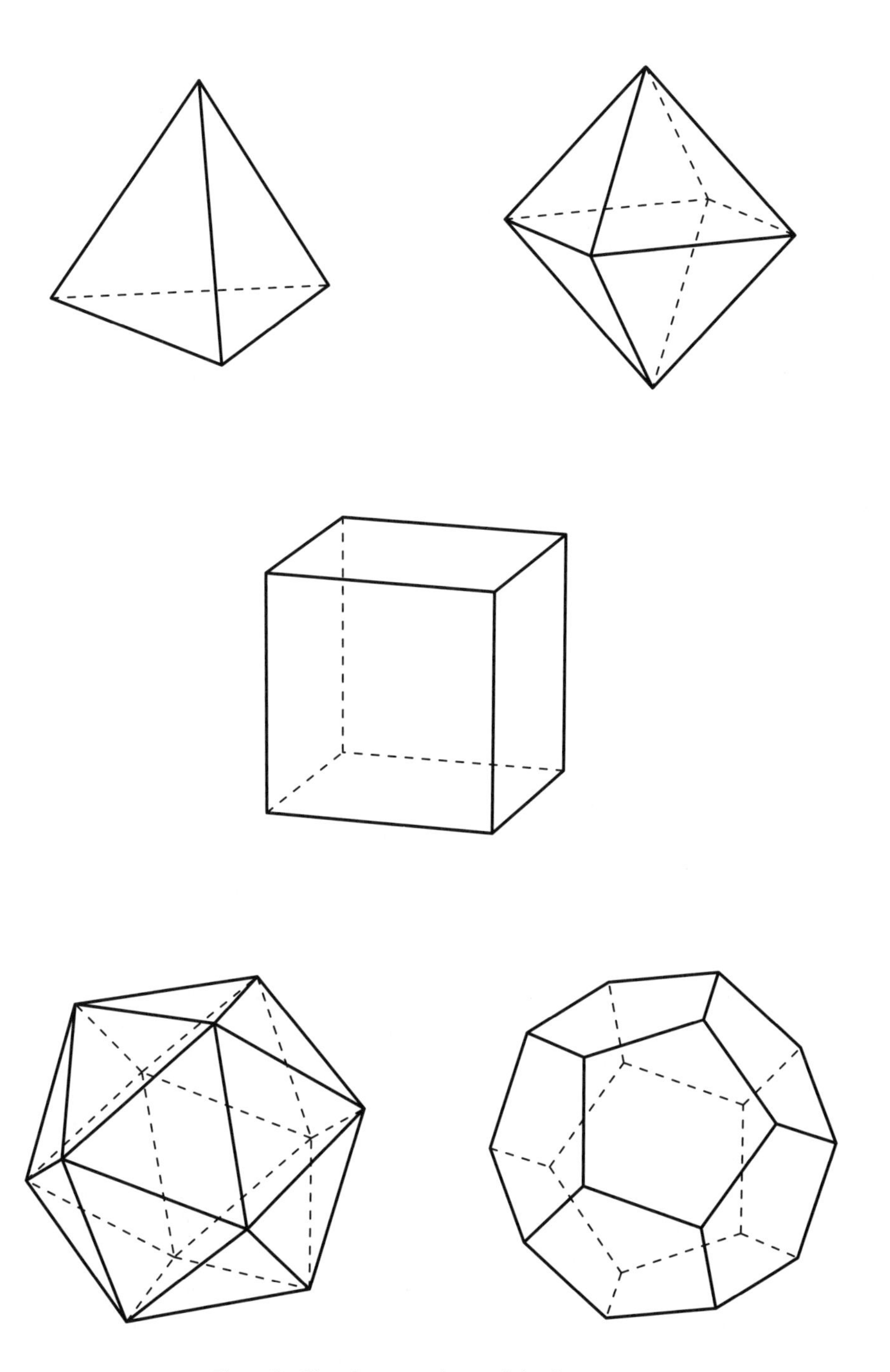

Plate 3. The five regular polyhedrons
(Platonic solids)

golden section as the essential ratio governing the linear proportions (planes or solids) inside each of these two solids as well as the proportions that connect the dodecahedron and icosahedron inscribed inside the same sphere or cube.

The same observation applies to the two star dodecahedrons obtained by lengthening the faces (or edges) of the dodecahedron or icosahedron,[4] and which in both cases constitute the expansion of the pentagram to three dimensions.

Every design, every projection representing these solids separately or combined will require the initial division of a segment according to the golden section. This was the particular case for the inscription of the

---

4. So while in the plane we have an infinite number of regular convex polygons and an infinite number of regular star polygons, in three-dimensional space we have only 5 regular convex bodies (the five Platonic polyhedrons: tetrahedron, cube, octahedron, dodecahedron, icosahedron) and two continuous regular star polyhedrons, which are precisely these two star dodecahedrons. (This is because Kepler's "stella octangula," an eight-point solid star produced by two distinct tetrahedrons cutting each other with an octahedron as their common nucleus, is only a "pseudo-star." Its correlation in the plane is the hexagram or "seal of Solomon," which is formed by two independent equilateral triangles placed head to tail on each other.)

The star dodecahedron of the first type (plate 5) is obtained by extending the faces (or the edges) of an icosahedron "nucleus"; the 20 points of the resulting solid star coincide with the vertices of an enveloping convex dodecahedron. The star dodecahedron of the second type (plate 6) is obtained by extending the faces (or edges) of a dodecahedron "nucleus." The 12 points of the resulting star coincide with the vertices of an icosahedron. It would therefore be tempting to call this second type a "star icosahedron," but the common name of star dodecahedron for both types is justified by the fact that each of them is clearly produced by the combination and fitting together of 12 plane faces (of the pentagrams) that intersect in space.

In these reciprocal and continuous generations by "budding" from a central nucleus, from the dodecahedron to the icosahedron, and from that to the dodecahedron, and so on, whose pulsing armature is produced by the alternating converging lines of the edges of the two star polyhedrons, and in which the growth of the radii, surfaces, and volumes is governed in geometric progression by the rhythm of the golden section, we have the ideal archetype of dynamic growth whose reflections, projections, and cross-sections we will find throughout this book. This is the reason I've provided illustrations here (plates 5 and 6). Plate 4, on the contrary, provides the "static" fitting together of the five Platonic solids inscribed within each other, and not connected by a polar projection or pulsation, but by the simple morphological correlations pointed out in the preceding note.

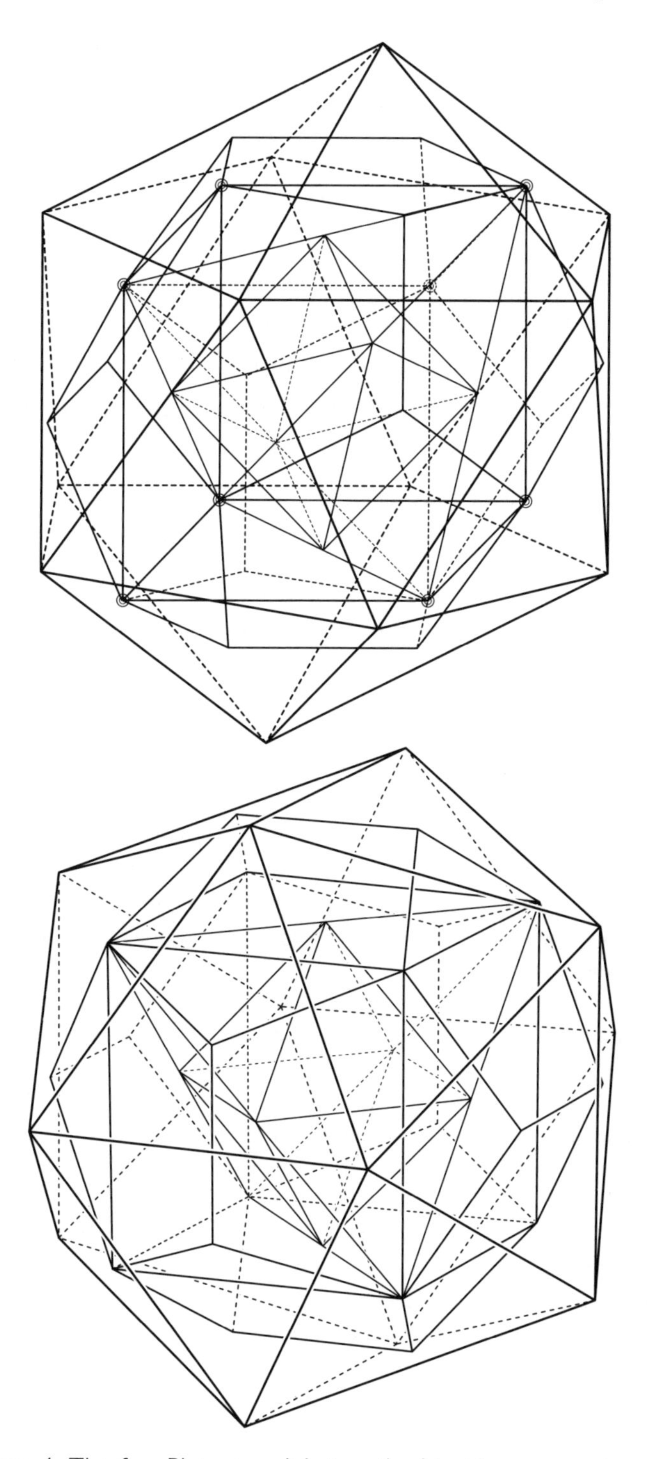

Plate 4. The five Platonic solids inscribed inside one another

dodecahedron in the sphere. It was for having revealed this "reserved" secret that Hippasus of Metapontum was excommunicated by the Pythagorean brotherhood.[5]

I showed in my *Esthétique des proportions* how the states of balance and configurations of all inorganic physicochemical material systems (which is to say, those that contain no living organisms) were strictly governed by the "principle of least action," or the Hamilton principle, which states that systems evolve from their least probable to most probable states (Boltzmann), and why, when the final states of balance produce regular geometric configurations (such as crystal formations), the symmetries created were always of a cubic type (square plane grids, cubic arrangements in space) or (even more often) hexagonal (triangular or hexagonal plane grids, cuboctahedral arrangements in space).[6] I also showed how, conversely, in the systems containing organic matter, containing life—which

5. Iamblichus, *Life of Pythagoras.* During this same period (450 BCE), Hippocrates of Chios was hit with a similar solemn expulsion, again for the revelation of a geometric secret concerning irrational proportions, and he has been credited with the invention of a special method for constructing a regular pentagon of a given side (cf. *Esthétique des proportions*), also based on the golden section. It is possible that it was precisely the disclosure of this "pentagram of Hippocrates" that was responsible for his excommunication, all the more as the pentagram, as we saw earlier, was the sign of secret recognition used by the Pythagoreans. Everything referring to its construction would have made it the mathematical secret "par excellence."

6. The tendency to level out, find balance, and establish an equipartition of energy leads to symmetry and to the equipartition of the plane or of space. The hexagonal and square symmetries are made necessary by the fact that the only regular polygons that can "fill" the plane (without interstices) are the square, the equilateral triangle, and the hexagon. The sole regular polyhedron that can fill space (through its repetition) is the cube. Two semi-regular polyhedrons also permit the equipartition of space: the hexagonal regular prism and Lord Kelvin's Archimedean polyhedron (8 hexagonal faces, 6 square faces, 24 vertices, 36 equal edges). Lastly, in the same way that the ideal isotropic system of points in the plane is provided by the centers of the "compact" arrangements of equal tangential circumferences (each circumference is tangential to the six that encircle it, the grid obtained this way being equivalent to the one that provided the vertices of the triangular equipartition, the vertices and the centers of the hexagonal equipartition of the plane), the grid of ideal isotopic points is provided by the "compact" arrangements of tangential equal spheres (each sphere is tangential to the twelve surrounding it, the 12 points of contact being the vertices of a cuboctahedron inscribed

can break from the principle of least action[7]—we often find forms based on pentagonal symmetry, that is to say, on the asymmetrical theme of the golden section (flowers, marine organisms, the human body).

The phenomenon that gives rise to the asymmetry here is the growth of living beings, a growth that works from the inside out, as if by "imbibition," a swelling, not by "agglutination" as is the case in crystals, and this living growth tends to produce successive "homothetic" forms, which is to say, they remain "self-similar." Here we find exhibited again the essential difference between the hexagonal symmetry that corresponds perfectly to inert equilibrium (whose ideal culmination is filling the plane or space, isotropy, static periodicity, juxtaposition of the same interchangeable motif, without any preferred direction), and the pentagonal symmetry that can come into play both in the plane (extending the lines of the pentagon produces pentagrams whose dimensions increase in geometric progression) and in space (generation, proliferation of alternating star polyhedrons out of a dodecahedron nucleus), a pulsation in geometric progression, a truly rhythmic dynamic periodicity, that

inside the central sphere). This isotropic system of points in space also falls under the heading of hexagonal symmetry of which it is a three-dimensional expansion. We should recall that the cuboctahedron (semi-regular Archimedean polyhedron with 12 vertices, 8 triangular faces, 6 square faces, 24 equal edges) has its edges equal to the radius of the circumscribed sphere. This polyhedron and the hexagonal regular prism are both the expansion in space of the plane hexagon.

7. This is the principle of least action in mathematical physics, or the Hamilton principle (also known as the principle of stationary action), a tendency to consume, to waste existing potential energies, and to level out, which strictly predetermines and governs the evolution of all "closed" physicochemical systems (which is to say, lifeless, as life acts in physics as an "external" force). Sometimes in this book there is mention made of a psychological law whose similar name can lead to confusion: this would be the hedonistic principle of the least effort, which instead of being a tendency to waste is a principle of economy with regard to mental or nervous energies. German aestheticians use it to explain the preference granted by the organism to certain colors and certain forms (the rectangle of the golden section, for example), and the "why" of the agreeable, harmonious corresponding sensation. Akin to this hedonistic principle of the least effort is the principle of the economy of concepts (*entia non sunt multiplicanda*), the Occam's razor of the English logicians, a tool of control, of "Taylorization," and reasoning, which permitted us to establish the golden section a priori (in chapter 1).

Plate 5. Star dodecahedron with twenty vertices
(obtained by two polar pulsations from a dodecahedron nucleus)

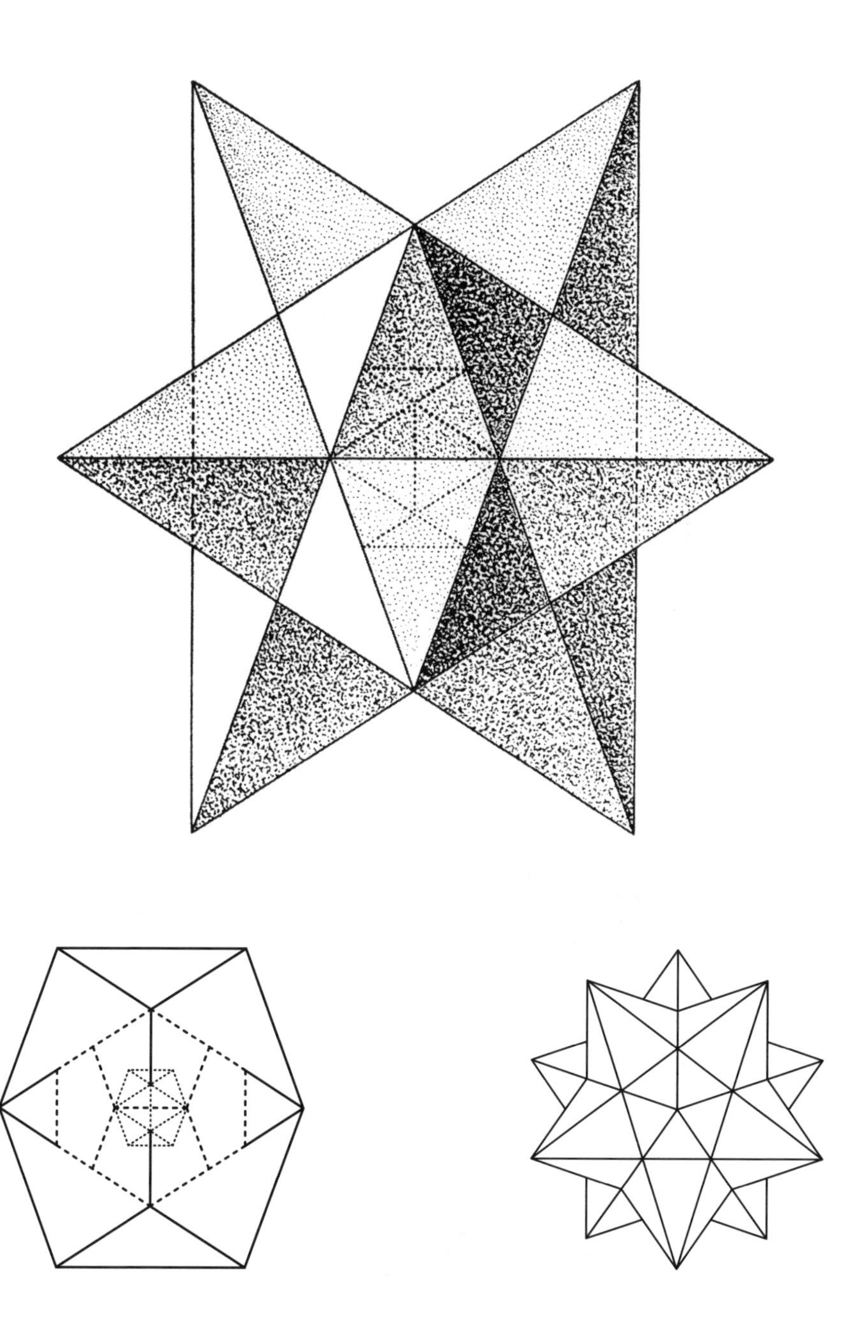

Plate 6. Star dodecahedron with twelve vertices
(obtained by two polar pulsations from an isocahedron nucleus)

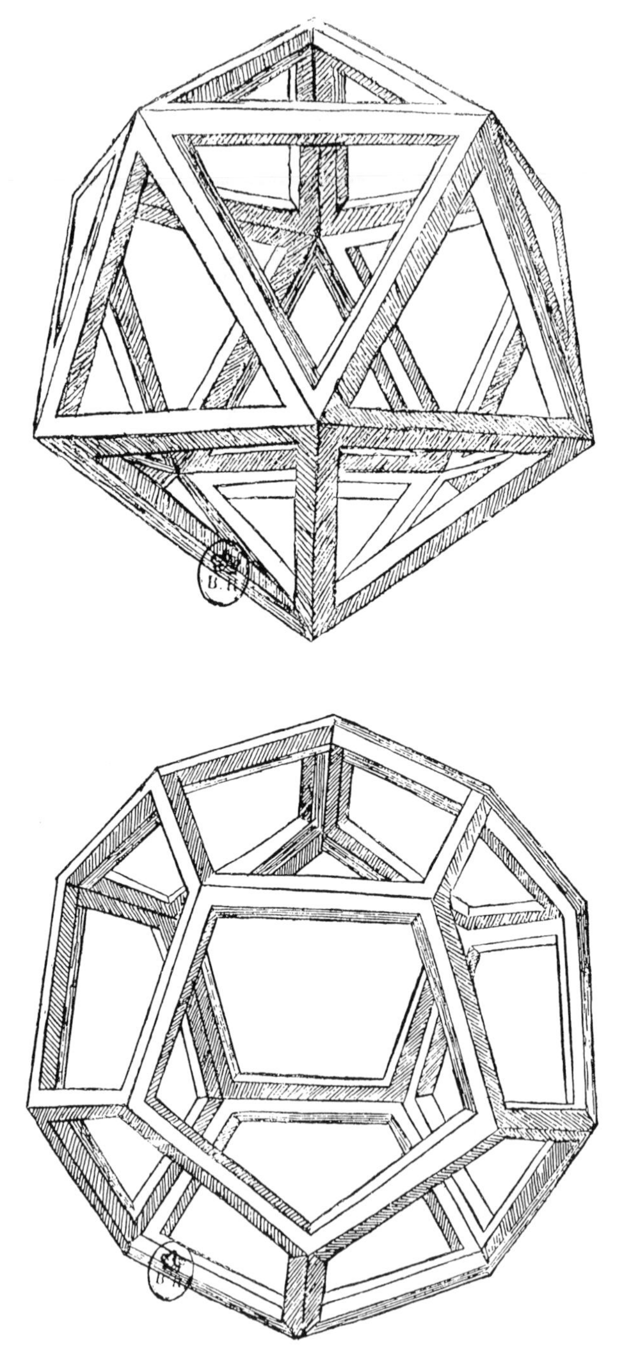

Plate 7. Icosahedron and dodecahedron drawn by Leonardo da Vinci for Fra Luca Pacioli's *De Divina Proportione*

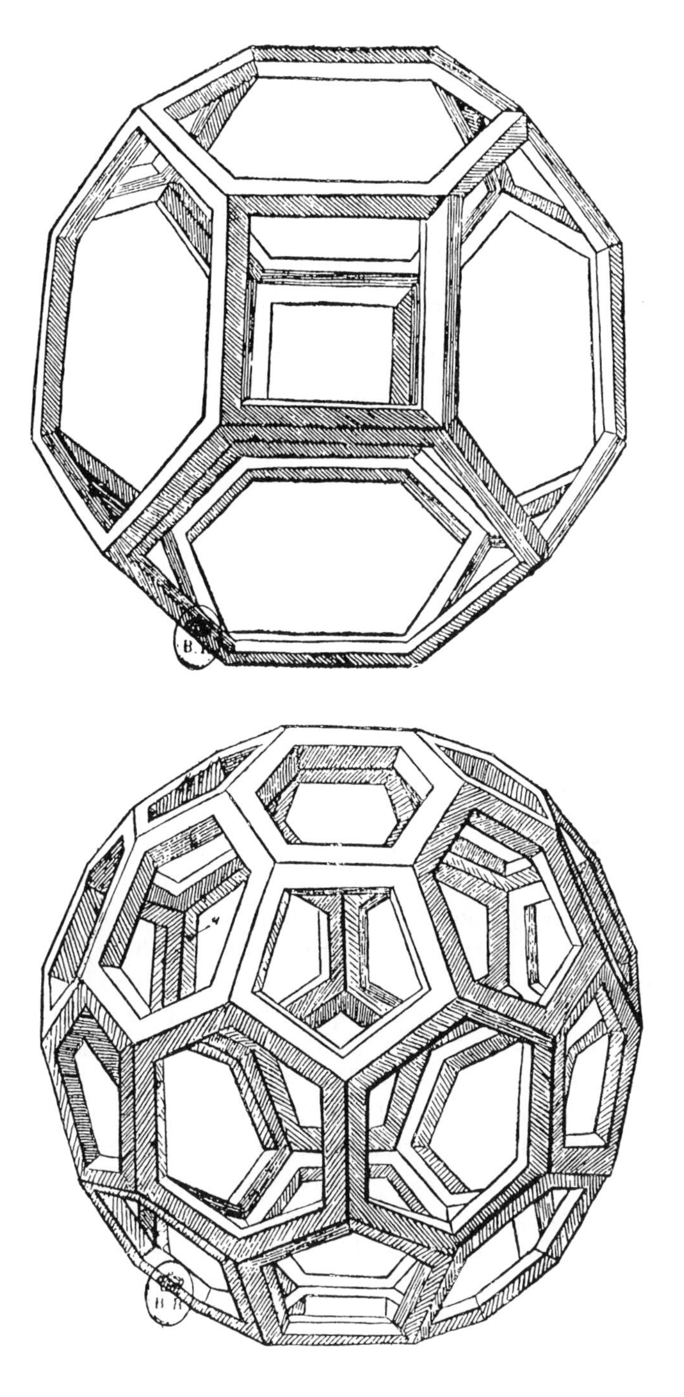

Plate 8. Semi-regular polyhedrons
drawn by Leonardo da Vinci

not only corresponds to growth of all kinds but to perfectly homothetic growth. This is because every pulsation in a geometric ratio can be envisioned as the schematic drawing of a logarithmic spiral, the ideal curve of "analogical" homothetic growth.[8]

This is all the more true as, among homothetic growth patterns, the one that resolves the problem of being both additive and geometric is governed by the spiral with a Φ quadrantal pulsation[9] and a directing rectangle with Φ as the module (because the Φ series is the only one that is an additive series at the same time it is a geometric progression, which is to say, a continuous sequence of proportions), and as this ratio is precisely the characteristic ratio of pentagonal growth and symmetries, we have an additional reason for the presence of pentagonal forms and symmetries in living organisms. We find in these organisms not only the elements of segments and surfaces that are proportional to the terms of the Φ series (as in the proportions of the human body), but

---

8. For more on the logarithmic spiral, discovered by Descartes, see my *Esthétique des proportions.* I will quote here several passages from the book by D'Arcy Wentworth Thompson, *On Growth and Form* (Cambridge: Cambridge University Press, 1917; reprint, New York: Dover, 1992), on the mathematical laws of the growth of living organisms: "And this remarkable property of increasing by terminal growth, but nevertheless retaining unchanged the form of the entire figure, is characteristic of the equiangular [or logarithmic] spiral, and of no other mathematical curve. . . . Any plane curve proceeding from a fixed point (or pole) and such that the vectorial area of any sector is always a gnomon to the whole preceding figure, is called an equiangular or logarithmic spiral" (758, 763).

(Thompson here reintroduced the word *gnomon* with the meaning it held in the Pythagorean theory of figurate numbers. This is a number or figure that when added to a given number or figure does not change its "form." The figurate numbers of a same series are all similar, and the successive differences are gnomons.) "If a growing structure be built up of successive parts, similar . . . and similarly situated, we can always trace through corresponding points a series of equiangular spirals" (763).

9. The quadrantal pulsation of a logarithmic spiral is the constant ratio between the lengths of two consecutive perpendicular radii; this is sufficient to define the spiral. By extending one of these radii from the other side of the pole until it meets the same volute of the spiral, we obtain a third point on the curve that gives us the third vertex of the main rectangle, the one for which the module (ratio between its larger and smaller sides) is equal to the quadrantal pulsation. Similarly, every rectangle has a corresponding logarithmic spiral passing through three of its vertices (plate 10).

also especially in botany (for example, in phyllotaxis, which studies the arrangement of branches, leaves, seeds), the numbers of the Fibonacci series, 1, 1, 2, 3, 5, 8, 13, 21, 34, 55, 89, 144, ... , which is an approximation of the whole-number terms of the Φ series.[10] This is a good time to note that this Fibonacci series corresponds to the tenth and last type of the proportions listed by Nicomachus of Gerasa (see chapter 1), the one whose equation (if *a, b, c* are three magnitudes obeying this proportion) is $\frac{c-a}{c-b} = \frac{b}{a}$. In fact, (by multiplying and subtracting) we get $c = a + b$, which gives (by starting from $a = 1$) the Fibonacci series.[11]

We have seen in chapter 1 that the last four kinds of proportion, thus this one (the tenth) in particular that we just examined, were discovered or made public by the neo-Pythagoreans of the Alexandrian school. I have also reminded the reader of the major place that the geometric outlines and numbers connected to the golden section (pentad, decad, pentagram, dodecahedron) held in Pythagorean mysticism. The role of the golden section in the proportions of the constructions or projections related to the inscription of the five Platonic solids in the sphere was not forgotten in the Middle Ages, as we know from a phrase by Campanus (thirteenth century) that was used anew by Luca Pacioli di Borgo, author of the treatise *De Divina Proportione* (1509) illustrated by Leonardo da Vinci.

10. Especially the series of pairs (ratios):

$$\frac{1}{2}, \frac{1}{3}, \frac{2}{5}, \frac{3}{8}, \frac{5}{13}, \frac{8}{21}, \frac{13}{34}, \ldots \text{ (approaching } \frac{1}{\phi^2}\text{) and}$$

$$\frac{1}{1}, \frac{1}{2}, \frac{2}{3}, \frac{3}{5}, \frac{5}{8}, \frac{8}{13}, \frac{13}{21}, \frac{21}{34}, \frac{34}{55}, \frac{55}{89}, \frac{89}{144}, \ldots \text{ (approaching } \frac{1}{\phi}\text{).}$$

Kepler (*De Nive Sexanbula*) also noted the kinship between the Fibonacci series and the golden section, as well as their presence in botany.

This role played by the Fibonacci series in botany was explained by the work of Braun, Church, and Bravais. In 1875, Professor Wiener found that the angle 137°30'28" recurred often in phyllotaxis in the constant angular separation (helicoidally) of branches and stems, which satisfied the equation $\frac{a}{360° - a} = \frac{360° - a}{360°}$ or $a = \frac{360°}{\Phi^2}$, and corresponds to the rigorous mathematical solution of the problem of optimal exposure (maximum in temperate climates) of leaves to vertical (or axial) light. He called this angle $a = \frac{\pi}{\Phi^2}$ the "ideal angle."

11. If one does not posit the condition that the terms *a, b, c, ...* , be whole numbers, and if one adds that they form a geometric progression $\frac{b}{a} = \frac{c}{b}$, we are brought back to the Φ series. We can say that the Φ series is the continuous algebraic *archetype* of the F series or the (discontinuous, in whole numbers) Fibonacci series.

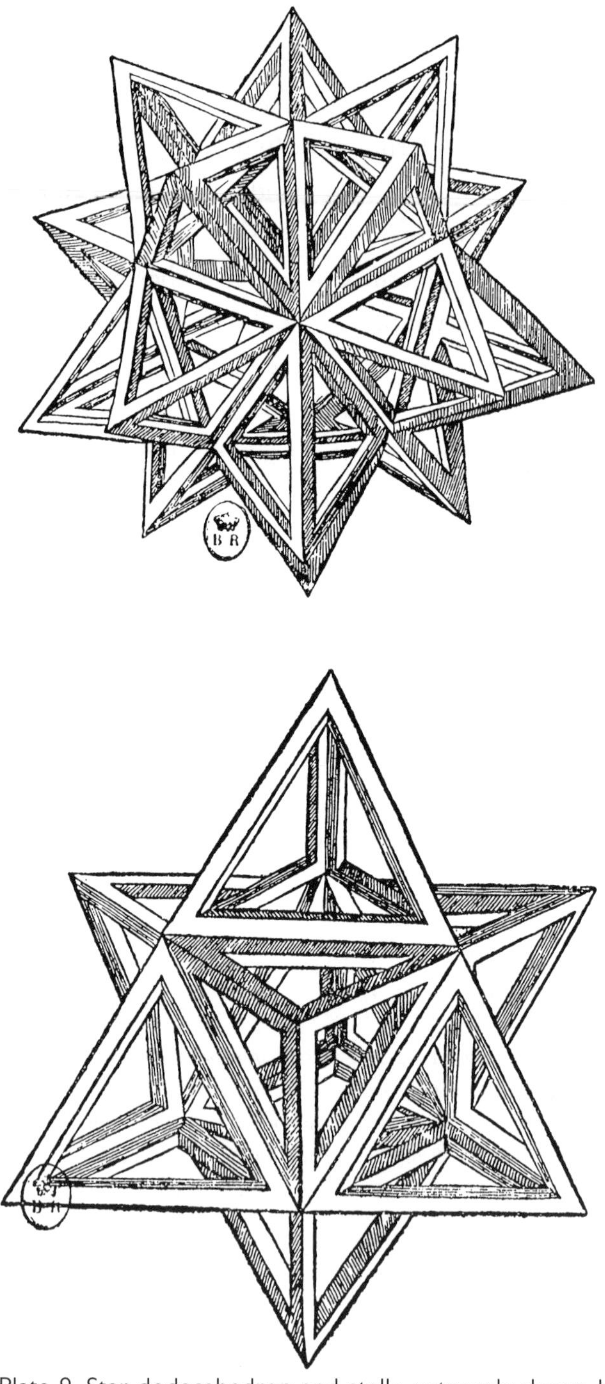

Plate 9. Star dodecahedron and stella octangula drawn by Leonardo da Vinci

The divine proportion or golden section was mentioned again by Kepler as one of the two "jewels of geometry,"[12] the other being Pythagoras's theorem on the square of the hypotenuse (which made it possible using a rope or chain divided into twelve equal parts to draw a right angle on the ground). It was then completely forgotten until the moment it was rediscovered and highlighted as a governing morphological principle by the German Zeysing (circa 1850). In chapter 3, we shall see the preponderant role it played in Egyptian, Greek, and Gothic architectural designs. It is not only the consequence of these architectural plans of decagons and pentagons inscribed inside the circles of orientation, or even the conscious use of volumes and proportions resulting from the inscription of the icosahedron or dodecahedron in the sphere, but also of the fact that during the great epoch of Greek architecture the human body was considered to be the most perfect example of symmetry and eurhythmy, thus serving the architect as an essential source of inspiration, if not more precisely a model, for the composition of his blueprints.

This point of view, moreover, is no more than the transposition into the domain of geometric form of the concept of the correspondences between the macrocosm (the universe) and the microcosm (the human being), of which the *Timaeus* gives us a metaphysical version (with even a triple play of correspondences between the human body, the human soul, and the world soul). The correspondence between the form of the temple and the universe is one that had already been mentioned in Egypt,[13] but the idea of realizing this goal, by taking as an intermediary

12. In his *Mysterium Cosmographicum de Admirabili Proportione Orbium Caelestium*, published in 1596.

Vitruvius, whose work contains no personal innovations, but is a presentation of the already five-centuries-old tradition of Greek architecture, emphasized this at length. When he discussed columns, he compared the proportions of the Doric column (a module of $\frac{6}{1}$ between the mean diameter and height) to that of the male body, while those of the Ionic column (module of $\frac{8}{1}$) evoked the more graceful female body, and those of the Corinthian column were comparable to the more willowy bodies of virgins.

13. In an inscription seen at the Cairo Museum by Princess Bibescu that I cite as the epigraph for chapter 3.

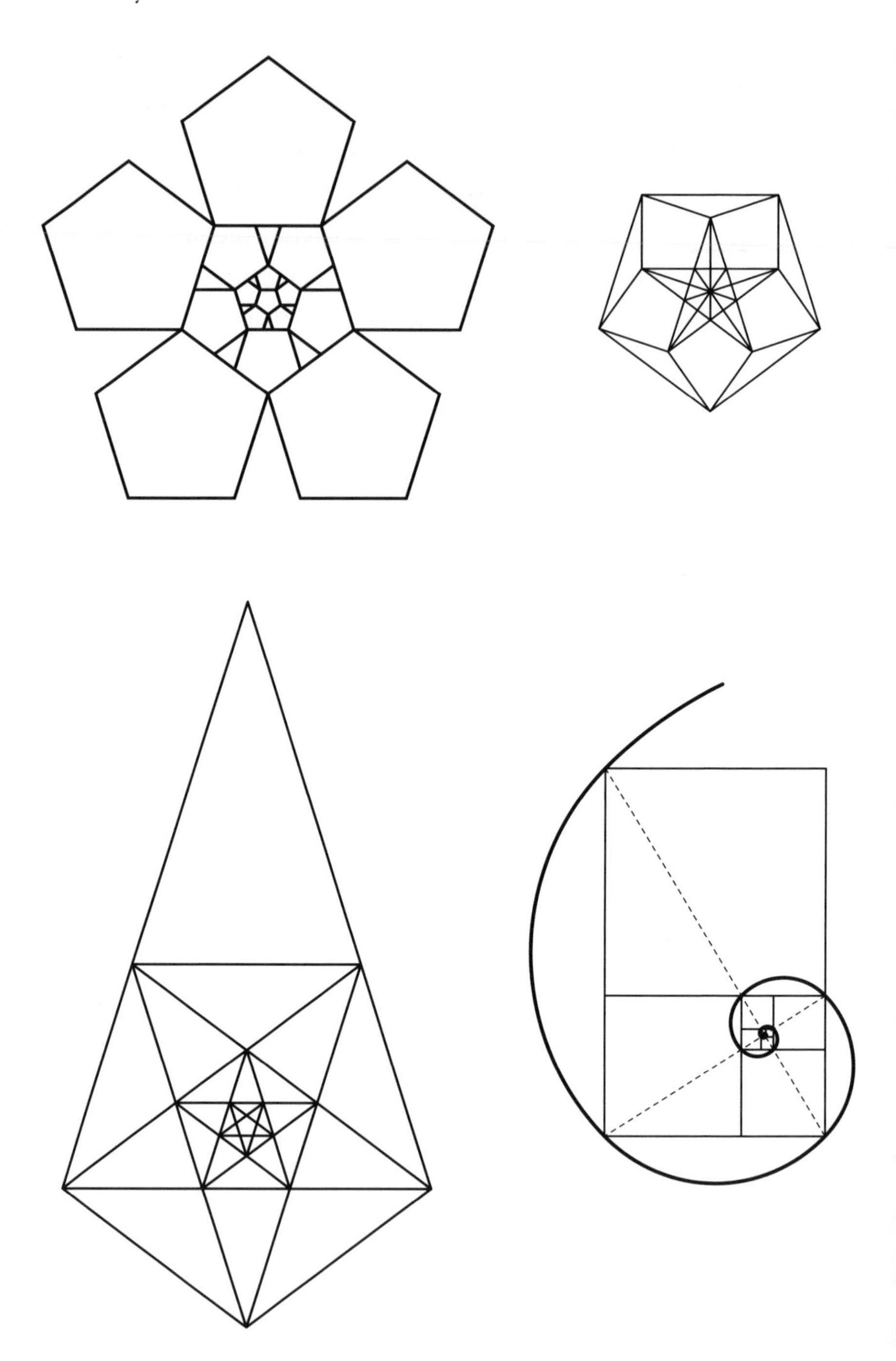

Plate 10. Pentagonal symmetries:
Φ rectangle and spiral of harmonious growth

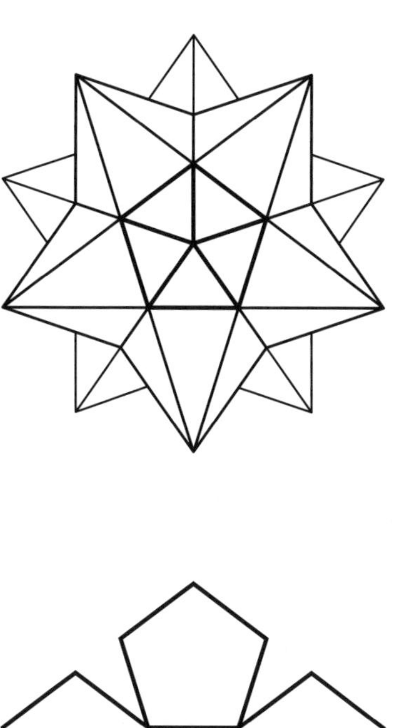

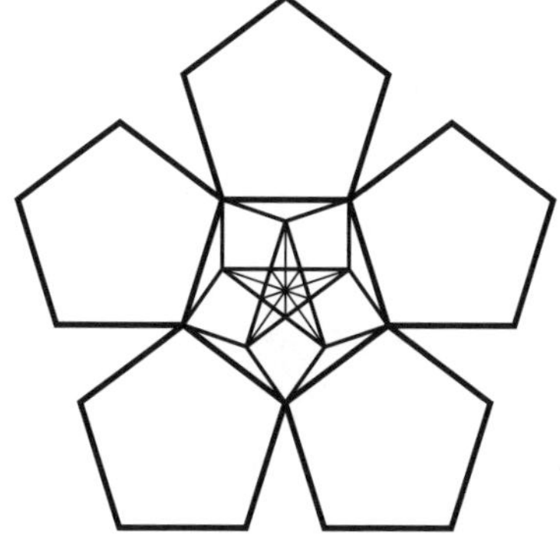

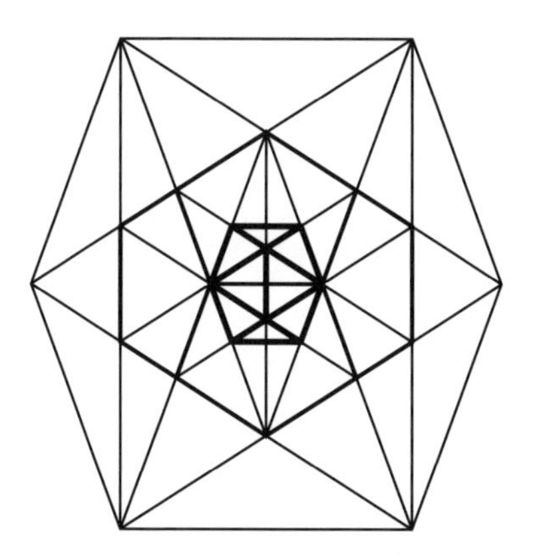

Plate 11. Three growth patterns governed by the golden section

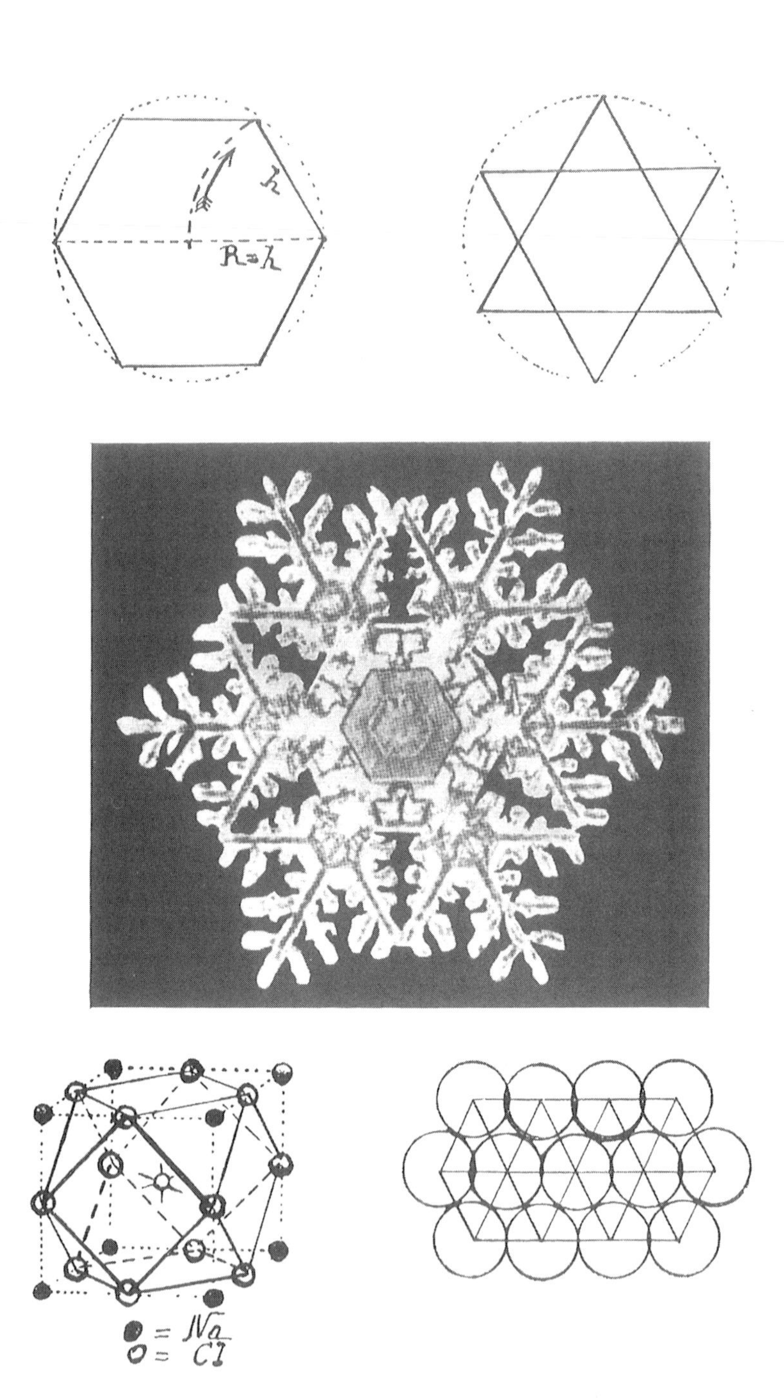

Plate 12. Hexagonal symmetries (crystals); in the center, a snowflake (photo P. La Cour)

model not the human form but the very subtle play of harmonies and proportions that can be seen there, appears specifically Greek.

We have seen that the Pythagoreans chose the pentagram, a symbol of living harmony and health,[14] as their rallying and recognition sign. We find it again among the Kabbalists, magicians, and alchemists of the Middle Ages and Renaissance as a symbol of the "microcosm," that is, man both physical and astral—to use the modern term that translates fairly well the ideas of occultists of all eras on the intermediary fluidic aura that they claim lies between the spiritual essence, the "nous," and the body. The best known of these representations of the man-microcosm, with arms and legs spread wide in such a way as to depict with the top of the head the five points of the pentagram, is that of Agrippa von Nettesheim in his treatise *De Occulta Philosophia* (see plate 17).

Furthermore, in the thirteenth century Villard de Honnecourt, in the famous sketch album housed at the Bibliothèque Nationale, sometimes used the pentagram as the governing outline for the head and human body as well as for leaves, as Leonardo da Vinci would later do in diagrams of flowers.

The ideas of the ancients and their spiritual heirs on the subject of these correlations between the human body and the pentagram would be curiously confirmed, in a dynamic rather than static fashion, by the experiments of Rudolf von Laban, director of one of Germany's most famous rhythmic choreography institutes. Von Laban observed that all the movements of the dancer (in the three dimensions) gave the extreme angular movements of 72°, and that the various directions in the space corresponding to these movements could be expressed by the radii of a circumscribed icosahedron ($72° = \frac{360°}{5}$ is the angle at the center of the pentagon).

Von Laban now uses this directing icosahedron in his courses (cf. his book *Die Welt des Tänzers*) (see plate 17).

---

14. I mentioned earlier that Lucian's well-known passage on this subject was corroborated by the Ptolemaic alabaster mold in the Basel Museum pointed out and reproduced by Perdrizet.

Plate 13. Pentamerous flowers
(photo E. Wasmuth, Berlin)

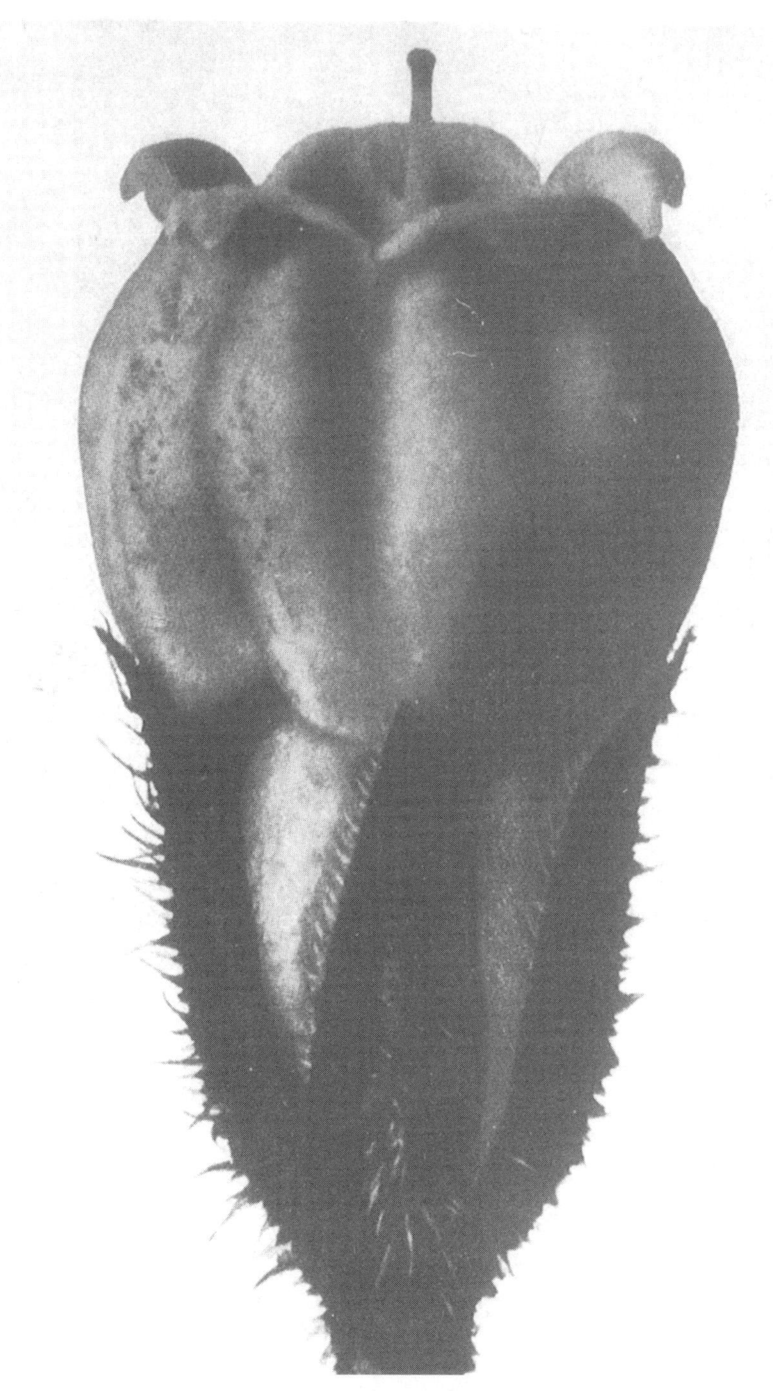

Plate 14. Pentamerous calyx (*Symphytum officinale*)
(photo E. Wasmuth, Berlin)

Plate 15. *Cardium pseudolima* and *Solarium perspectivum* (photo Wendington)

I would like to refer back to my *Esthétique des proportions* and specifically to plates 19 and 22, which concern the rigorous establishment of a canon of human proportions governed by the golden section and its related themes in accordance with the ideas of Zeysing, Sir Theodore Cook, and Hambidge. The conclusion that Hambidge came to was that there was an individual theme specific to each skeleton, obtained by analyzing not only its linear proportions but the surfaces (of the three principle projections) broken down into harmonious rectangles using the procedure that served him well in his harmonic analyses of Greek temples and vases[15]—a conclusion I feel worth keeping. It clearly emerges from several passages in the work of Vitruvius[16] that Greek sculptors and painters had studied the question of the proportions of the human body with all possible care and that they were no more satisfied than the architects with an arithmetical canon, a numerical grading scale, but instead applied what Vitruvius calls, in opposition to the "arithmetic" or static "symmetry" obtained by a simple scale of whole coefficients or fractions, "geometric symmetry." This geometric symmetry, in other words, is the establishment of proportions by means of a graphic method, surfaces

---

15. In the next chapter the reader will find a brief reminder of Hambidge's procedures. What is interesting in his method is that it does not set up a rigorous single canon directly based on the golden section, but it offers an infinite number of themes related to this proportion. An ideal canon clearly explaining the golden section appears, conversely, as the average resulting from a large number of observations; for example, the observation first stated in our era by Zeysing, that the navel divides the (adult) human body in accordance with the ratio $\Phi = 1.618$ ("Fibonaccian" approximations, framing Φ, $\frac{8}{5} = 1.6$ and $\frac{13}{8} = 1.625$) is exact as the resulting statistical average. The search for an aesthetic canon of the proportions of the human body is more than ever on the agenda. I can cite the interesting studies in Italy by Professor Paolo Cipriani of Rovigo, *Geometria del corpo umano* (Rovigo, 1928). Professor Umbdenstock of the École Polytechnique and the École des Beaux-Arts of Paris and R. Musmeci-Ignis (Rome) have announced works on the same subject.

16. *"Reliqua quoque membra suos habent commensus proportionis, quibus etiam antique pictores et statuarii nobiles usi magnas et infinitas laudes sunt assecuti. Similiter vero sacrarum aedium membra ad universam totius magnitudinis summam ex partibus singulis convenientissimum debent habere commensuum responsum. Item corporis centrum medium naturaliter est umbilicus."* The navel was already designated as the center of symmetry.

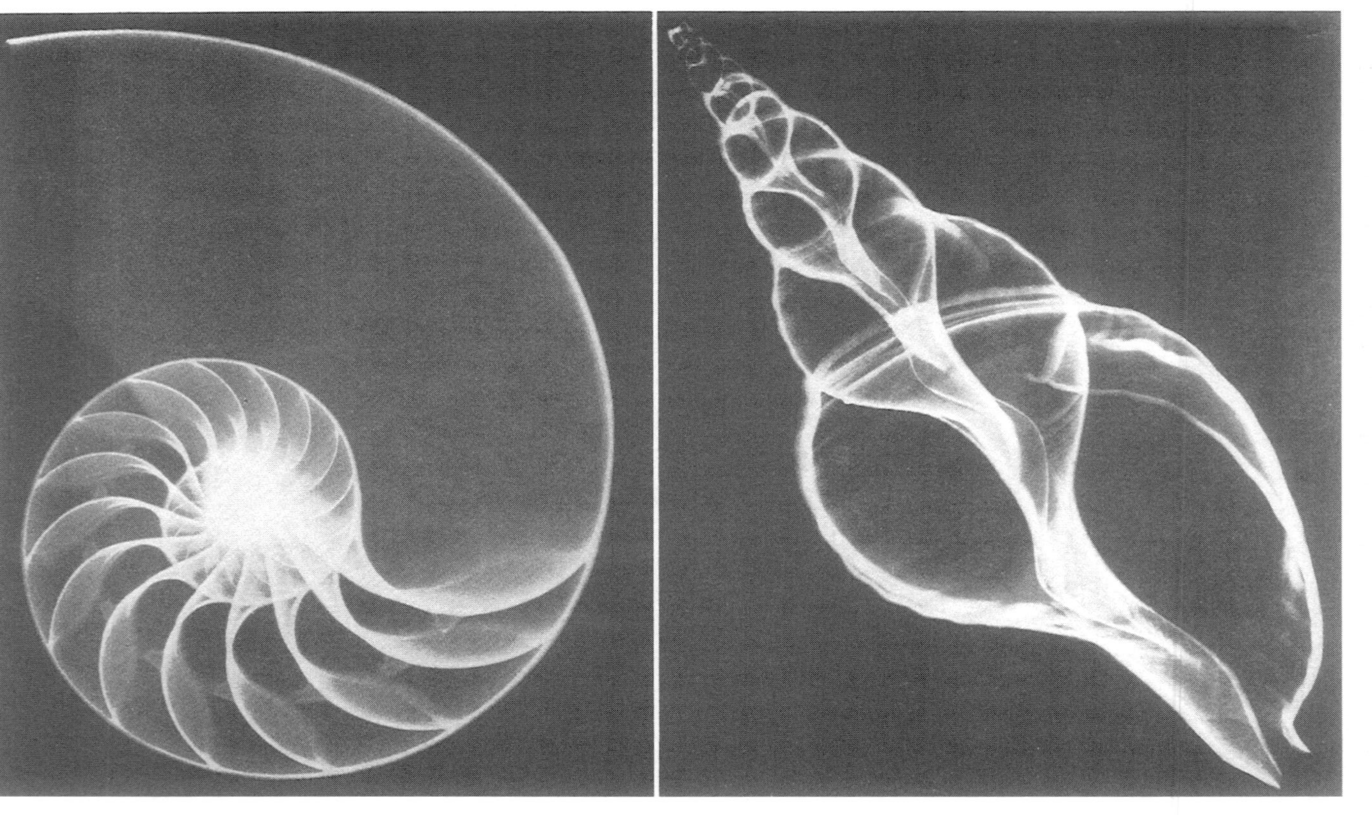

Plate 16. *Nautilus pompilius* and *Triton tritonis* (X-rays)

whose linear dimensions can present irrational ratios that in accordance with Plato's expression in the *Theaetetus* are "potentially commensurable." It is Hambidge who deserves the credit for finding an entirely satisfying key to Plato's "potential commensurability" (dynamic symmetry) through identifying it with Vitruvius's "geometric symmetry." The word *geometric* here has the same acceptance as it does in the expression *geometric proportion* or *geometric mean*. It concerns the continuous geometric proportion or analogy and "irrational proportions" (like the golden section) applied to surfaces and volumes. This graphic treatment precisely forms one of the mathematical secrets of the Pythagoreans, a secret partially divulged by Hippocrates of Chios (with his pentagram that we looked at earlier), and then by Theaetetus, Eudoxus, and Plato.

In conclusion, it is permissible to believe that Greek painters and sculptors had established for the proportions of the human body:

1. A practical "arithmetic" canon with whole and fractional coefficients, whose elements can be found in Vitruvius and which has been handed down to us by Pacioli, Leonardo da Vinci, and the painter-geometers of the early Rennaisance;[17]
2. An ideal "geometric" canon based on the golden section, like the reformulated one based on the *Doryphorus* of Polykleitos;[18]

17. The ratio $\frac{8}{5}$, or $\frac{10}{6} = \frac{5}{3}$, between the total height and the distance of the navel from the ground; $\frac{1}{8}$ as dimension of the head (in proportion to total height); $\frac{1}{10}$ for the face (from the chin to the hairline); $\frac{1}{6}$ for the foot, $\frac{1}{10}$ for the hand, and so forth can be found in Vitruvius.

18. Cf. *Esthétique des proportions,* plate 9 of the "ideal" woman's body as proposed by Sir Theodore Cook (*The Curves of Life*). In this ideal canon the height of the navel divides the total height exactly in accordance with the golden section. The same proportion is determined (but with the lower number on the bottom) by the level of the middle fingertips when the arms are hanging in a vertical position.

We even have the Φ ratio between the height of the face (up to the hairline) and the vertical distance of the arch of the eyebrow from the chin, and between the distance from the base of the nose to the chin and the distance of the corner of the lower lip to the chin.

The three phalanges of the middle or ring finger give three consecutive terms of a

3. A graphic method making it possible to adjust the variants of the ideal canon, which probably used procedures that were identical or analogous to those of Hambidge for the harmonic composition or decomposition of surfaces and volumes.

I will permit myself to offer some verifications of the ideal canon here.

The profile in plate 21 is that of Isabella d'Este, drawn by Leonardo at the time when his friend Luca Pacioli, the monk "drunk on beauty," was giving conferences at the court of Ludovico il Moro, Duke of Milan (and brother-in-law of Isabella), on the "divine proportion" illustrated by the magnificent sketches of Leonardo. This brings to mind this phrase by Paul Valéry: "He worships the human body, male and female, which measures and is measured by all things. . . . And the face, that enlightening and enlightened thing—of all things the most visible, the most magnetic, the most difficult to regard without studying it—the face haunts him."[19]

The photograph in plate 18 of Miss Helen Wills (Mrs. F. Moody) possesses the rare property of not only providing a theme "related" to the golden section but also an "ideal" model of the canon that it strictly follows (plates 19 and 20).

It is not hard to find again in the living "microcosms" of plates 22 and 23, in the features of the Olympic tennis champion there depicted, the Platonic symphonies resulting from the inscription in the sphere of

(cont. from p. 59) Φ series. We should recall here that a characteristic property of the Φ series (increasing, for example) is that each term is equal to the sum of the two preceding terms (because $\Phi^2 = \Phi + 1$). This "additive" property introduces in all patterns with a Φ rhythm, in addition to subtle asymmetrical proportions, simple whole-number ratios and even symmetrical division into two equal parts. It is this unique property that thereby makes it immediately possible to pictorially construct a Φ series as extensive as one could wish through elementary compass manipulations when starting from two elements in Φ ratio. Mario Meunier tells me that Plutarch mentions the famous canon of Polykleitos as being established in accordance with Pythagorean ideas.

19. Paul Valéry, *Introduction to the Method of Leonardo da Vinci*, in *Paul Valéry: An Anthology*, ed. James Lawlor (Princeton: Princeton University Press, 1956), 66–67.

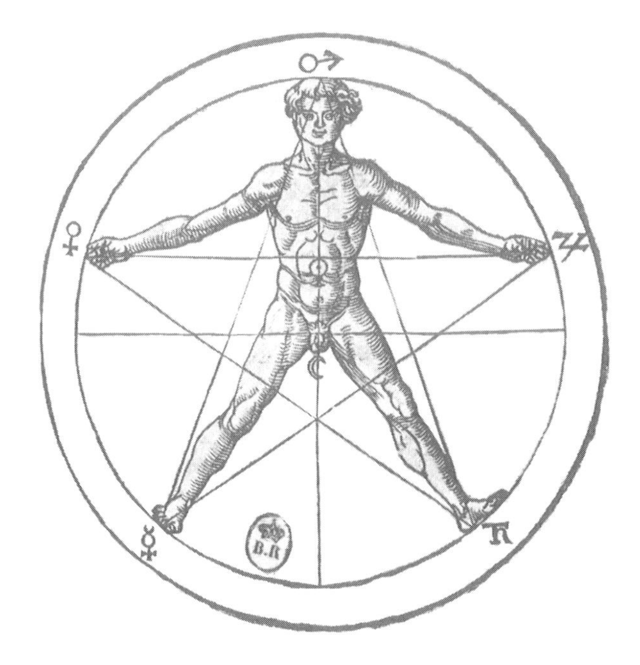

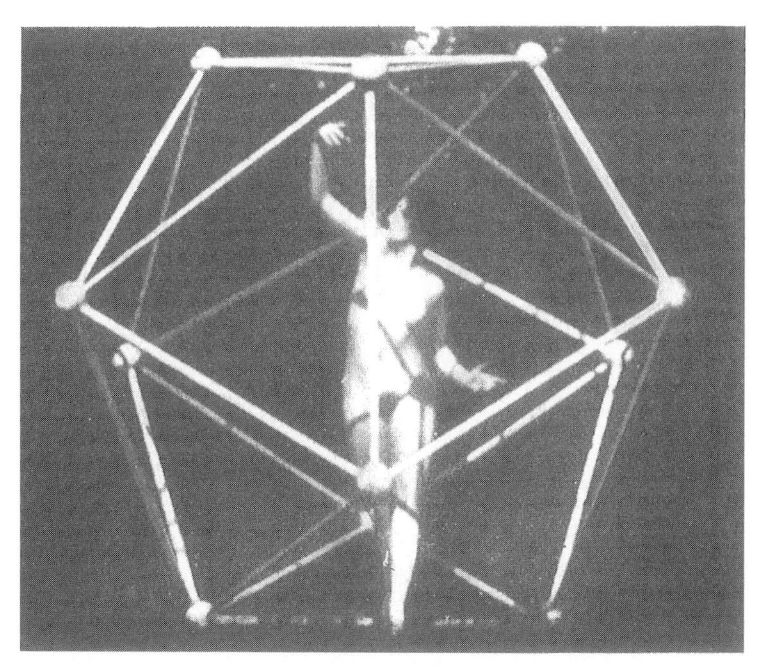

Plate 17. *Top,* the man-microcosm according to Agrippa von Nettesheim; *bottom,* the directing icosahedron in the choreography manual by R. von Laban, *Choreographie* (Jena: Diedrichs, 1926)

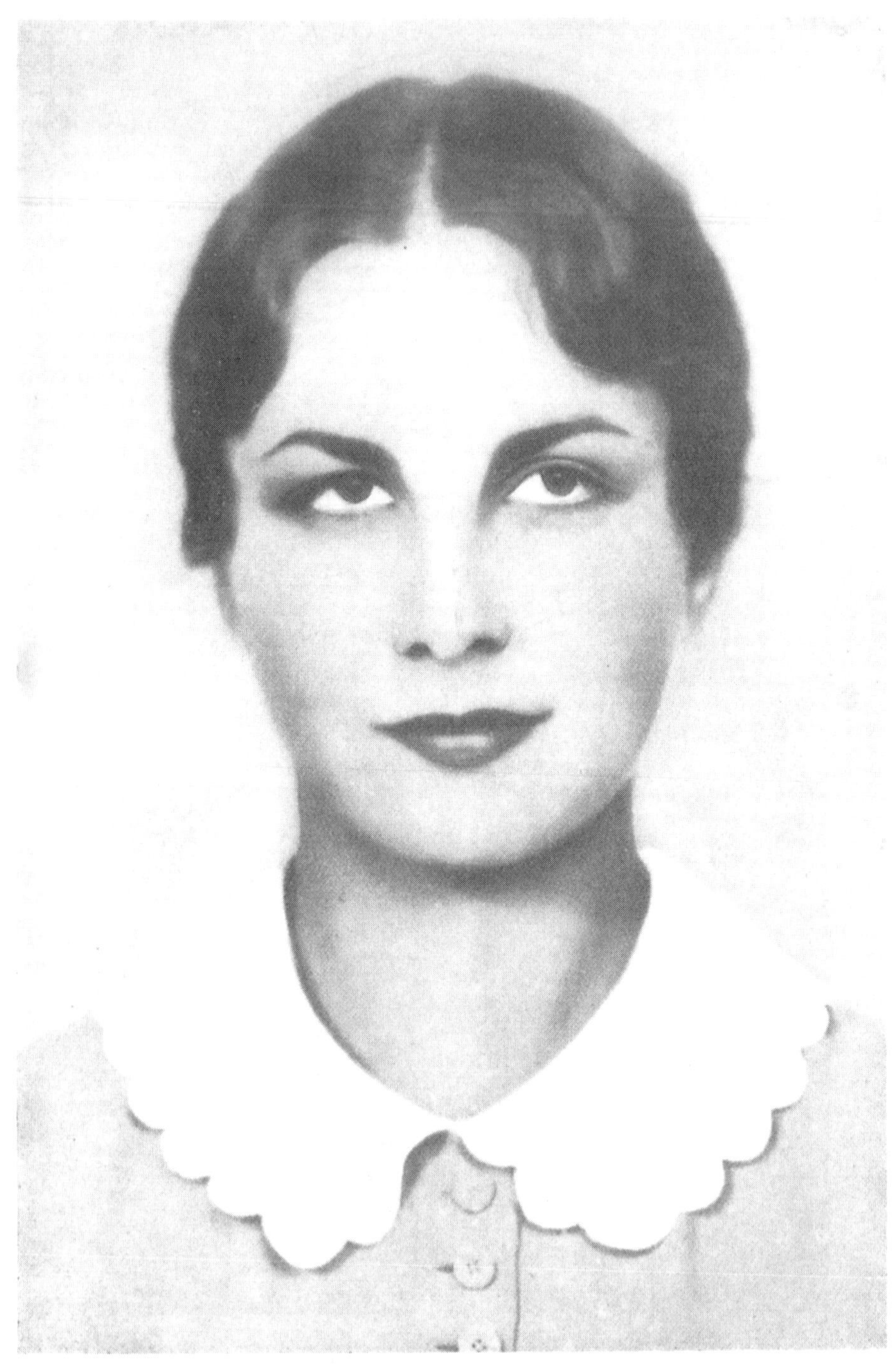

Plate 18. Miss Helen Wills (Mrs. F. Moody)
(photo Dorothy Wilding, London)

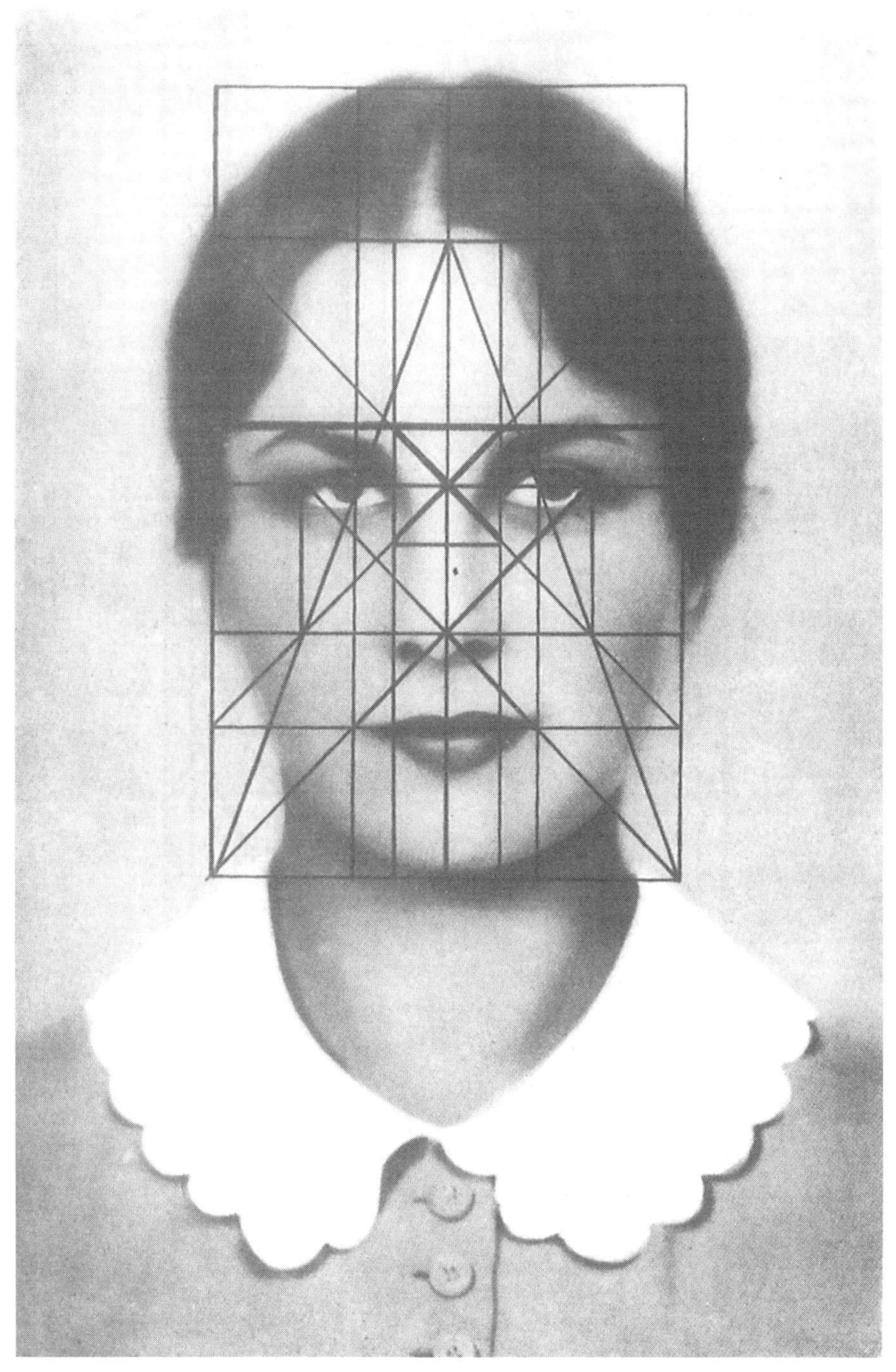

Plate 19. Harmonic analysis of the previous photograph (strictly applying the golden section)

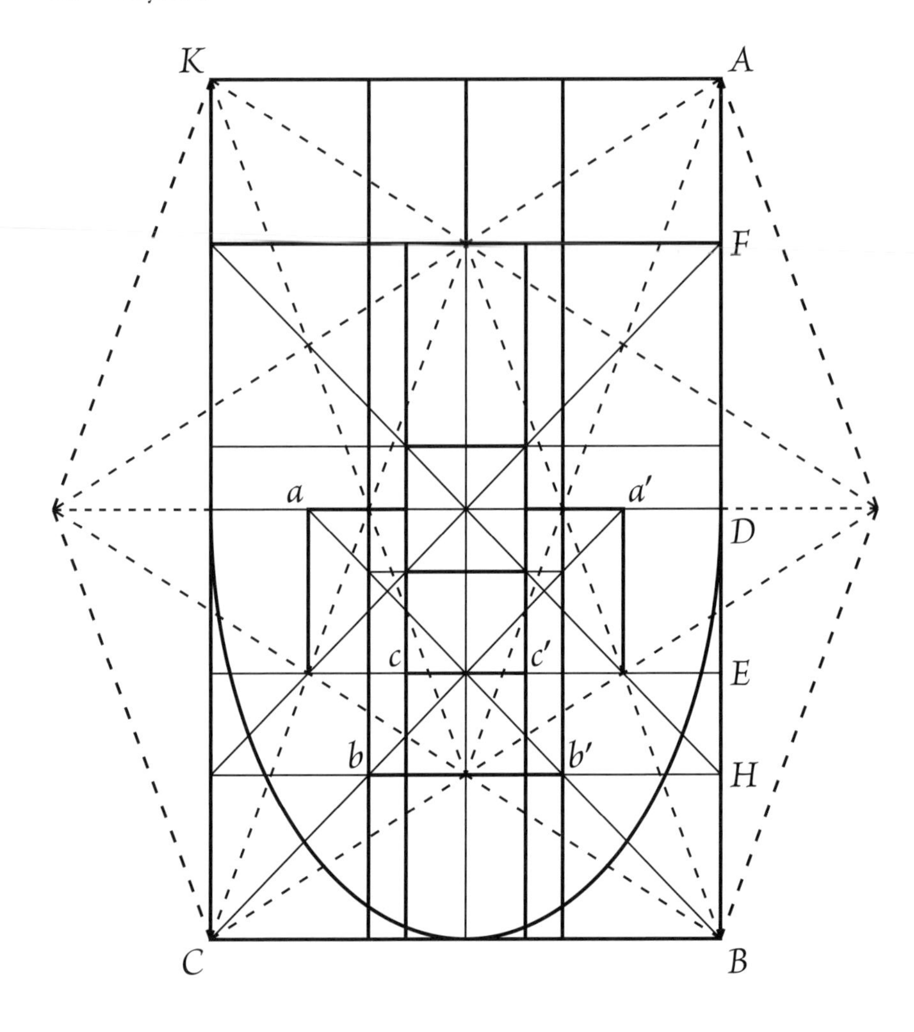

$$\frac{AB}{BC} = \frac{AD}{FD} = \frac{DB}{EB} = \phi = \frac{\sqrt{5}+1}{2}$$

$$\frac{FD}{DE} = \frac{DH}{DE} = \frac{EB}{HB} = \phi$$

$$\frac{CB}{a\,a'} = \frac{a\,a'}{b\,b'} = \frac{b\,b'}{c\,c'} = \phi$$

Plate 20. Explanation of the diagram in plate 19 (Helen Wills)

regular polyhedrons, and from the pulsing, alternating gemmation of the star polyhedrons derived from a dodecahedron (nucleus or enveloping), the dodecahedron of *Timaeus,* the geometric paradigm of the harmony of the cosmos.

We will find these patterns again in the designs of the temples that we will analyze in chapter 3, thereby illustrating the words that the author of *Eupalinos* put in the mouth of the ancient architect: "There where the passer-by sees but an elegant chapel . . . there have I enshrined the memory of a bright day in my life. O sweet metamorphosis! This delicate temple, none knows it, is a mathematical image of a girl from Corinth. . . . It reproduces faithfully the proportions that were particularly hers."

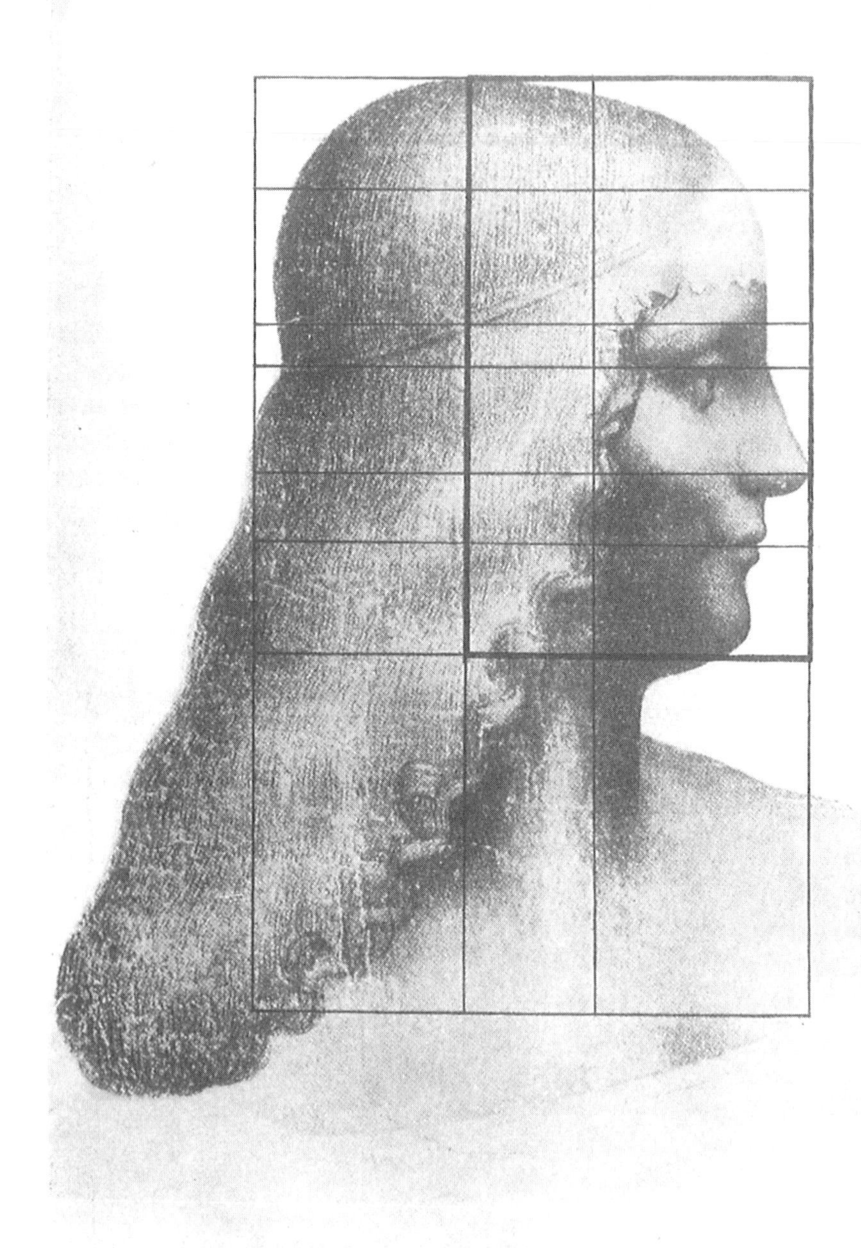

Plate 21. Isabella d'Este, by Leonardo da Vinci
(photo Alinari)

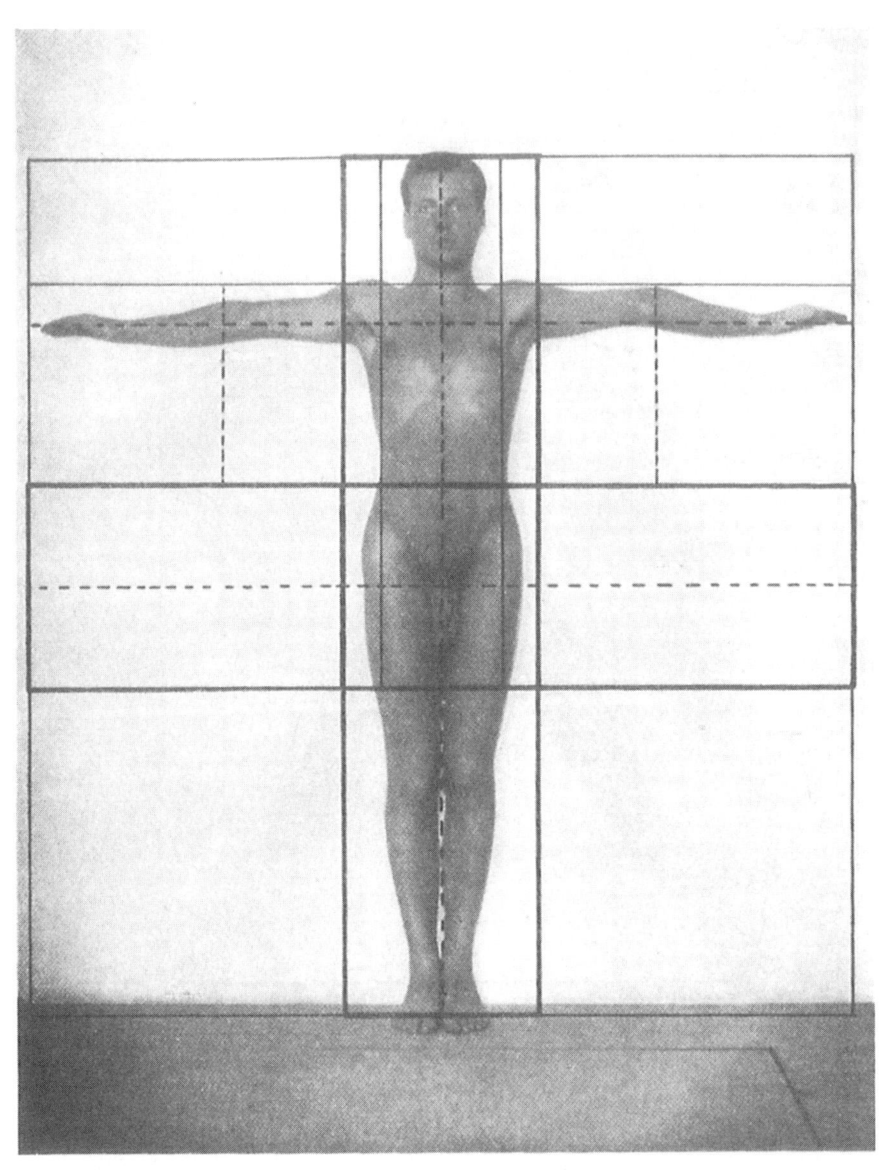

Plate 22. Male nude, harmonic analysis (square and golden section) (photo Manassé)

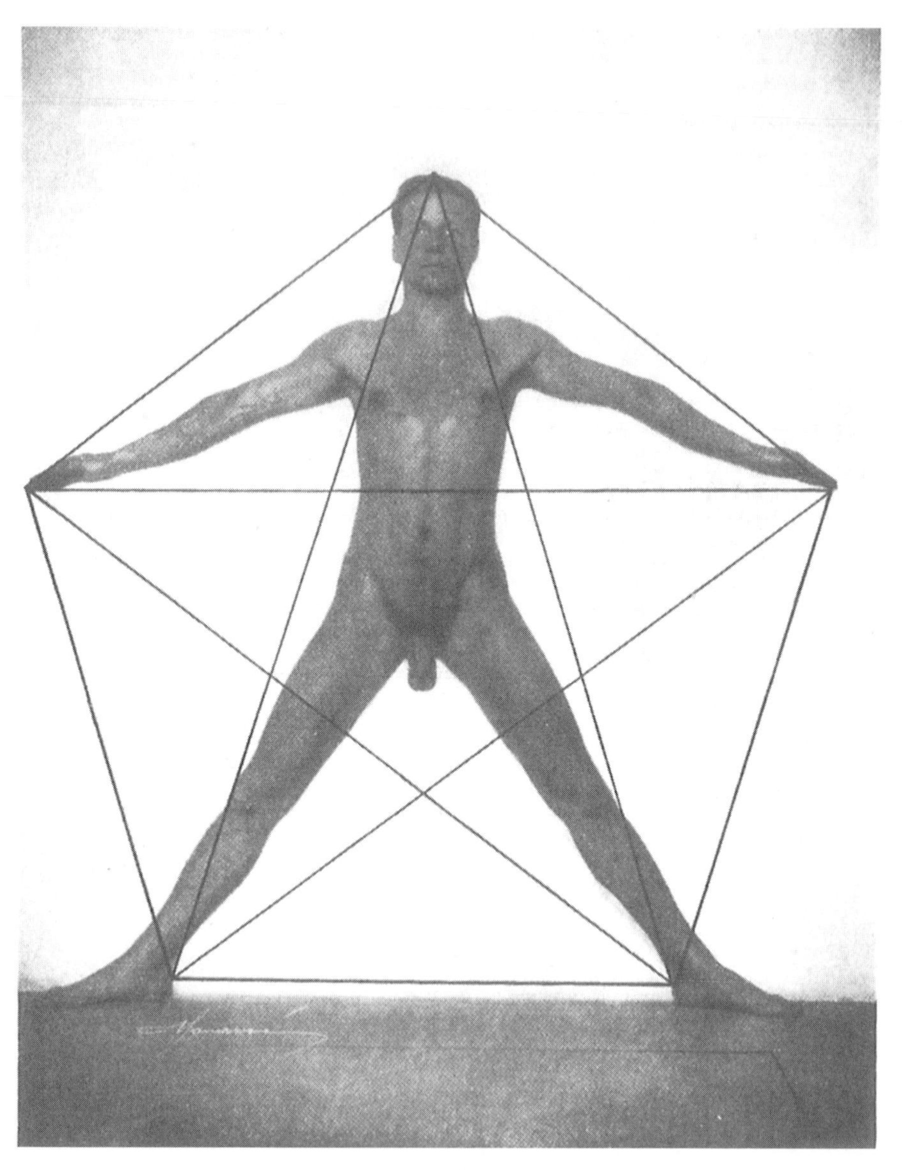

Plate 23. The "microcosm"

Plate 24. Hellenistic bas-relief (Rome); frame and composition in $\sqrt{2}$, the proportions of the bodies in Φ (golden section) (photo Anderson)

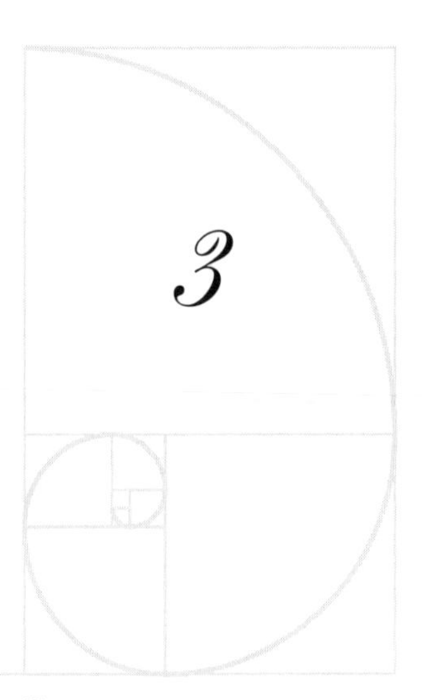

The Geometric Canons of Mediterranean Architecture

This temple is like unto the heavens in all its arrangements.

Inscription on a fragment from the temple of Ramses II, Cairo Museum

Among the various systems or hypothetical canons of establishing proportion suggested for decoding the complex geometry of Greek, Egyptian, and Gothic architecture, I examined two of the more recent systems in detail in my previous book. The first was that of the American Jay Hambidge,[1] called "dynamic symmetry," and at whose foundation we find:

1. The predominant use, for the overall framing surfaces and surface elements, of a certain number of rectangles of the type Hambidge calls "dynamic," which is to say, their modules (the ratio between the lengths of the larger and the smaller sides is sufficient to characterize the form of a rectangle) were not rational ratios like $\frac{4}{3}, \frac{4}{1}, \frac{3}{2}, \frac{3}{1}$ (the name "static symmetry rectangles" or "static rectangles" is reserved for these), but simple incommensurable ratios like $\frac{\sqrt{2}}{1}, \frac{\sqrt{3}}{1}, \frac{\sqrt{5}}{1}, \frac{\sqrt{5}}{2}$, and $\frac{\Phi}{1} = \frac{\sqrt{5}+1}{2} = 1.618...$

1. This theory was presented by Hambidge in *Dynamic Symmetry,* and by Lacey D. Caskey, Ph.D., curator of Greek antiquities at the Boston Museum of Fine Arts (*Geometry of the Greek Vase*), as well as in the review *Diagonal,* published by Yale University.

(ratio of the golden section, algebraically and geometrically akin to the themes of $\sqrt{5}$); the rectangles $\frac{\sqrt{4}}{1} = \frac{2}{1}$ and $\frac{\sqrt{1}}{1} = \frac{1}{1}$, which is to say, the double square and the square, are part of the series of dynamic rectangles as well as that of the static rectangles; and

2. The "harmonic" subdivision of these rectangles that frame rectangular surfaces of different sizes connected to each other by a continuous chain of proportions.

   The ingenious method that produces these "harmonic decompositions" is based on the recurrent creation from within the enclosed surface and its primary subdivisions of similar ("reciprocal") or related surfaces through the simple drawing of diagonals and perpendiculars dropped onto them from the vertices of the various given or progressively obtained rectangles (see plate 25).

The reader will especially note the harmonic divisions of the Φ rectangle, whose module is equal to the ratio of the golden section. It possesses the remarkable property that the elementary harmonic subdivision (obtained by dropping a perpendicular from a vertex onto the opposite diagonal) determines a square, in addition to a rectangle with a Φ module (or rather $\frac{1}{\Phi}$, similar to but arranged perpendicularly to the first inside it).[2] As this diminishing subdivision can be repeated indefinitely, Hambidge gave the name "rectangle of whirling squares" to the rectangle of the golden section. This diagram of whirling squares has a directing spiral that is the "curve of harmonious growth," the logarithmic spiral with the quadrantal pulsation Φ, the ideal envelope for "pseudognomonic" growth outlined by the F or Fibonaacci series (1, 1, 2, 3, 5, 8, 13, 21, 34, 55, 89, 144, ...) that we encountered in chapter 2 as the

2. This square is the "gnomon" of the small rectangle $\frac{1}{\Phi}$ since, when added to it, it reproduces a similar figure. The Pythagoreans studied this homothetic growth by successive gnomons (which graphically depicts our "calculus of finite differences") with regard to polygonal figurate numbers. Descartes adopted this method and also applied it to "solid" figurate numbers.

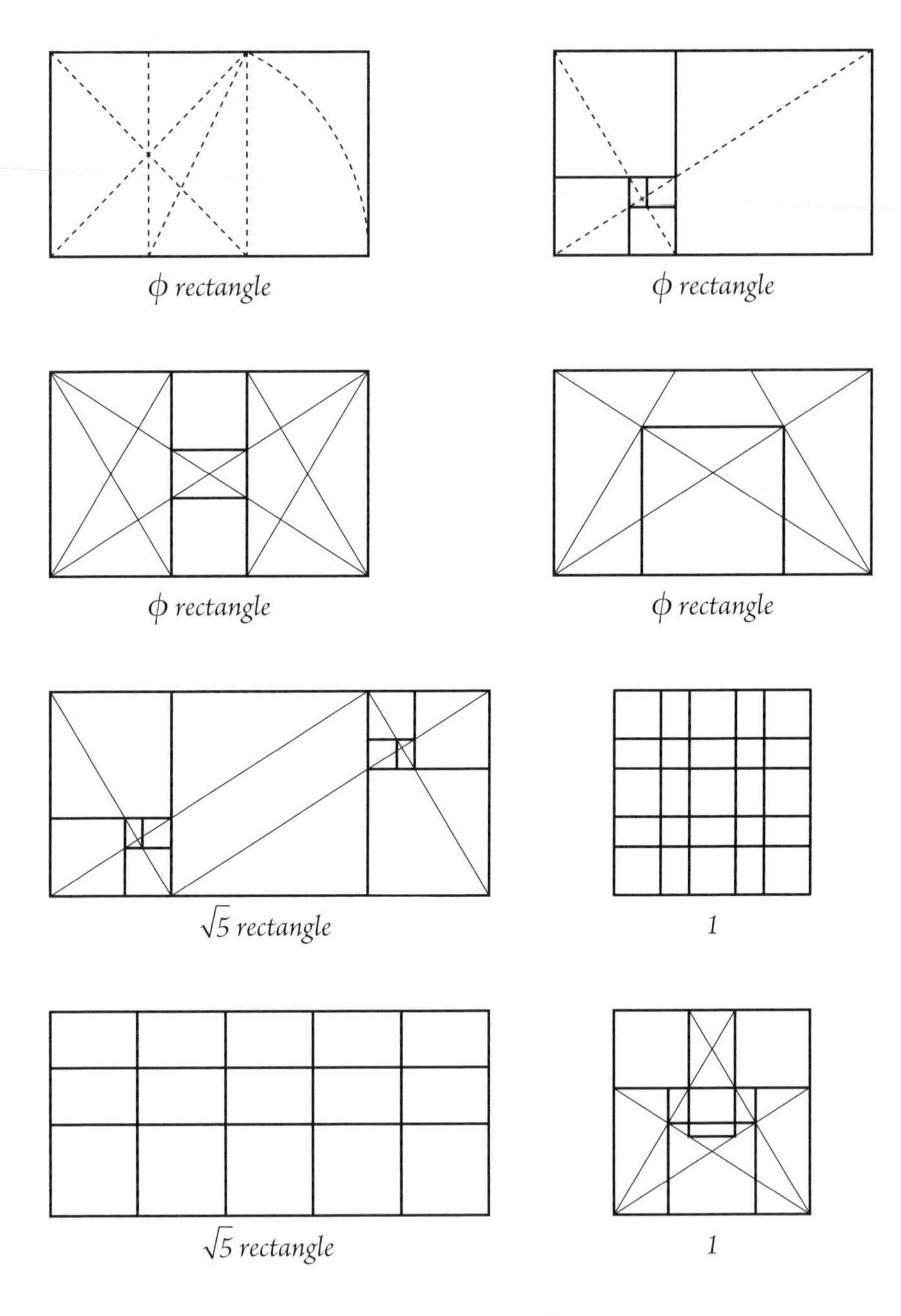

Plate 25. Harmonic rectangles (Φ and √5) based on Hambidge

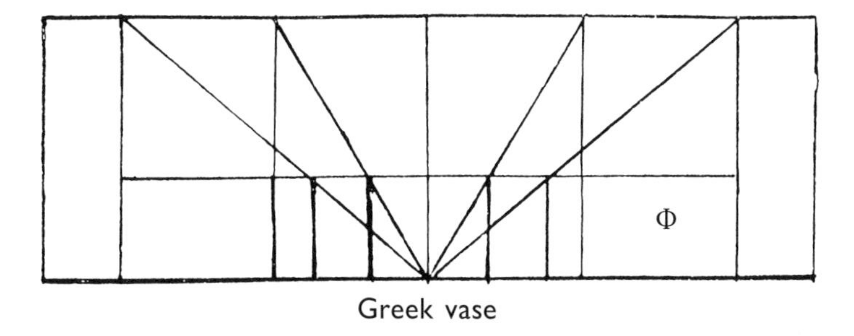

Greek vase

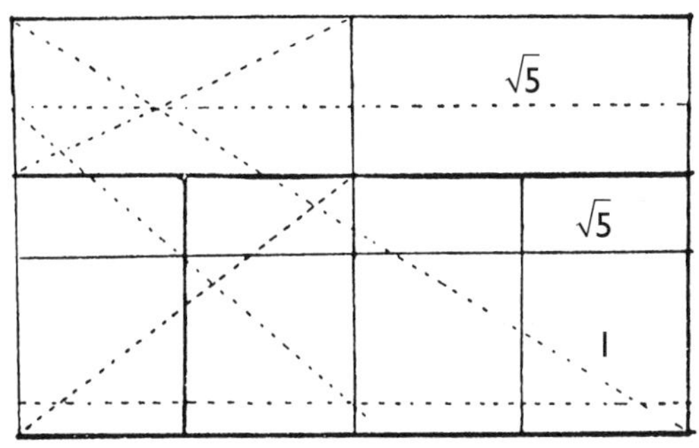

Façade of the Parthenon

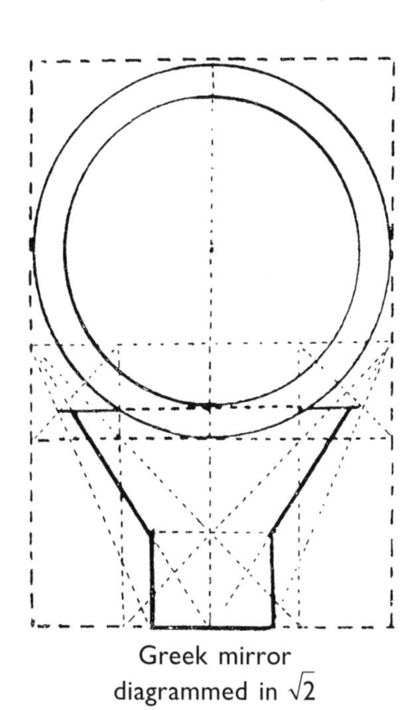

Greek mirror
diagrammed in $\sqrt{2}$

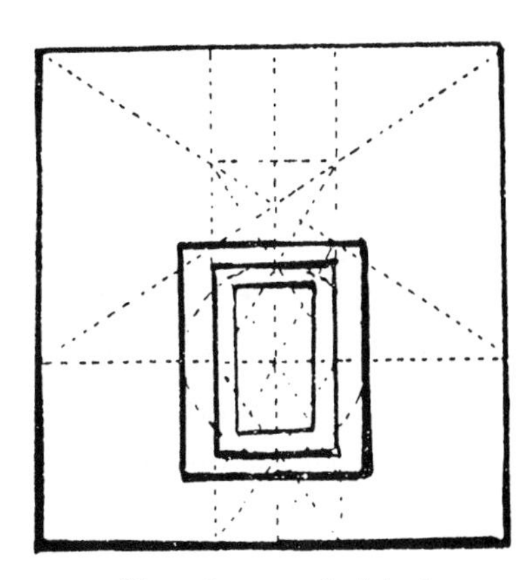

Egyptian tomb (plan)
diagrammed in Φ inside
a square

Plate 26. Harmonic decomposition of drawings
according to Hambidge's method

natural discontinuous approximation of the ideal continuum (Φ series).

I have already noted that the primary idea of this method of composition and analysis of rectangular surfaces was suggested to Hambidge by several passages in Plato's *Theaetetus,* most specially by the expression *δύναμει συμμετρόι,* numbers (or ratios) that are "potentially commensurable." It applies perfectly to those combinations of rectangular surfaces that, although derived from framing rectangles with incommensurable modules (but easy to construct with a rule and compass), $\sqrt{2}$, $\sqrt{3}$, $\sqrt{5}$, Φ, and so forth, are not only (the surfaces obtained by this "harmonic" decomposition) commensurable among themselves, but always form a graduated sequence, a series or progression of surfaces connected (among themselves and with the surface of the whole) by one same proportion, exactly as Vitruvius demanded for the surface or volume elements connected "by symmetry based on the proportion that the Greeks called analogy." And the *Theaetetus* and the *Timaeus* show us, moreover, that when the ancients spoke of analyzable surfaces among "plane numbers," it was always "rectangular numbers" (of the type $a \times b$, product of two component numbers), which is to say, the rectangular surfaces that they compared.[3]

It is this study of the proportions among rectangular numbers ($a \times b$, $c \times d$, or a, b, c, d were the origin of whole numbers) that led directly to the study of irrational proportions (but "potentially commensurable") from the time an attempt was made to place a "geometric mean" between two plane numbers. Plato personally discussed the corresponding problem in three dimensions (see in chapter 4 "Plato's theorem" on the two geometric means that can be set between two cubes, and also the related problem of the duplication of the cube, which leads

3. And not the other plane or polygonal numbers strictly speaking, which form the series of triangular, pentagonal, and similar numbers, with which, incidentally, Nicomachus and Theon of Smyrna were concerned. Just as the "solid numbers" that Plato and Nicomachus discussed from the point of view of their proportions are the numbers of the form $a \times b \times c$, representing right rectangular parallelepipeds (and cubes) and not the other polyhedral, pyramidal, dodecahedral, and similar numbers.

to not only incommensurable means, but ones that cannot be constructed by euclidean methods). Until Hippocrates of Chios's "betrayal," this question of incommensurable proportions, with the connected question of the construction of the pentagram and dodecahedron, was the mathematical secret reserved for Pythagorean initiates alone.

With respect to the practical application of establishing proportion in architectural designs, especially those for religious buildings, this secret appears to have been one of the confidential teachings that were handed down in families of architects and builders' guilds. Just as in religion and philosophy, the teaching of trades in antiquity had an esoteric basis, and this applied as much to the architect as it did to the sculptor and the doctor. As we shall see in the second part of this book with regard to the royal art of architecture and the geometry that formed its essence, this state of mind, with its related initiatory ritualism, was handed down unbroken to the builders' guilds of the Middle Ages.

If there are many obscurities or, rather, omissions in Vitruvius, they are, generally speaking, deliberate.

The now-vanished figures that accompanied his treatise would probably offer us the keys to its obscurities. In their absence, Hambidge's interpretation of "dynamic symmetry" provides us a fully satisfactory hypothesis for explaining what the Latin text refers to as the famous trinity of eurhythmy, symmetry, and analogy, and especially to the distinction between arithmetic and "geometric" symmetry that makes it possible to establish proportions for surfaces.

The proof of the importance of "symmetry"[4] as the master discipline of the architectural science of antiquity (corresponding to harmony and counterpoint in the curriculum of a music composer today) can be found precisely in the work of Vitruvius where this term pops up repeatedly as a leitmotif and as a summary of the essence of architecture.

4. I would again like to remind the reader that for Plato, Vitruvius, and all the architects of the early Renaissance, the word *symmetry* meant "commensurability" among all the elements of a whole, and between each of these elements and the whole.

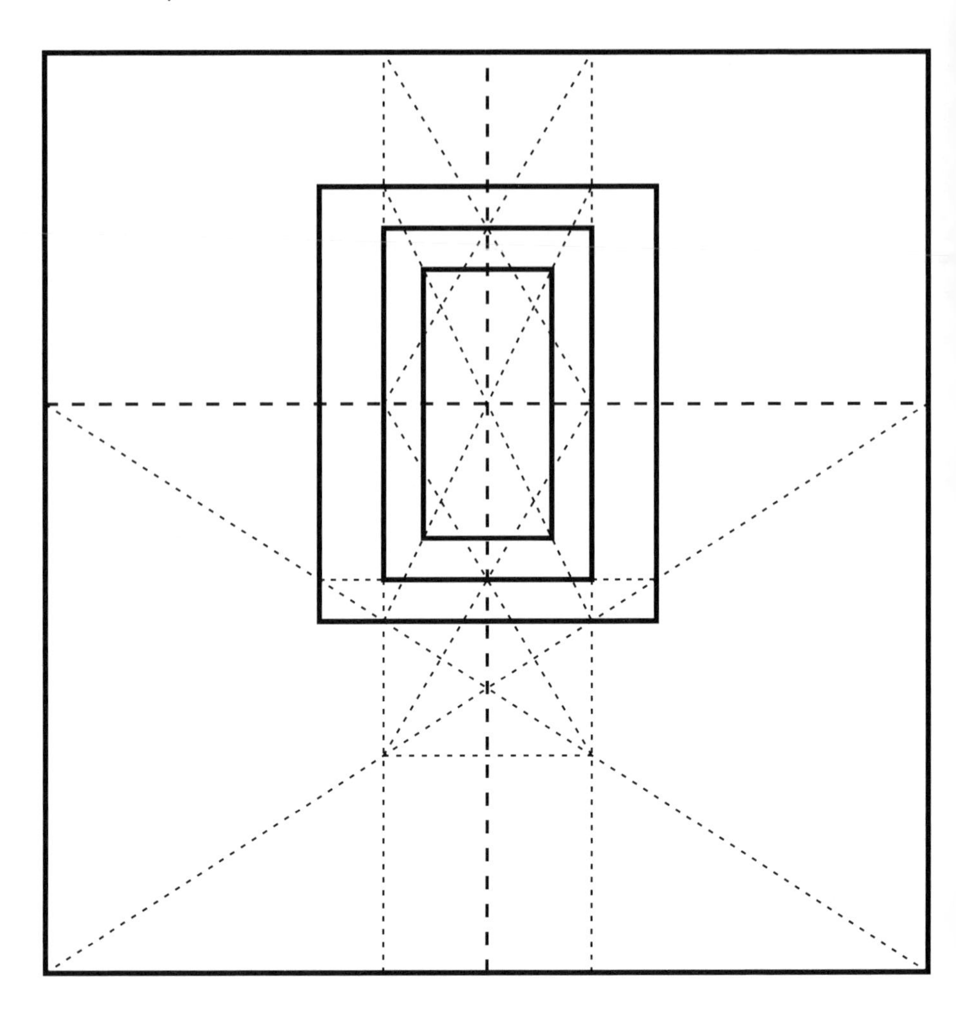

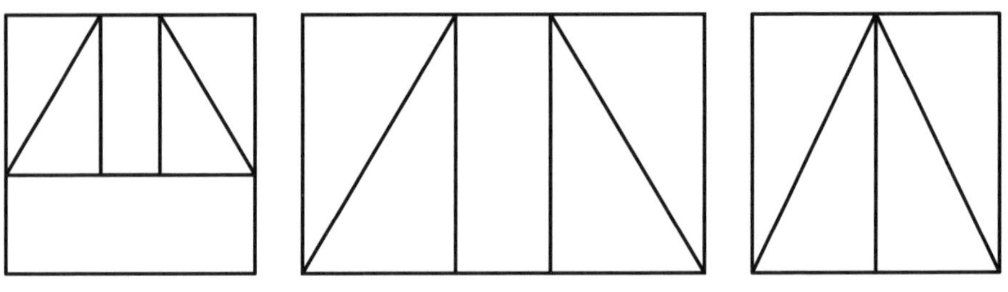

Plate 27. Egyptian harmonic drawings

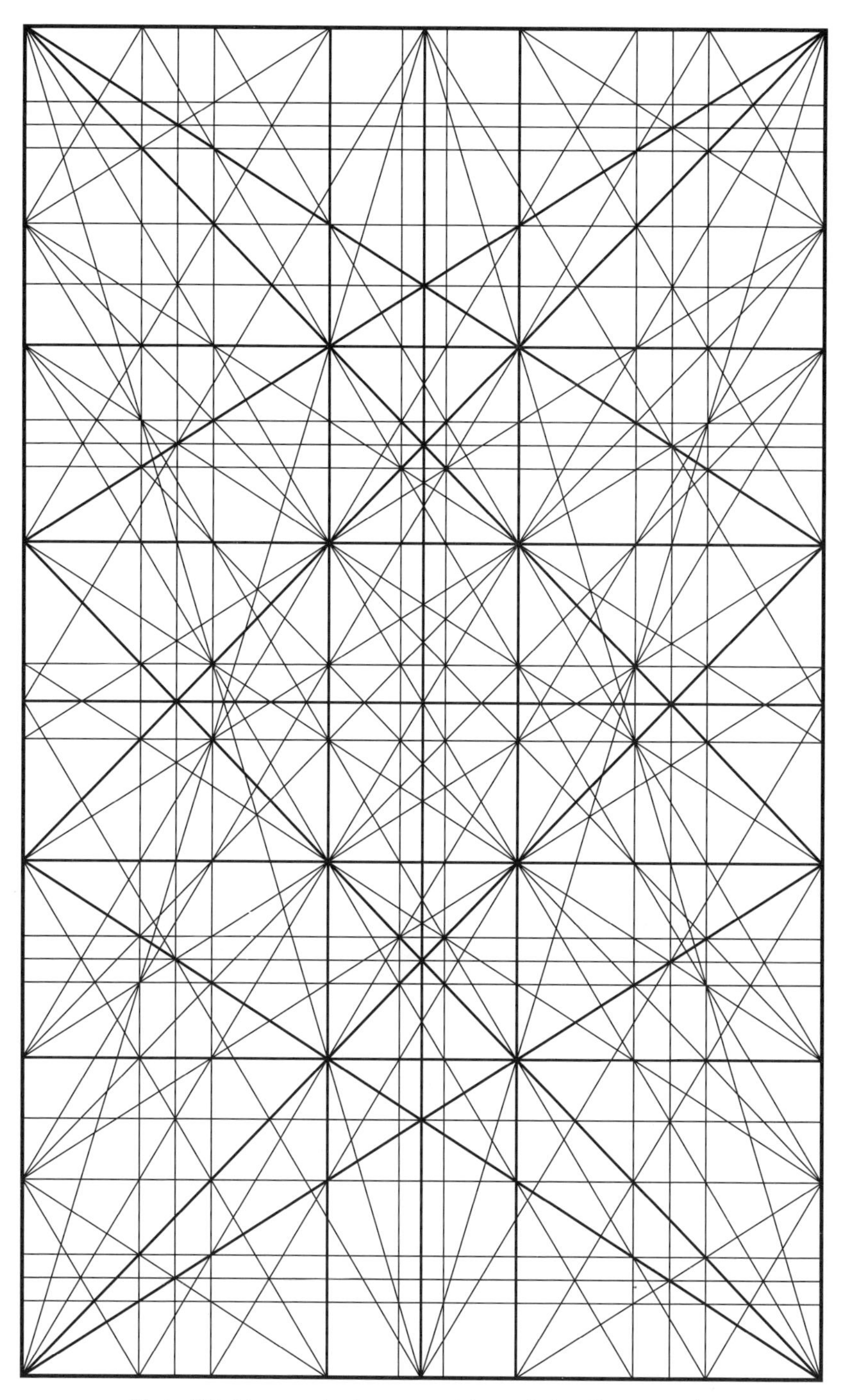

Plate 28. Harmonic decomposition of the Φ rectangle
(D. Wiener)

The Greek or Latin treatises (alas, now all lost) that he cites as architectural classics are almost all titled *Treatise on Symmetry.*[5]

The role of this *commodulatio* symmetry, the commensurability of the parts to each other and of the parts to the whole, sometimes with regard to the linear elements, sometimes with regard to surfaces ("dynamic" symmetry), was perfectly understood by the first commentators of Vitruvius, architects and mathematicians, whose works, thanks to the invention of the printing press, we still have. This includes architects working in the Gothic tradition (like Caesar Caesariano, who in 1521 chose to illustrate the text by Vitruvius with plans for the cathedral of Milan, mathematicians like Luca Pacioli,[6] or early Renaissance masters like Alberti.

Without the invention of printing, architecture would have

---

5. Examples: "*Silenus de symmetriis Doricorum edidit volumen.*" And "*Philo de aedium sacrarum symmetriis et de armamentario quod fecerat Piraei in portu.*"

Vitruvius again cites Argelius as author of a book on symmetry in the Corinthian order, and nine other, less important authors of treatises on "symmetries."

The mention of Philo needs to be pointed out, because an inscription found in Greece gives all the sides and measures for a pediment of the Piraeus arsenal specifically featured here, and which Auguste Choisy, thanks to these specifications, succeeded in reconfiguring as an "organic" design in which everything was linked together geometrically.

Choisy, moreover, in his excellent commentary on Vitruvius, concludes that the ratios of commensurability (the "commodulations") from which *symmetria* arises are not necessarily arithmetic (static) but can be purely geometric. They are precisely the "*geometricae rationes*" discussed in the Latin text. He notes that in the plan of the Greek theater described by Vitruvius (as in the proportions of the human body), the two rhythms—modular and geometric—are combined. And he highlights the phrase with which Vitruvius opened his discussion that the delicate questions associated with symmetry are resolved by geometric ratios and methods. (*Difficilesque symmetriarum quaestiones geometricis rationibus et methodis inveniuntur.*)

6. Pacioli specifies (*De Divina Proportione,* chap. 5) that architects should make use of all the symmetries, "even the irrational ones that, like the ratio between the diagonal and the side of the square, cannot be specified by whole numbers and their aliquot parts." And much later (in chap. 20) he revisits this idea: when one cannot employ simple symmetries like $\frac{1}{2}$, $\frac{1}{3}$, $\frac{3}{4}$, $\frac{2}{3}$, and so on, but must use irrational proportions, the points instead of being determined by numbers will be determined by lines or surfaces "because proportion has much greater use in the domain of continuous quantities than in that of whole numbers . . . and geometry is as much concerned with the irrational domain as it is the other," and so forth.

remained an esoteric discipline, for in the absence of printed texts accessible to everyone, the transmission of principles and procedures by professional "initiation" (and this point is very important for understanding this book) was not the result of an inexplicable or puerile love of the occult, but a necessity.

Certain passages by Alberti furnish a full confirmation of this method.

> Harmony is the consonance of several sounds that are pleasing to the ear. . . . Architectural harmony, meanwhile, consists of the use of simple surfaces by architects, as these are its elements, not in a random or indiscriminate manner but by ensuring they correspond to one another through harmony or symmetry. For example, if someone wanted to erect walls around an area whose length was double its width, it would not be suitable to use triple harmonies but only double harmonies. (*De Re Aedificatoria,* 1485)

Then comes the famous passage in which Thiersch accurately discerns the law of analogy, the repetition of the fundamental form, which is the key to architectural harmony: "*Lineamenta sentiamus ubi una atque eadem in illis spectatur forma.*"

I refer the reader back to my *Esthétique des proportions* for more detailed accounts of the remarkable results obtained by Jay Hambidge in his analysis of Greek temples and vases (see plate 26) and for the analysis of his method applied to the human body. I would again add here as additional confirmation of the "Greek" character of his constructions that, as he intuitively deduced, the diagonals in fact do play a major role in the graphic manipulations of the Greek geometers; Eratosthenes's method for solving the duplication of the cube by shifting the diagonals brings the problem squarely back to one of proportions.[7]

I would like to remind the reader here that of all the dynamic

7. Cf. chapter 4, Auguste Choisy, who wrote his commentaries on Vitruvius fifty years before Hambidge discovered his theory of "dynamic symmetry," noted that in an atrium

"themes," those based on $\sqrt{5}$ and Φ (in other words, the golden section) offer a much greater variety and flexibility than the two other simple dynamic themes ($\sqrt{2}$ and $\sqrt{3}$) and can be found with greater frequency than these latter (the "crystalline" themes, of inorganic matter), in Greek and Egyptian architectural and decorative schemas.

The second system is that of the Norwegian archaeologist F. Macody Lund who, in addition to Greek temples, gave special study to Gothic designs.[8]

Generally he found, on a grid of double squares, "starry" designs [pentagrams] having as an asymmetrical pole (often coinciding with the center of the high altar on the plan and with the center of the main rose window on the elevation) the center of a pentagon or pentagram.

Readers can also find in my *Esthétique des proportions* a more detailed examination of this theory and several graphic examples reminiscent of the abstract diagrams of pentagonal polarity in chapter 2 (and, for example, the construction that I reproduce in plate 35 of this book, a vertical section that relates the transverse diameter of a Gothic nave to the diameter of the pillars).

Like Hambidge, Lund was set on his path and guided by passages from Plato.

He traced, through the ages, the influence of the ideas expressed in the *Timaeus* regarding the importance of the five regular polyhedrons and their structure; quite naturally, he happened upon the planar diagrams obtained by the projections of polyhedrons inscribed within a single sphere. The major role Plato attributed to the dodecahedron

---

(cont. from p. 79) plan described in detail by Vitruvius the (irrational) ratio of the diagonal to the side of the rectangle governed the symmetry of the plan.

Theon of Smyrna cites among figurate numbers "diagonal numbers" and "side numbers" whose series provide rational approximations of $\sqrt{2}$, and Proclus (commentary on Plato's *Republic*) mentions the Pythagorean origin of these "diagonal numbers."

8. F. Macody Lund, *Ad Quadratum*, 2 vols. (Kristiania, Norway, 1919). The English edition was published by Batsford in London; the French by Albert Morancé. Lund has also published a sequel to his first work with the title *Ad Quadratum II* (Farsund, Norway: Aktieselskabet det Lundske Forlag, 1928).

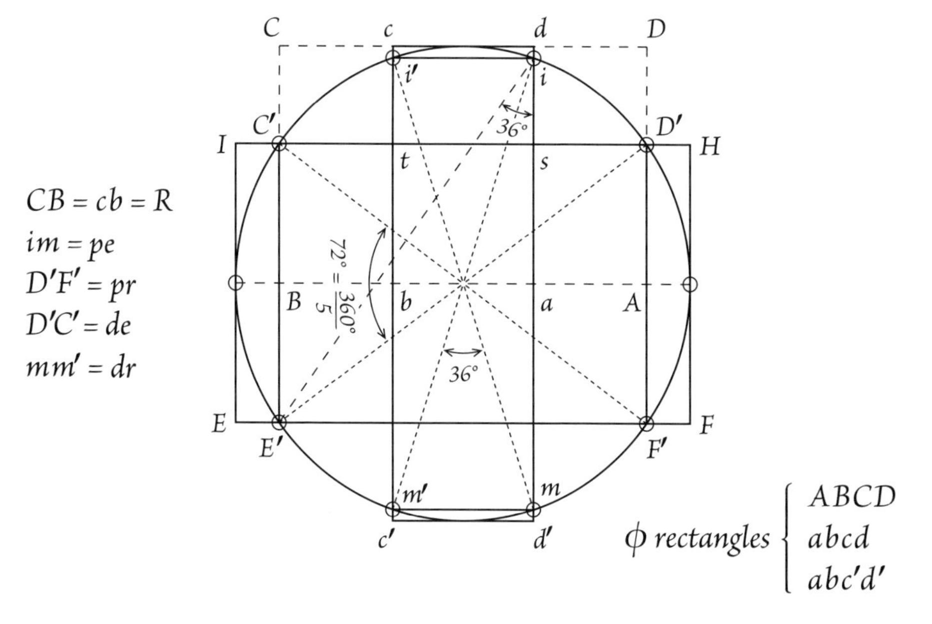

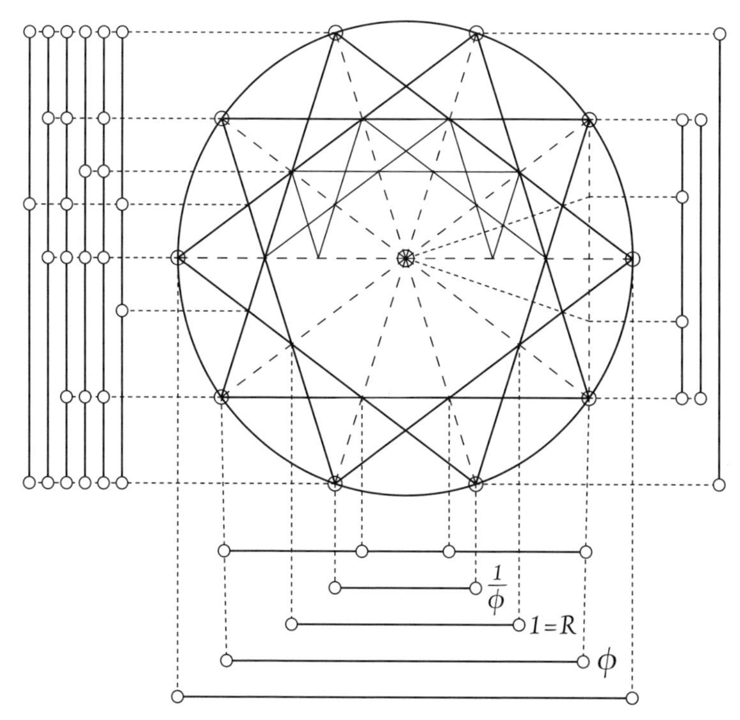

Plate 29. Systems of proportion obtained by the polar segmentation of the circle

(geometric symbol of the cosmos and the element ether, the "fifth essence"), the importance of the pentagram, and thus the golden section, guided Lund's discovery in a "Gothic" text of Campanus of Novara's (thirteenth century) homage to the golden section (*proportionem habentum medium duoque extrema*) as being the proportion that "in an irrational symphony" (in other words, in a "dynamic" symmetry with irrational ratios only "potentially" commensurable) bestows "in the most rational manner" (here in the sense of "logically harmonious"[9]) the proportions of the Platonic solids.

This phrase, which was not forgotten—since I found it triumphantly cited by Pacioli[10] more than two hundred years later (his treatise on divine proportion, although printed in 1509, was written in Milan before 1500, at the same time Leonardo created the magnificent plates that illustrate it)—shows how the people of the Gothic era valued the "eurhythmic" properties of the golden section.

Lund, who unearthed many other interesting references on this subject, demonstrates how the grid of large and small double squares, which he could find in almost all Gothic diagrams, only provides the elementary framework, articulated or draped like cloth over the armature of the design. However, the "section," the principal rhythm of this armature, is almost always a theme that is independent of this grid and whose principal elements, for both the horizontal plan and the elevations, are often provided by a large pentagon and the decreasing series of pentagrams that are naturally inscribed within it.

---

9. The value of these observations by Campanus resides precisely in the apparent spiritual antithesis (*Rationabiliter . . . irrationali symphonia*) between "*irrationali*" used in the strictly mathematical sense of incommensurable, and "*rationabiliter*" used in the general sense. Campanus of Novara was a chaplain of Urban VIII and canon of Paris. This passage can be found in his commentary on a translation of Euclid.

10. "This prerogative . . . that our (divine) proportion has certainly received from the invariable nature of the higher principles (and whereby), as said by the great philosopher Campanus, our extremely famous mathematician, is the one that through an irrational symphony harmoniously aligns the solid bodies (the five Platonic solids) from the perspective of their volumes with regard to the number of their faces, as well as their forms." *De Divina Proportione.*

Since the publication of my *Esthétique des proportions,* I have become aware of a third hypothesis, one that has only been presented recently but is founded upon a long series of observations and comparisons. It is the hypothesis of Professor Mössel[11] of Munich.

Mössel is an architect who, having decided that the question that prevailed over all others in architecture was that of proportion, devoted a good part of his life to measuring or collating, from the perspective of their lengths, surfaces, and volumes, the dimensions and proportions of all the Egyptian, Greek, and Gothic buildings for which we have exact plans. He is not beholden to any theory or a priori synthesizing notion, but the comparison of these hundreds of drawings gradually allowed him to see the emergence of analogies and similarities, if not outright identical features. In the thousands of numerical ratios established this way, certain numbers consistently recurred, as did their powers and the series of these powers in ordered progressions.

The geometric diagrams could all be reduced—for the horizontal plans as well as the elevations and vertical sections—to the inscription inside a circle, or several concentric circles, of one or more regular polygons.

Sometimes it involved the division on the horizontal plane of the directing circle into 4, 8, or 16 equal parts, and the various combinations of squares and rectangles suggested by the points and lines obtained in this way provide the skeletal structure of the plan. In this case, a simple idea became quite evident: this directing circle was derived from the building's circle of orientation that was itself first drawn on the ground, an idea in keeping with the almost religious importance attributed to the orientation of temples among the Egyptians, and then by the Greeks and Romans.

Vitruvius describes the procedure quite clearly. On a large circle traced on the ground, the minimum shadow cast (corresponding to the

11. *Die Proportion in der Antike und Mittelalter.*

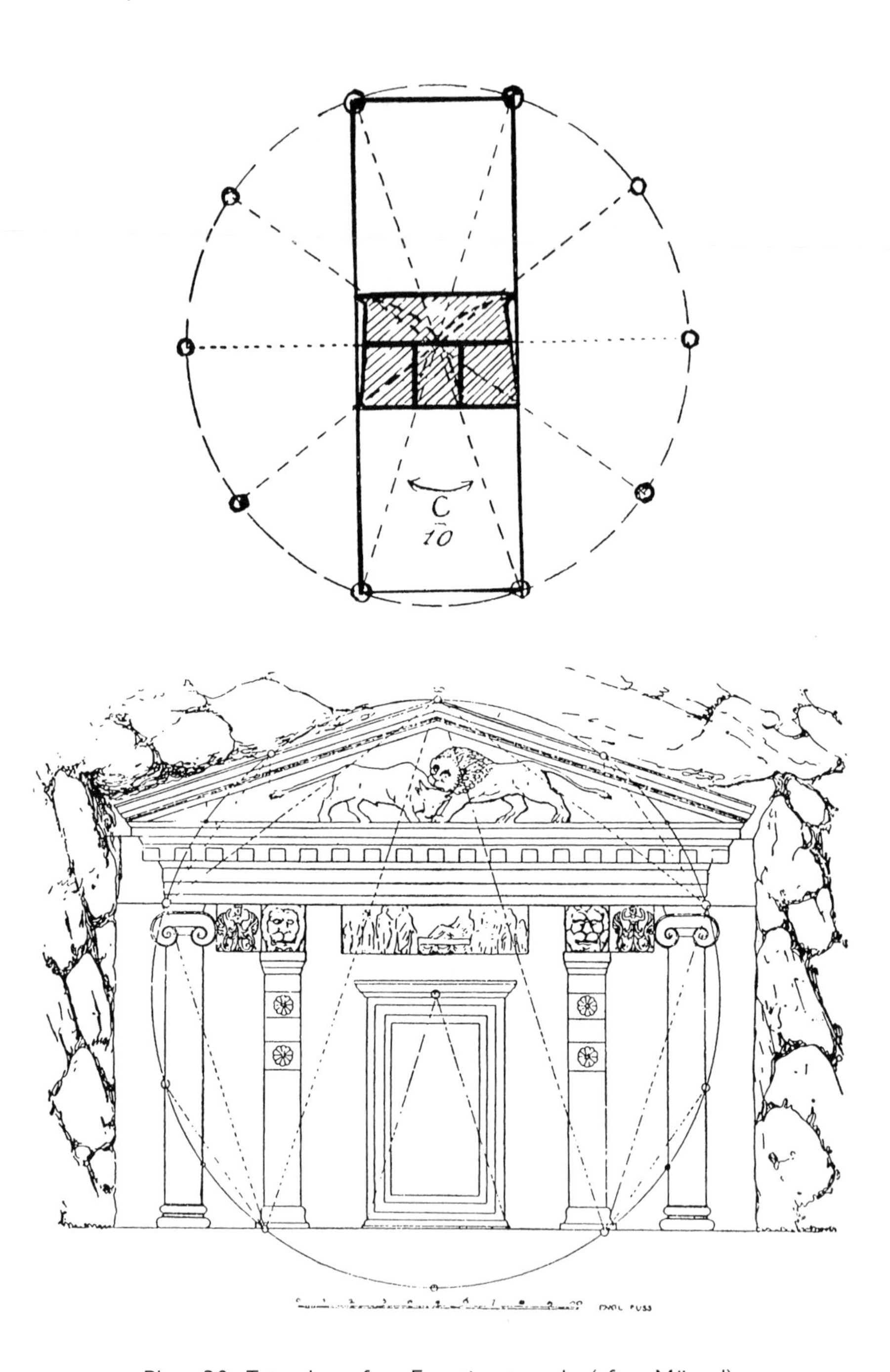

Plate 30. *Top,* plan of an Egyptian temple (after Mössel); *bottom,* rock temple of Mira (Asia Minor), from Mössel, *Die Proportion in der Antike und Mittelalter* (Munich, C. H. Bock, 1926)

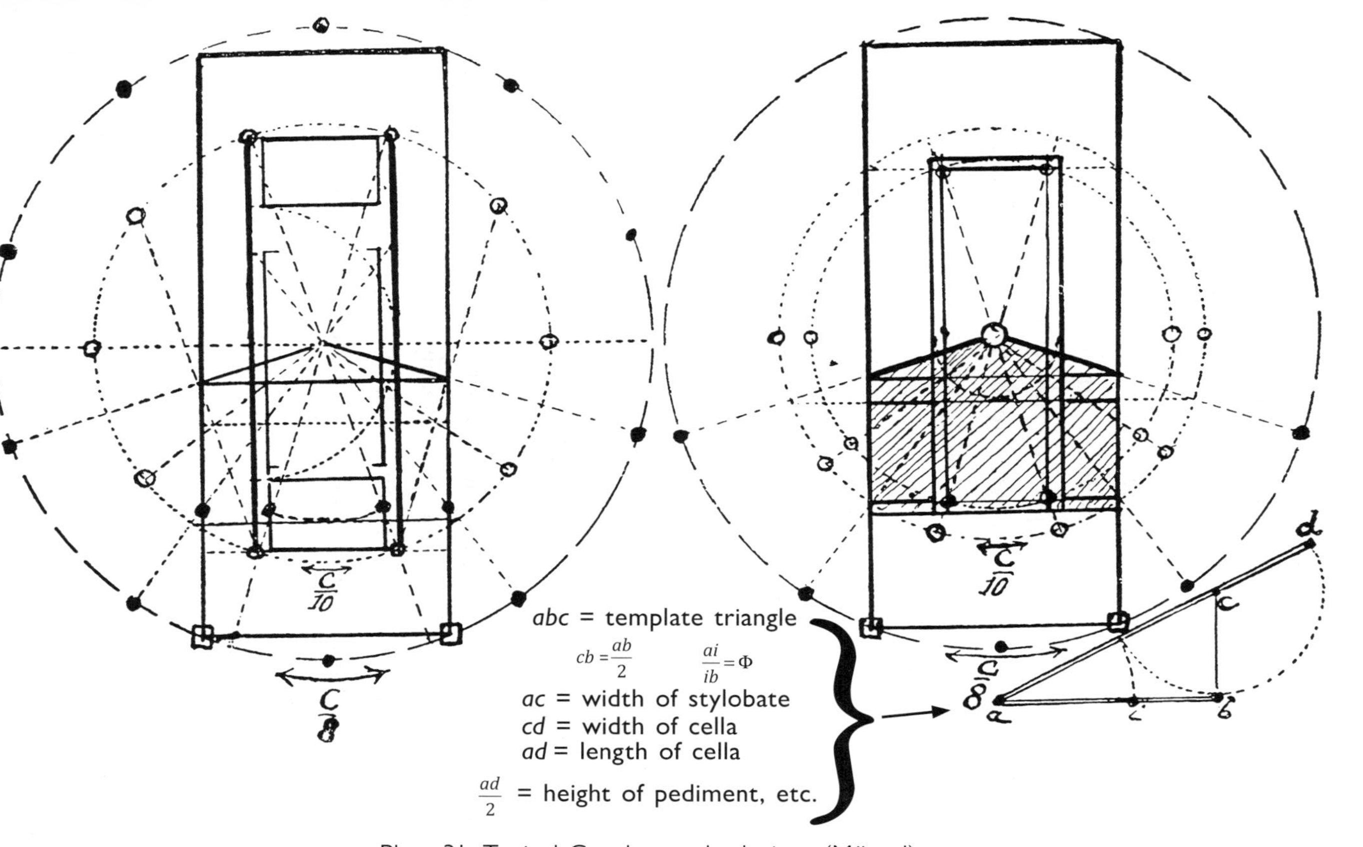

Plate 31. Typical Greek temple designs (Mössel)

maximum height of the sun above the horizon at "true noon") by a pole placed in the center of the circle (a gigantic sundial) provides a precise reading of the north-south axis; the perpendicular diameter will reveal the east and west points on the circle. We know that by means of a rope divided by knots into 3 + 4 + 5 = 12 equal segments, the *arpedonaptes* [rope stretchers] or geometer-surveyors of antiquity were able, by drawing a Pythagorean triangle on the ground with the help of three stakes, to determine a precise perpendicular line, and the extraordinary importance the ancients attributed to Pythagoras's discovery of the square of the hypotenuse stems in large part from the possibility it gave of actually constructing the right angle.[12] It is obvious, moreover, that the Egyptians already knew many particular cases of the theorem of Pythagoras, the 3-4-5 "sacred triangle" in particular. It was the opinion of the Greeks that Mediterranean geometry was born from the practices of these geometer-surveyors of ancient Egypt. All know the famous phrase in which Democritus of Abdera boasted of "not having found anyone who has surpassed me in the art of drawing lines in figures, and of demonstrating their properties, not even the Egyptian *arpedonaptes.*"[13] This same Democritus (450–360 BCE), who like Thales and Pythagoras spent long periods of time in Egypt and studied mathematics and the "natural sciences" there, was the originator of the atomic theory. He knew Philolaus and was probably initiated by him into the philosophical-musical speculations of the Pythagoreans. In any case, he was the first philosopher (to our knowledge) to use the expressions *microcosm* and *macrocosm.*

Regarding the rest, Mössel notes that the largest number of drawings were not produced by this natural "astronomical" subdivision of

12. This is certainly how Pacioli describes the importance of the discovery of the 3-4-5 right triangle by Pythagoras, who, he reminds us, sacrificed one hundred cattle in its honor. Pacioli also calls this fortuitous find "the discovery of the proportions of the right angle" and says that the disciples of Pythagoras called the right angle, the "angle of equity." According to Diogenes Laertius, tradition also attributes Pythagoras with the discovery that the angle inscribed inside a half-circle is a right angle.

13. Clement of Alexandria attributed these words to him.

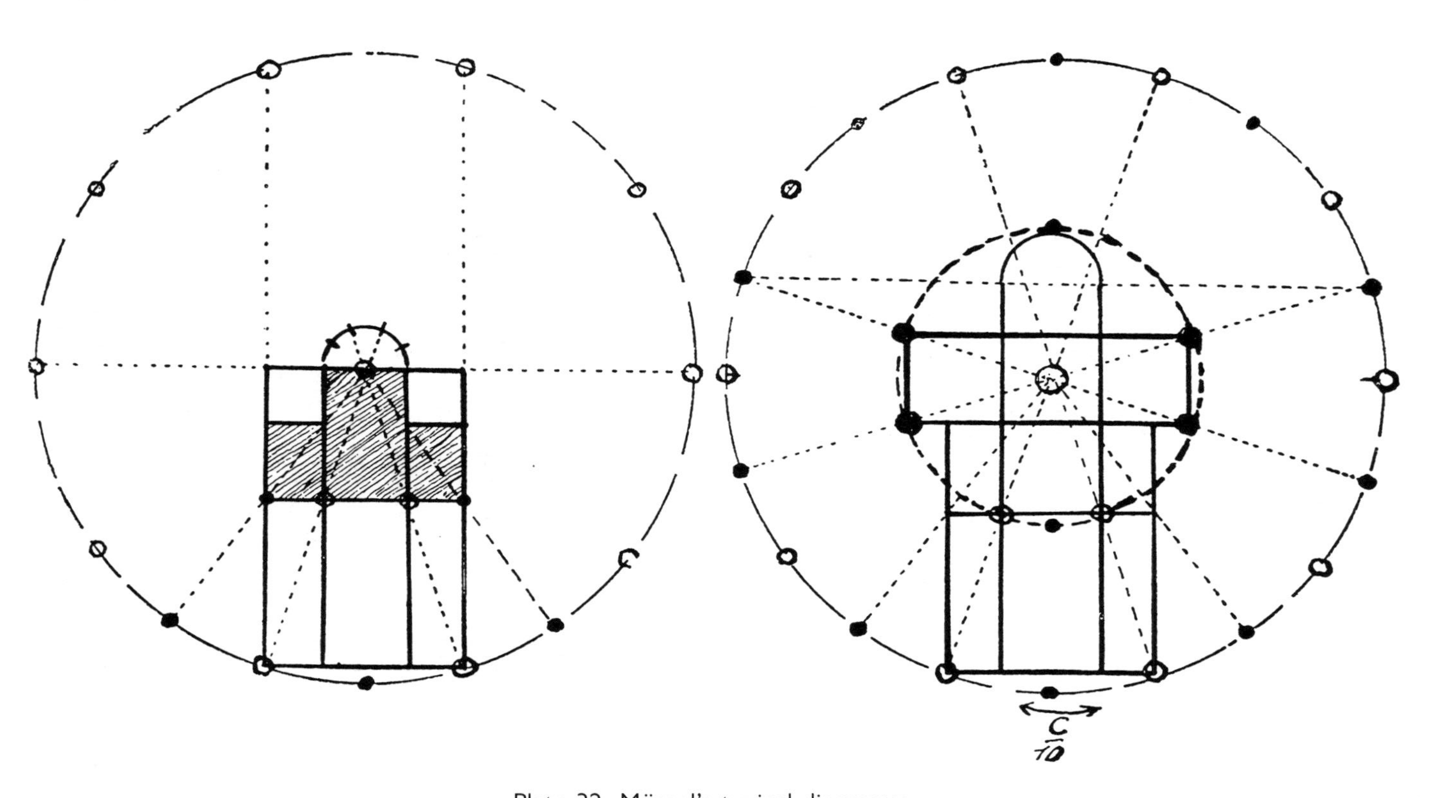

Plate 32. Mössel's typical diagrams:
Left, early Christian basilica; *right,* Gothic church

the circle of orientation into 4, 8, or 16,[14] but by its more subtle division into 10 or 5 parts, that is, by the inscription inside this circle, which then becomes the directing circle of a life-size template, of a regular pentagon or decagon. This applies equally to Egyptian drawings and to those from the great era of classical Greece, and takes us directly to Lund's Gothic drawings with their starry plans and pentagonal polarity.

So it should come as no surprise to note that the ratios as well as their powers that continually appear in the numerical tables of proportions calculated by Mössel, independently of his drawings, are the ratios $\Phi = 1.618...$, or the golden number, its powers Φ^2, Φ^3, and so on, the diminishing powers $\frac{1}{\Phi}, \frac{1}{\Phi^2}, \frac{1}{\Phi^3}, \ldots$, as well as $\sqrt{5} = 2\Phi - 1 = 2.236...$, as we know that arising from this ratio of the golden section are the harmonics and correlations of everything drawn from a pentagonal or decagonal base (plate 29).

Mössel could therefore classify the diagrams of almost all the monuments he analyzed into a certain number of specific types based on what he called the *Kreisteilung* or polar segmentation of the directing circle. This holds true for the elevations as well as for the horizontal plans, which for each building can be set based on a single diagram, with the elements and wholes of these projections linked both vertically and horizontally by chains of proportions in which the recognized themes of the golden section recur like a leitmotif. Thus there is a point in common here with the systems of Hambidge and Lund (of which Mössel had no knowledge when he first published the summary of his conclusions in 1926). We should immediately point out that while Hambidge's method imposed what I would call the "law of non-blending of themes," Mössel found in some cases two concentric directing circles, of which the larger, corresponding to the outer design of the given building, is divided into 8 or 16 parts (octagonal symmetry, that is, square, with a module of $\sqrt{2}$),

14. We can see in Vitruvius that this segmentation of the orientation circle according to the compass rose was used very much in the urban planning of antiquity, even for drawing the plans of cities themselves and determining the direction of all their major arteries.

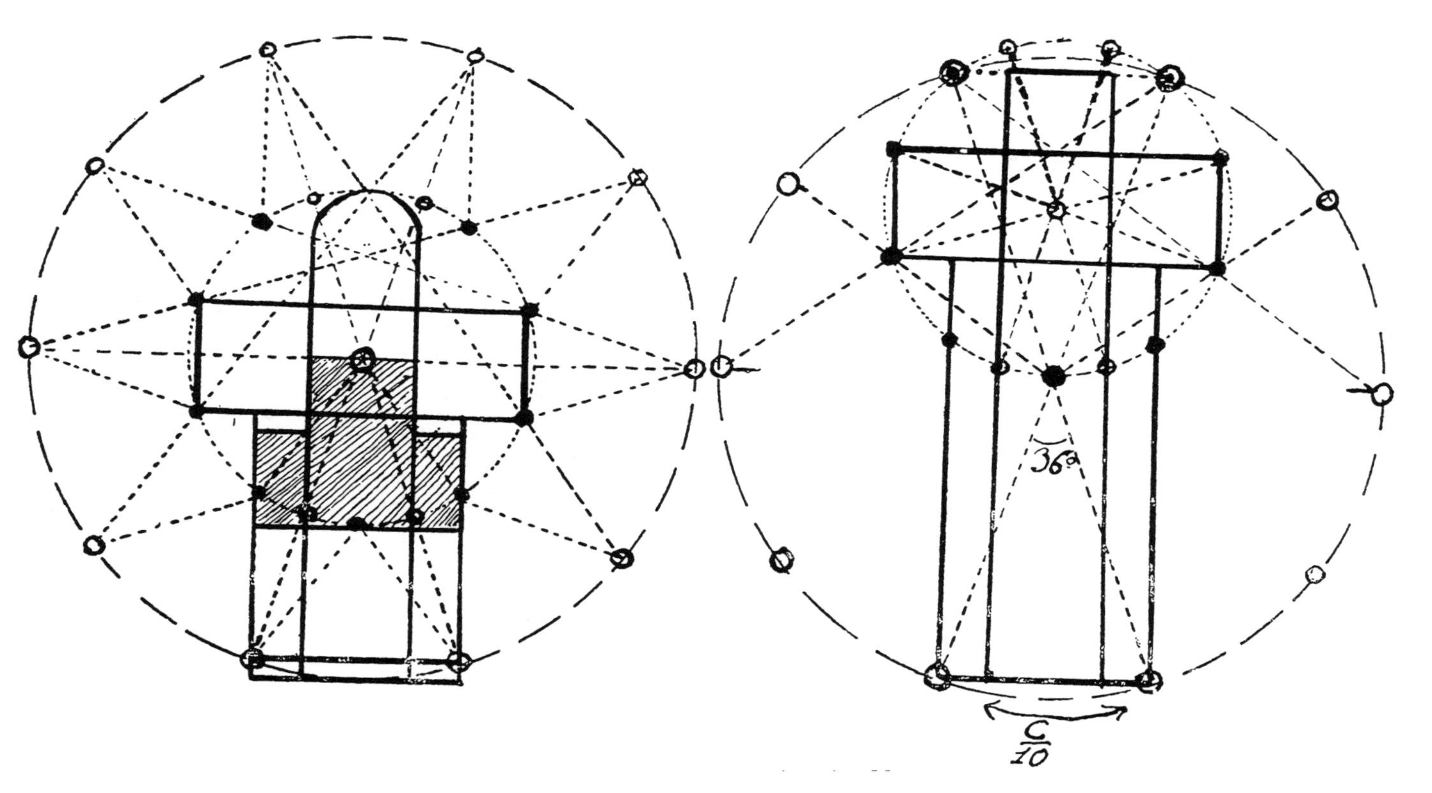

Plate 33. Typical Gothic designs according to Mössel
(showing decadal segmentation of the directing circle)

whereas the other, corresponding to the nucleus (the naos, cella, and so on) is divided into 5 or 10 parts (pentagonal or "golden" symmetry with a module of Φ or $\sqrt{5}$). The vertical outline (façade or transverse section) of the building, deriving from the same diagram, is in this case governed by the inner directing circle (with pentagonal symmetry), but one of its linear elements is equal to an element supplied by the other circle, which creates an organic bond, a "concatenation" of all the elements despite the presence of two different themes.

It suffices to examine the overall types of the diagrams that I have reproduced from Mössel's book (plates 30–34) to see that his system brings together, and to some extent combines, Lund's starry diagrams and Hambidge's dynamic rectangles, while suggesting a reasonable starting point: the manipulations and segmentations of the orientation circle, taken from the start as a large directing circle on a life-size template, traced right on the ground.

This starting point allows one to imagine that, once architects, through the practice of graphic manipulation of the "golden" series stemming from the division of the circle into 10 or 5, perceived the flexibility of these modulations, an extremely rapid evolution toward a symphonic subtlety of rhythms conforming to the "harmonic" ideal of the Pythagorean school as described by Plato took place. All one needs to do is compare the text of Vitruvius with the *Timaeus* to see that this rigorous mathematico-musical aesthetics dominated the architecture of antiquity.

The general theory of proportions, including the harmonic and geometric proportions associated with the decad and the tetractys, the study of proportions among the volumes of the five regular solids, the astronomical and biological rhythms that we find mentioned in the *Timaeus* and the *Republic* (the number of the world soul, the nuptial or marriage number, and so on), all combined with the earlier Egyptian notion of the desirable correspondence between the temple and the universe, and the correlation between the living universe and the human being (macrocosm-microcosm), necessarily culminated in the architectural technique of designs based on the subtle eurythmic correspondences among

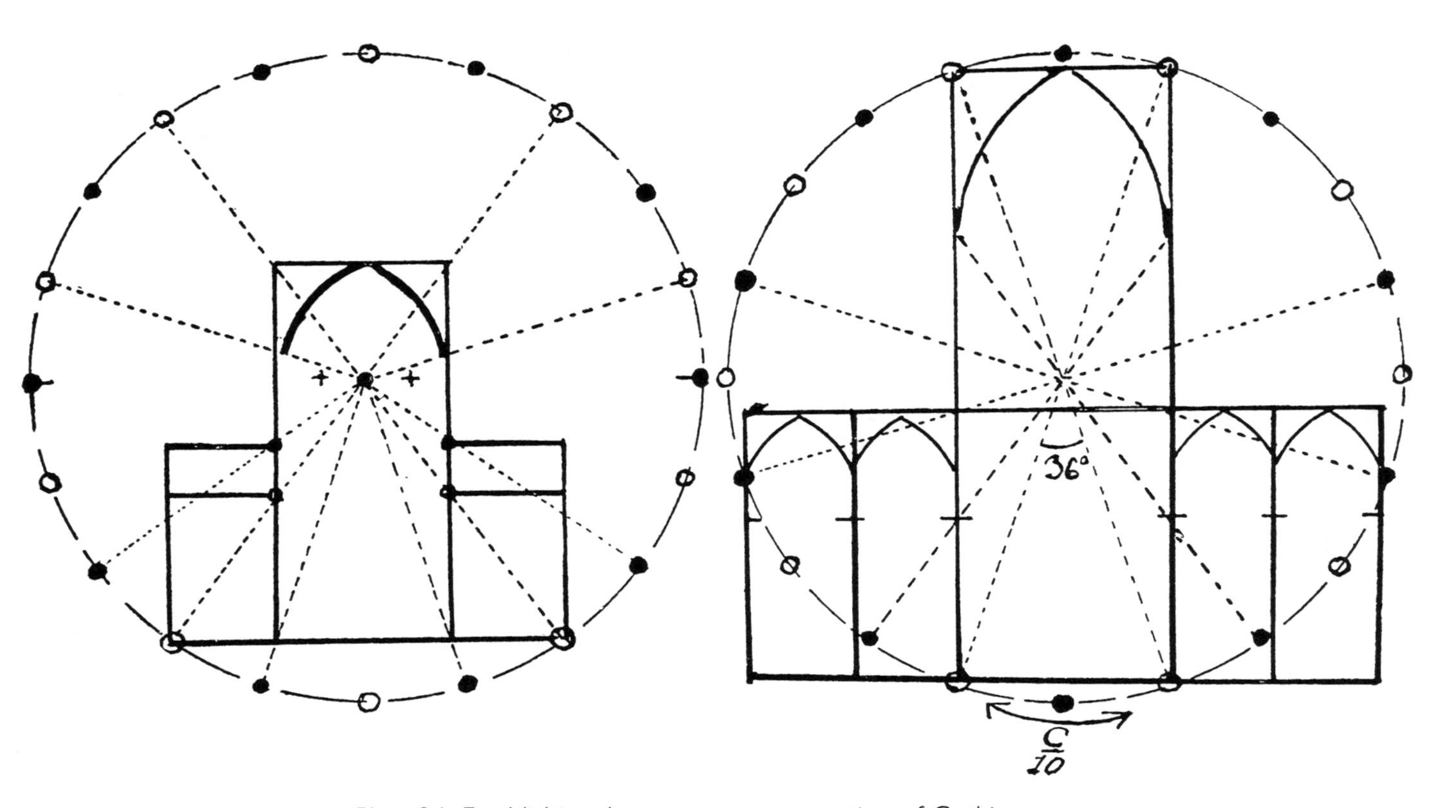

Plate 34. Establishing the transverse proportion of Gothic naves
(after Mössel)

lengths, surfaces, and volumes, which through the difficulty required to decipher them surely deserve the label of esoteric. The contemporary architecture of Pythagorean mathematics and the religion of Eleusis, like all those of a ritual and initiatory nature, and this tradition of secrecy covering all things concerning the sacred, also came from Egypt.[15] The Greeks added correlations to it that were not only harmonic but explicitly musical,[16] and, pushing it to the limits, developed a metaphysical concept of number and its emanations: proportion, rhythm, form.

15. It is worth mentioning here a very important, practical reason that compelled Egyptian architects to secrecy with regard to their methods and even their designs. The tombs of the pharaohs and other high figures held as great a place in the constant output of the architect as the building of temples as such, and the discovery of Tutankhamen's burial chamber revealed what great treasure hoards accompanied the important dead into their tombs. Hence the imperious necessity to protect these riches against grave robbers whose industrious ways have been with us since the most remote past, by limiting to a strict minimum the number of people who knew the plans for the burial chambers and the means to gain access to them. The hereditary succession of the vocation of architect in certain families was one of the surest safeguards (in theory) against the danger of revelation, and this hereditary transmission can be seen next in Greece and Rome. The "Hippocratic oath" tells us the same held true for physicians. We also know that the torchbearers (*dadouchos*), heralds (*keryces*), and other officiants of the mysteries of Eleusis were theoretically recruited from certain important families of Athens. In all cases adoption made it possible to bend the rules while following them to the letter.

16. It can be asked if in this perception of harmonic correlations between architecture and music, Greek architects, who were undaunted by any numerical subtleties, did not attempt, in addition to symmetries and eurhythmy of a purely spatial nature, to intentionally introduce into their designs not only analogical reflections of their musical theories, but even proportions and rhythms that rigorously reproduced the mathematical elements of these theories. This must have been all the more tempting as these elements were not frequencies but the lengths of vibrating strings [of the lyre] (inversely proportional to the number of vibrations per second characterizing the sounds), and the casual way Vitruvius displayed his knowledge of the mathematical theory of the diatonic scale and related matters, by citing Philolaus and Aristoxenus of Tarentum, suggests that his Greek teachers, only whose most superficial secrets he reveals, *ad usum Cæsaris*, seems to show they were not, indeed, able to resist the temptation. It so happens that a Greek scientist, A. Georgiades (an engineer in Paris, and ex-departmental engineer for Attica) has studied the dimensions and proportions in the temples of Hellas, but from a very particular perspective. His conclusions, published in Athens in 1926 (under the title *Harmony in Architectural Composition*), are affirmative.

CANON OF PYTHAGORAS

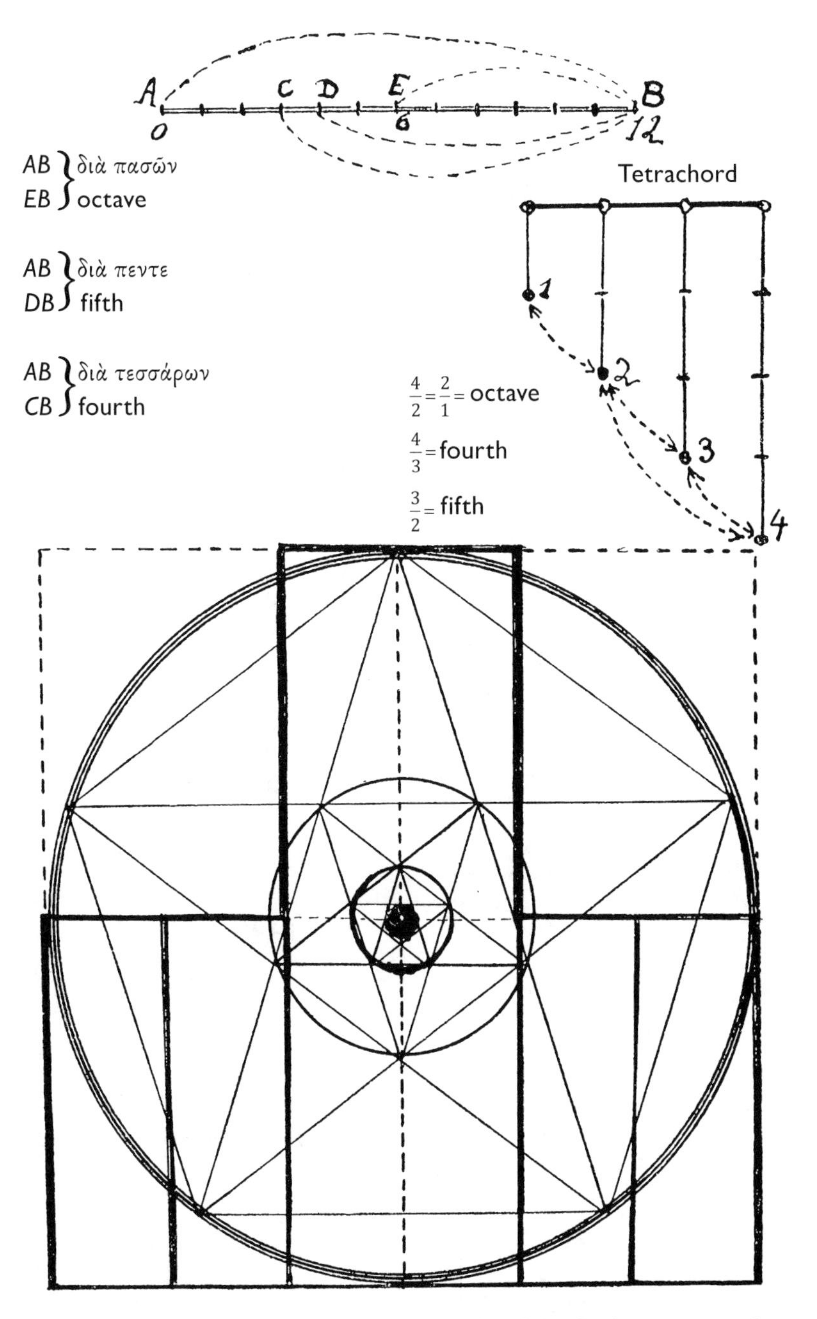

Plate 35. *Top,* Canon of Pythagoras: intervals of the Pythagorean scale; *bottom,* Gothic design (vertical section showing width of the nave and diameter of the pillars connected through the pentagram) (after F. M. Lund)

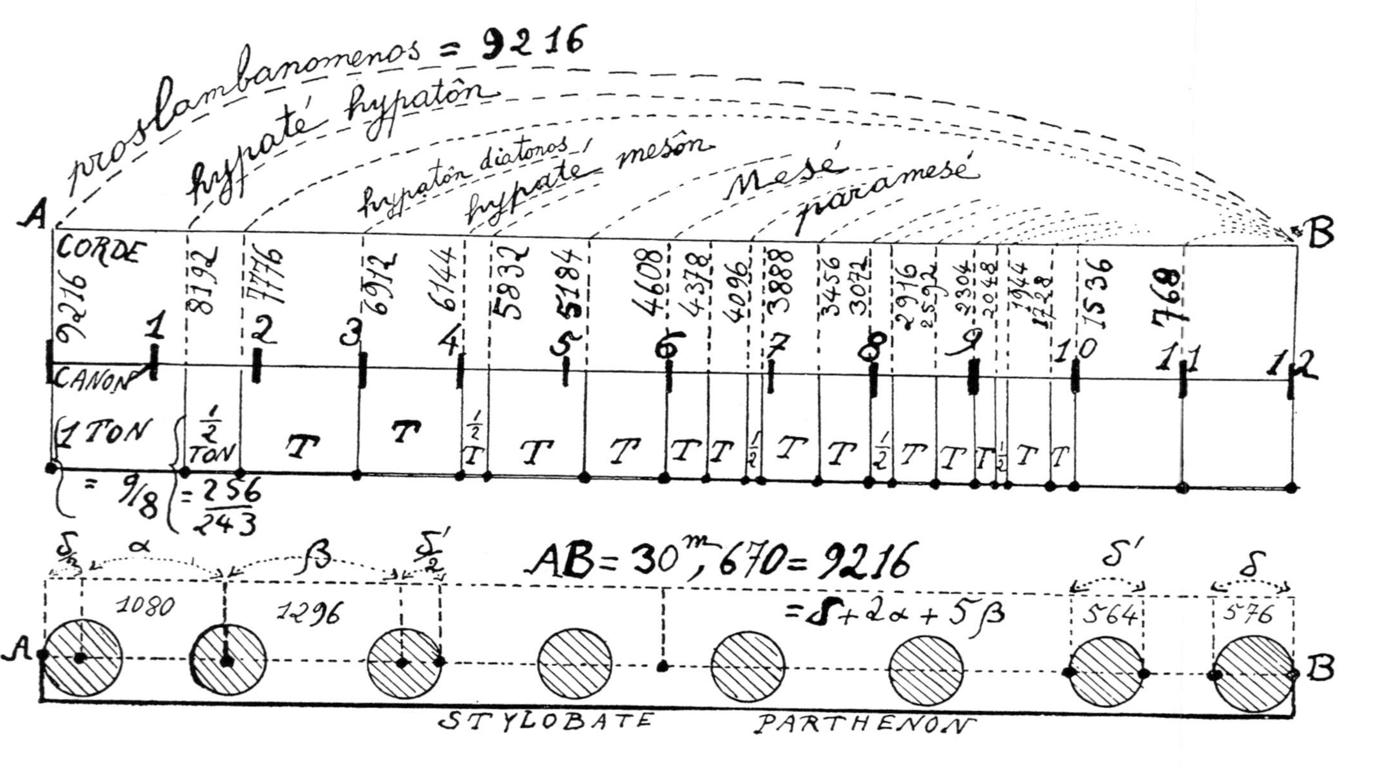

Plate 36. Relationship between the Pythagorean scale and the spacing of the columns of Greek temples (schematic design of the Parthenon) (after Georgiades)

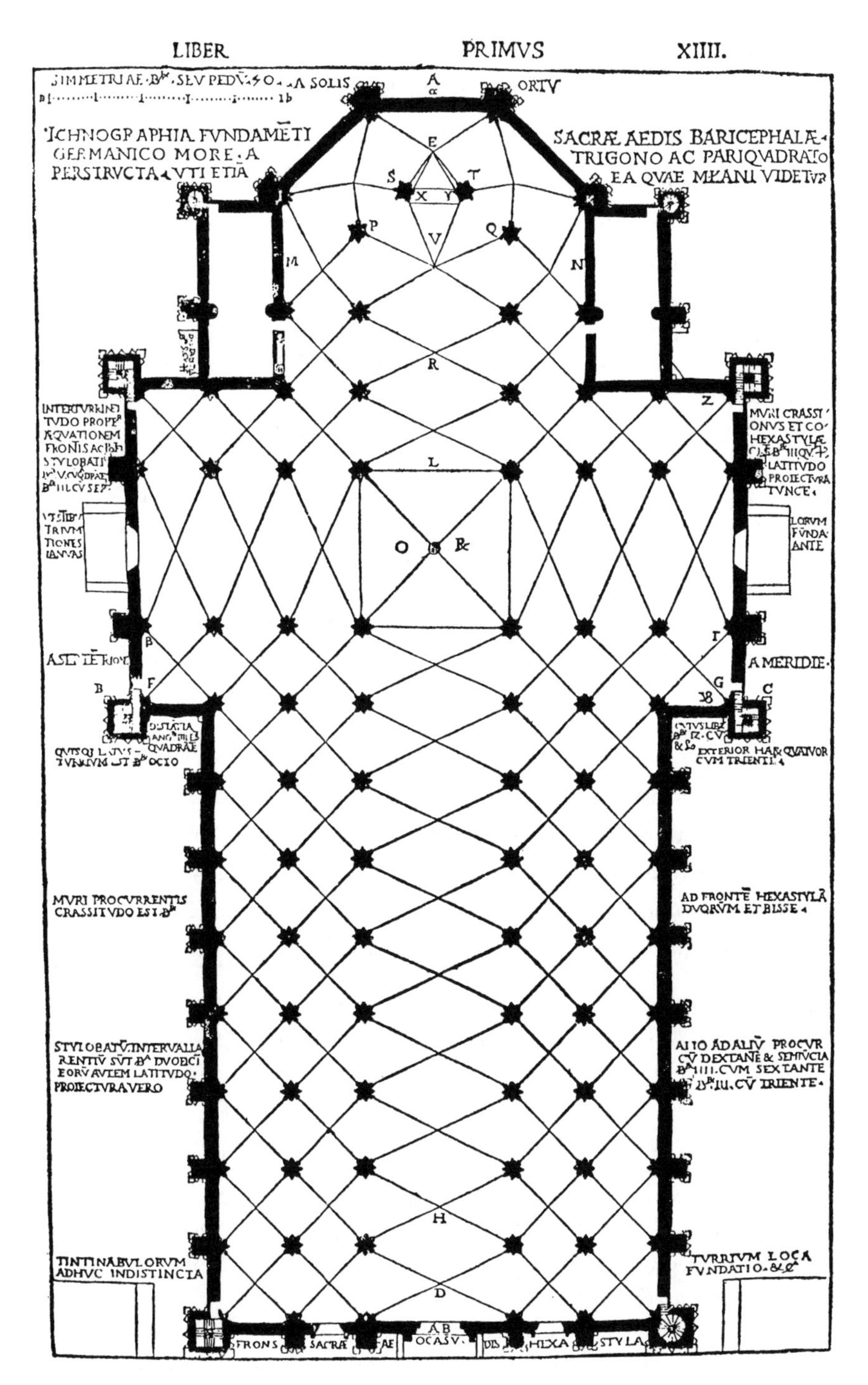

Plate 37. Milan cathedral; plan reproduced by Caesar Caesariano in his commentary on Virtruvius (1521)

The transmission of this esoteric notion of architecture to the Gothic builders was carried out by the builders' guilds and Neoplatonic philosophy; the Celtic-Nordic spirit enriched and renewed the classic theory of proportion by incorporating into it dream and the "Gothic forest" without taking anything from its rigor and geometric reliability.

Again, it was Plato who provided Mössel with the Ariadne's thread

---

(cont. from p. 92) We know that the measurements taken of Greek temples reveal, aside from the deviations or distortions obviously intended to produce what is called "optical corrections" (like the bulging of the columns, the inward inclination of the outer columns, the upward swelling of the cornices, and so on), other intentional irregularities that are harder to explain, especially in the diameter of the columns and the space separating them. It is in this arrangement of the columns on the stylobate that Georgiades discovered (for the Parthenon and the Propylaea, among others) numbers that were rigorously proportional to the elements of the Pythagorean scale when one takes the width of the stylobate as the "canon" (musical string whose length is altered to obtain the different intervals and chords) or *proslambanomenos* of 9,216 units.

For the Parthenon (see plate 36), for example, the lengths of the principal elements in meters on the one hand and in "diatonic" units on the other are:

	In Meters	In Diatonic Units
Width of the stylobate:	$\sigma = 30.670$	$9{,}216 = 12 \times 12 \times 64$
Interval between the corner column and its neighbor:	$\alpha = 3.594$	$1{,}080 = 12 \times 90$
Normal interval (between other columns):	$\beta = 4.313$	$1{,}296 = 12 \times 108$
Corner columns' diameter:	$\gamma = 1.916$	$576 = 12 \times 48$
Other columns' (normal) diameter:	$\delta = 1.875$	$564 = 12 \times 47$

Georgiades shows that these numbers α, β, γ, and δ, by virtue of their common factors and the proportions that connect them to the "canon" 9,216 (width of the stylobate), produce through simple combinations (additions, subtractions, ratios) all the intervals and chords corresponding to the diatonic scale. (For example, $\frac{\beta}{4(\beta-\alpha)} = \frac{1296}{864}$ produces the fifth $\frac{(9)^2}{(8)^2} \times \frac{32}{27} = \frac{3}{2}$; $\frac{3\gamma - \frac{\gamma}{2}}{2\gamma - \frac{\gamma}{2}} = \frac{3\gamma - \frac{\gamma}{2}}{\alpha} = \frac{1440}{1080}$ produces the fourth $\frac{9}{8} \times \frac{10}{9} \times \frac{16}{15} = \frac{4}{3}$; and so forth). For his demonstrations and reductions Georgiades used the famous "universal" proportion 6, 8, 9, 12 that Iamblichus attributes to Pythagoras himself and whose virtues were praised by Nicomachus (it contains, in fact, the three kinds of classic proportions, the arithmetic series 6, 9, 12, the harmonic series 6, 8, 12, because $\frac{12}{6} = \frac{12-8}{8-6}$, and the discontinuous geometric proportion $\frac{12}{8} = \frac{9}{6}$); he found rigorous harmonic compositions of this kind in the stylobates of twelve of the major Greek temples (including Eleusis). His findings also provide justification for the assertion made by René Praux (*Temps,* September 20, 1926) that "the entire temple was a musical

that led to his engaging synthesis—the *Timaeus* (as always) and this passage from the *Philebus:*

> But by beauty of shape I want you here to understand not what the multitude generally means by this expression, like the beauty of living beings and the paintings representing them, but something rectilinear and circular, and the surfaces and solids that one can produce from the rectilinear and the circular, with compass, cord, and set square. Because these things are not, like the others, conditionally beautiful, but are beautiful in themselves.

We know from Synesius that one of the higher initiatory grades of the Eleusinian Mysteries was called "initiation into the circle." The Pythagorean-Orphic inscription of Petilia has allowed us to approximately recreate the details of this ceremony[17] in which a circle traced on the ground plays a major role.[18]

We again find the importance of the circle and its center indicated

---

symphony in marble." It could also be said that the Pythagorean initiate contemplating the façade of a temple could not only see there "the proportions of a young woman he had happily loved," or those of his favorite athlete, but even the leitmotif—to the letter—of this or that Orphic hymn.

We should note that the number 576 also appears in the proportion of the Great Pyramid (it is equal to 4 × 144, the number of royal cubits from $a + c$, the total of the small side and the hypotenuse of the meridian half-triangle; for more see footnote 22 later in this chapter), and in the number of Plato's world soul (as well as, moreover, 1,296 and 9,216), which is also equal to 8 × 73 (72 being 360° ÷ 5, the angle at the center of the pentagon), and finally, that the number 108, which appears in various multiples in these diatonic elements of the stylobate of the Parthenon, is the vertex angle of the pentagon.

17. Victor Magnien, *Les Mystères d'Éleusis* (Paris: Payot, 1929).

18. This is the first of the upper grades, or the *holoclere* (complete) initiation, that follows the *epopteia,* which permits the soul to free itself from subjugation to its many desires and find its unity, in accord with the divine unity.

Paul Le Cour has shown me a photograph that he took at Eleusis of a circle with eight rays carved on a marble slab. He also photographed identical circles, but reduced to the size of masons' marks, on other stones in Eleusis.

in a mysterious medieval quatrain that was handed down by the master stone carvers of the Gothic period.

> *Ein punkt der in dem Zirkel geht,*
> *Der im Quadrat und Dreyangel steht,*
> *Kennst du den Punkt, so ist es gut.*
> *Kennst du ihn nil, so ist's umbsonst!*[19]

In part 2, I will provide a variation of this curious saying that takes on a very precise meaning when seen in the light of Mössel's diagrams: in the designs of the architect or master, as in the mason's mark that the journeyman or master entering a strange city had to use to prove himself (by placing it in its grid inside a circle), it always involves finding the directing circle and the pole of symmetry that governs all the geometry of the design and provides the key to its unity.

Vitruvius describes the classic design of Greek theaters in which three squares are inscribed in the directing circle, and the Roman design in which, conversely, four equilateral triangles come into play. In both cases the circle is divided in twelve segments, and the text is precise enough to allow the corresponding plans to be drawn. In these plans, as in Mössel's diagrams, not a single point is left to chance, but all are determined by the geometric symmetry of the concept.

Similarly, a very interesting plate in Caesar Caesariano's edition of Vitruvius (Como, 1521) depicting the façade of the Milan cathedral conforms equally well to both Lund's and Mössel's theories: the directing circle is not only drawn but mentioned explicitly in the commentary: "*ut possint per orthographiam ac scenographiam perducere omnes quascumquae lineas non solum ad circini centrum*" (see plate 38).[20]

---

19. A point that goes in the circle,
And that sits in the square and the triangle:
If you know the point, then all is good,
If you know it not, then all is vain!

20. The Germanic style of establishing proportion uses equilateral triangles. A sketch that dates from 1391 housed in the archives of the Milan cathedral shows the exact same design but without the circles that provide its key.

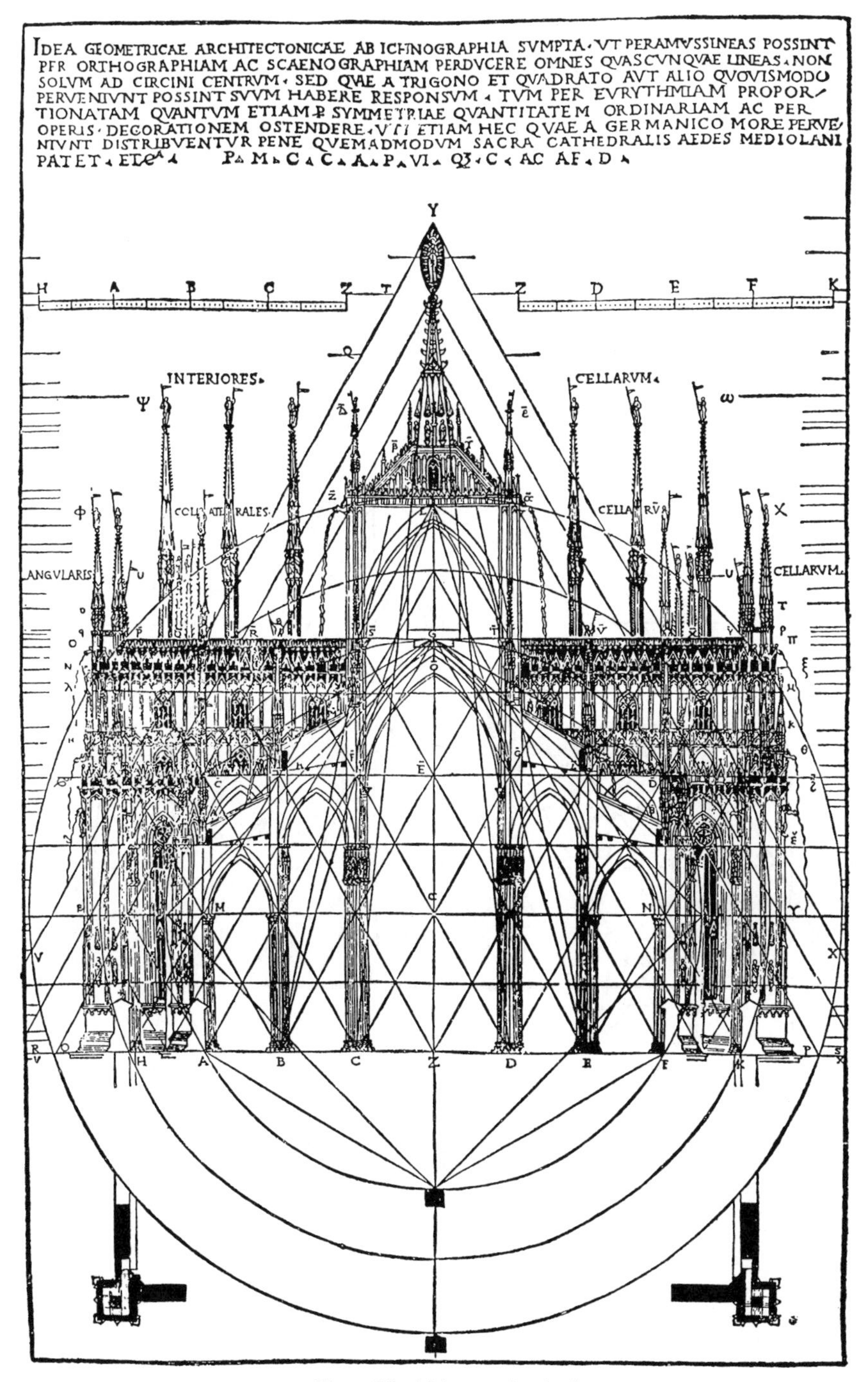

Plate 38. Milan cathedral
(elevation and vertical section by Caesar Caesariano, 1521)

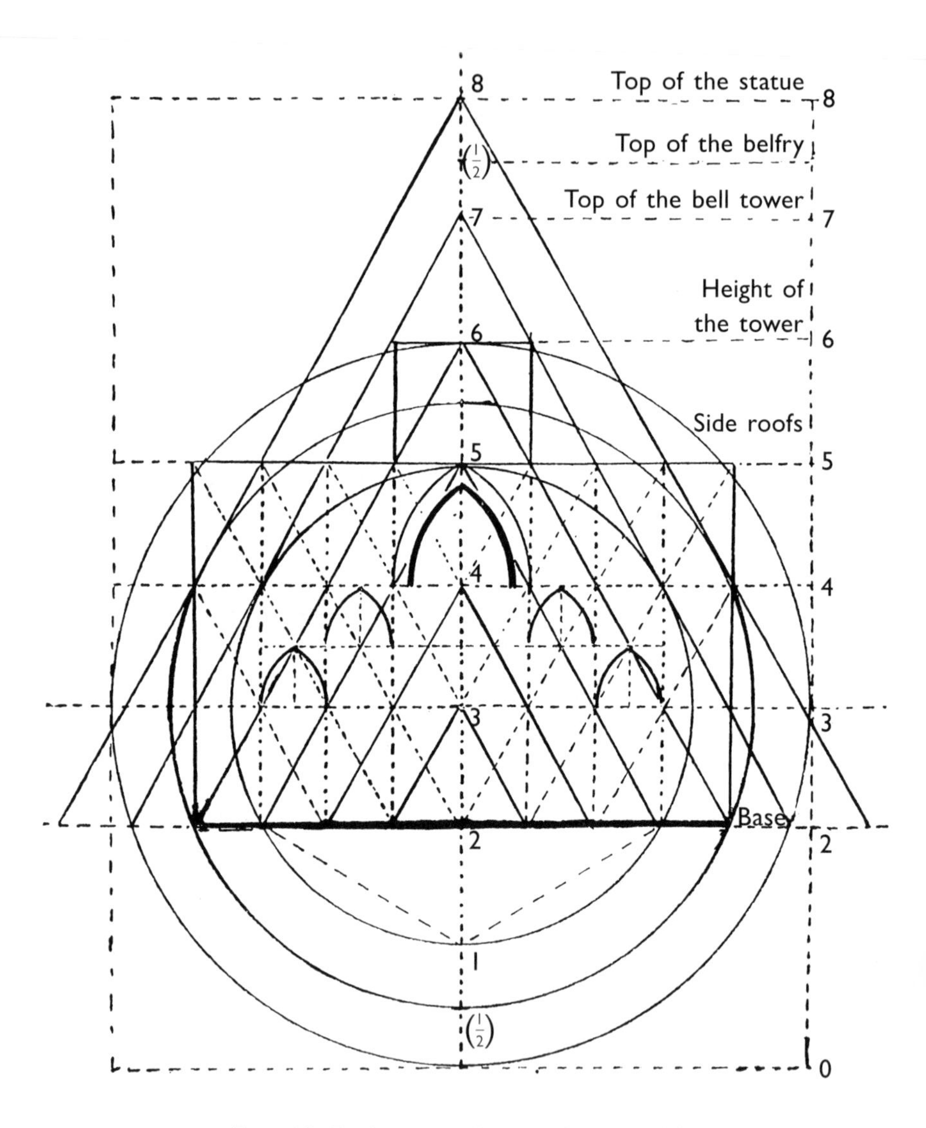

Plate 39. Explanatory diagram for plate 38
(D. Wiener)

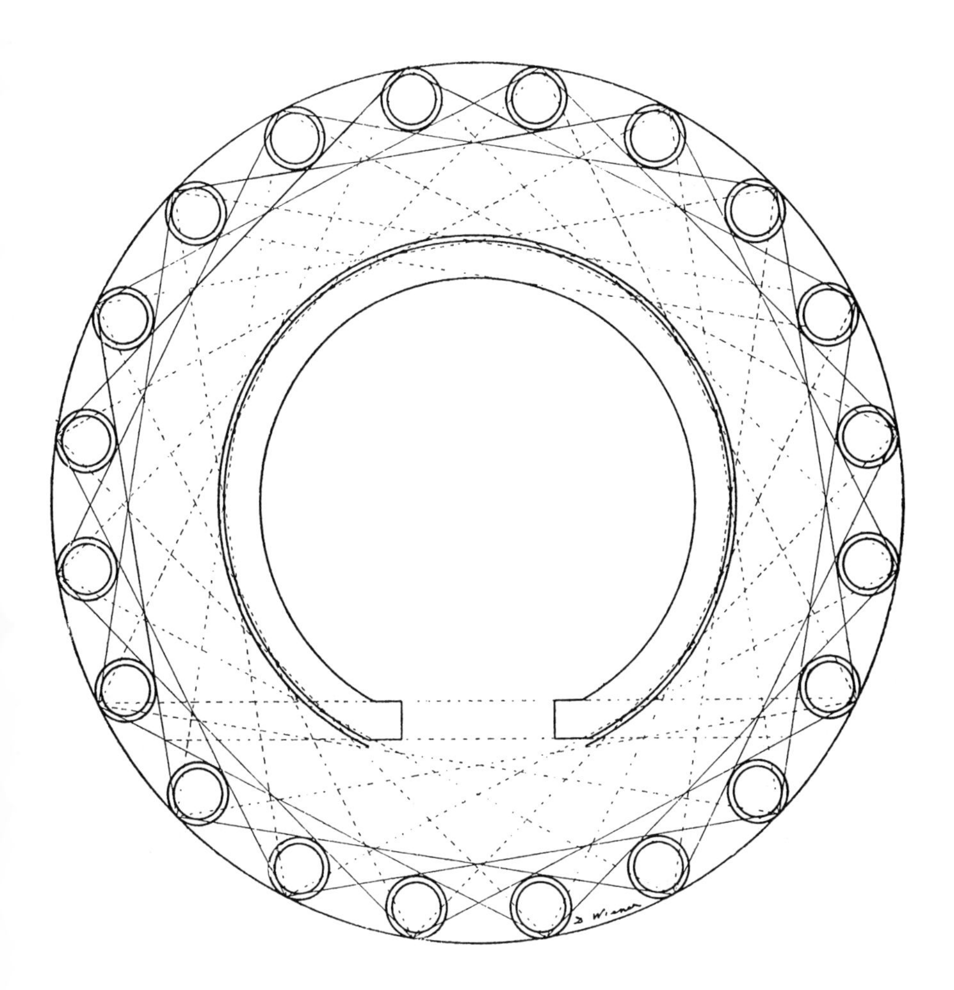

Plate 40. Small temple of Minerva in Rome, harmonic analysis (D. Wiener)

Moreover, I could not do better than to cite several passages in which Mössel has lucidly summarized his system.

> The composition of (architectural) plans since the beginning of Egyptian architecture until the end of the Middle Ages is in the vast majority of cases not arithmetical but geometrical. It derives from regular, angular segmentations of the circle. . . . From these various divisions of the circle arise systems of rectangles, triangles, and convex and star polygons, which represent grids having the shape and role of systems of coordinates. These geometric conformations are the foundation of the artistic compositions in architecture, painting, and bas-relief sculpture. This geometry that moves in the plan (horizontal contour and elevation) can be regarded as the projection of geometry in space. The specific divisions of the circle and the numerical ratios that characterize them appear in the planar projections of the regular solids inscribed in the sphere—tetrahedron, octahedron, cube, dodecahedron, and icosahedron. These "Platonic" solids are at play in all the theory and practice of antiquity and the Middle Ages, with an exceedingly important role as the starting point for cosmological speculations.
>
> The decadic division of the circle and its derivatives appears to be the system most often used by the old masters. Through the golden section (the preeminent continuous proportion[21]) that is the consequence of these systems, the elements of the buildings (or other artworks) are harmonized in increasing or decreasing chains from the measures of the outside dimensions of the plan or the elevation down to the smaller subdivisions of the component parts, and all in the simplest way possible.

21. German authors often quite simply called the golden section "the continuous proportion" (or "constant," *stetige Proportion*). It is possible that for Vitruvius as well, what he called analogy was not continuous geometric proportion in general, but the particular, characteristic case of the golden section, the preeminent engenderer of repeating analogies. This, moreover, was the interpretation of Pacioli and other Renaissance architects.

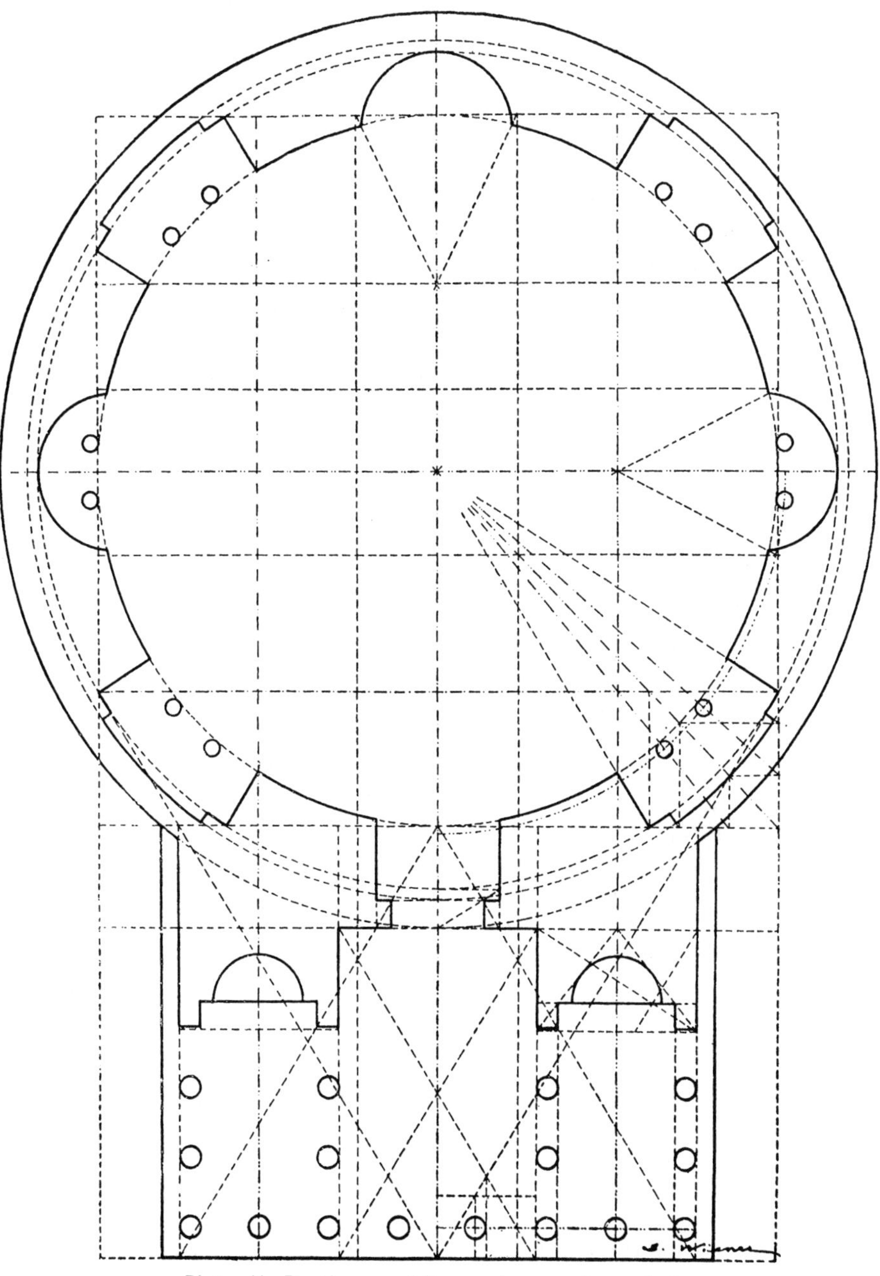

Plate 41. Pantheon of Rome, harmonic analysis
(D. Wiener)

Again, the best commentary on Ernst Mössel's theory is still provided by his diagrams and the specifications and numerical ratios connected to the dimensions of the buildings analyzed.[22]

22. These figures are all the more interesting as Mössel shows in juxtaposed columns exact measurements on one side (generally to the nearest millimeter) with mention of the sources from where they were collected, and on the other the theoretical figures that would result from a design that conformed strictly to the abstract diagram type from which the example appeared to arise. The slight gaps and deviations are just as impressive as the vast number of times the two columns rigorously coincide. The Fibonacci approximations 160, 100, 60 appear quite often (in the units of length employed in construction), and were already noted in the Pyramid of Cheops in my *Esthétique des proportions.* I provide several examples here of the numerical verification:

EGYPT

Tomb 87 at Giza

L = 9.52

W = 5.82

$9.52 \times 0.618 = 5.883$

Tomb 1 at Souat-el-Meitun

L = 10.32

W = 6.42

$10.32 \times 0.618 = 6.38$

Temple f at Naga

L = 16.30

W = 7.33

$7.33 \times 2.236 = 16.39$

L interior cella = 9.05

W interior cella = 5.63

$5.63 \times 1.618 = 9.11$

Tomb 105 at Giza

L = 4.75

W = 2.95

H = 1.80

$4.75 \times 0.618 = 2.94$

$4.75 = 0.618^2 - 1.81$

Great Temple of Philae

Interior volume of the *prosekos*

W = 9.53

Depth = 3.64

H = 5.90

$9.53 \times 0.618 = 5.89$

$9.53 \times 0.618^2 = 3.64$

L = length; W = width, H = height

The readings are in meters unless otherwise indicated.

$0.618 = \frac{1}{\Phi}$

$1.618 = \Phi$, ratio of the golden section.

$2.236 = \sqrt{5} = 2\Phi - 1.$

All the other dimensions in Philae are connected to the preceding ones, including those of the monolithic tabernacle, producing a Φ series diminishing to $9.53 \times \frac{1}{\Phi^6}$. It seems

Like Hambidge and Lund, Mössel noted that his diagrammed plans, with their polygons and pieces of polygons inscribed within a

---

that whole-number approximations (in Egyptian cubits or simple multiples of the cubit) have often been used in preference to the strict golden section. These approximations were invariably borrowed from the terms of the Fibonacci series, 1, 1, 2, 3, 5, 8, 13, 21, 34, 55, 89, 144, which also appears everywhere in nature as a very strict discontinuous approximation of the ideal Φ series. I noted in *Esthétique des proportions* (an observation made by Jarolinek and Klippisch) that the half-meridian triangle of the Great Pyramid (a rectangular triangle whose hypotenuse and small side are at first glance in the strict ratio of the golden section) appears to result, if one calls r (= 0.524m) the Egyptian royal cubit, from a very ingenious construction starting from $a + c = 144 \times 4r$; then $c = 89 \times 4r$, and $a = 55 \times 4r$ (55, 89, and 144 are precisely consecutive terms of the Fibonacci series, and 144 is also the square of 12). The height, h, of this right triangle and of the pyramid (146.6m) is approximately equal to $70 \times 4r$, because of the curious coincidence $55^2 + 70^2 = 7{,}925$ and $89^2 = 7{,}921$.

The length of the sides at the base of the pyramid in accordance with this system would theoretically be: $2a = 2 \times 55 \times 4 \times 0.524 = 230.560$m.

It so happens that the last measurements made on-site in 1925 produced as the average for the four base sides (with a deviation of 20 centimeters between the largest and smallest) the value 230.634m (plate 44; another average figure provided by Borchardt: 230.36m).

Ernst Mössel noted similarly in the dimensions of the west temple of Philae expressed in Egyptian cubits the series 8, 13, 21, 34.

Lastly, a temple façade depicted on a fresco in the hall of pillars of the temple of Khonsu in Karnak is a rigorous Φ rectangle harmoniously subdivided. The specifications here are in millimeters:

Total width: 329	$329 \times 0.618 = 203$
Total height: 203	
Pylon width: 127	$329 \times 0.618^2 = 126$

These strict correlations and slight deviations can be found again in the dimensions of Greek temples and the Roman and Gothic constructions analyzed by Mössel.

For example, in the church of the Maria Laach Abbey, we have:

Total length: 216.8	$100 \times \Phi 2 = 216.8$
Height of the tower: 100	$100 \times \frac{1}{\Phi} = 61.8$
Length of the nave: 99.20	
Width of the nave: 61.42	

The numbers are given in Rhenan feet of 0.315m (the unit of length used during the time of its construction). The reading of 261.8 for the total length, whereas $\Phi^2 = \Phi + 1 = 2.168$, could not be any more suggestive.

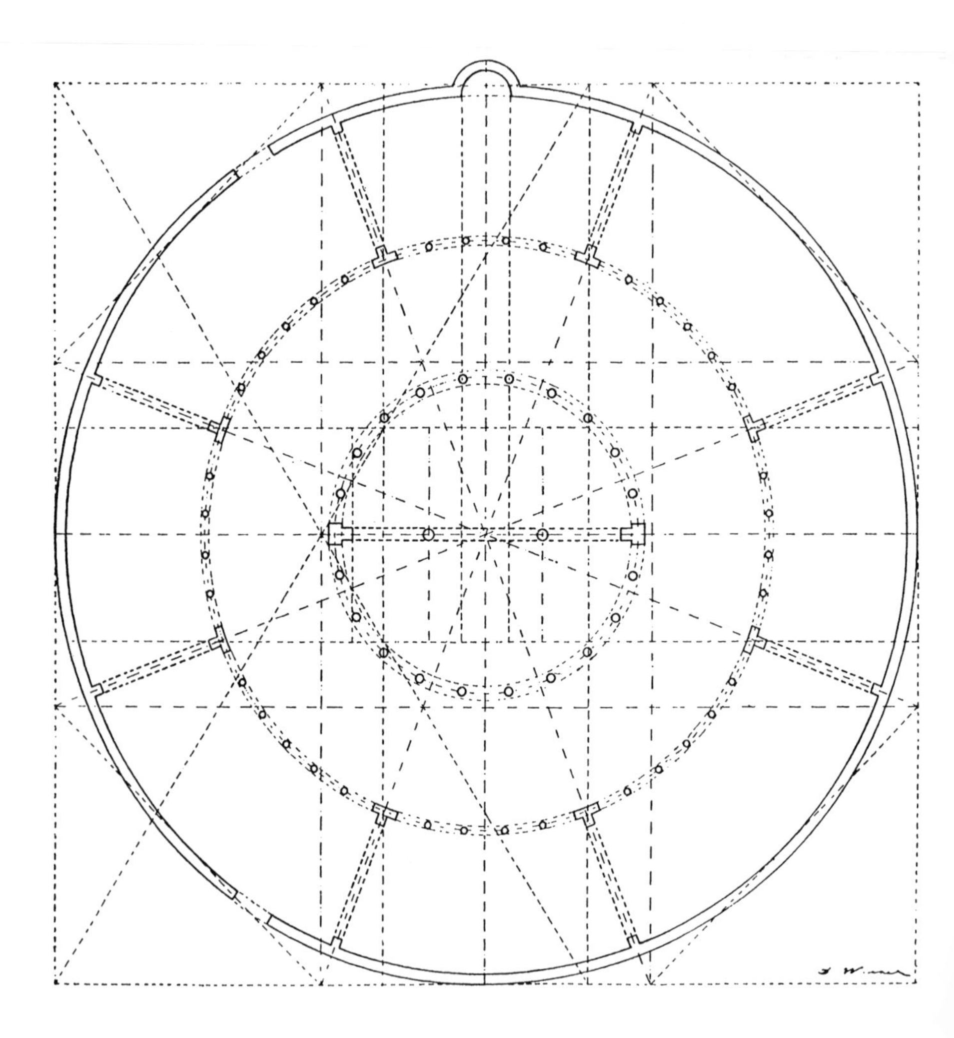

Plate 42. San Stefano Rotunda in Rome, harmonic analysis (D. Wiener)

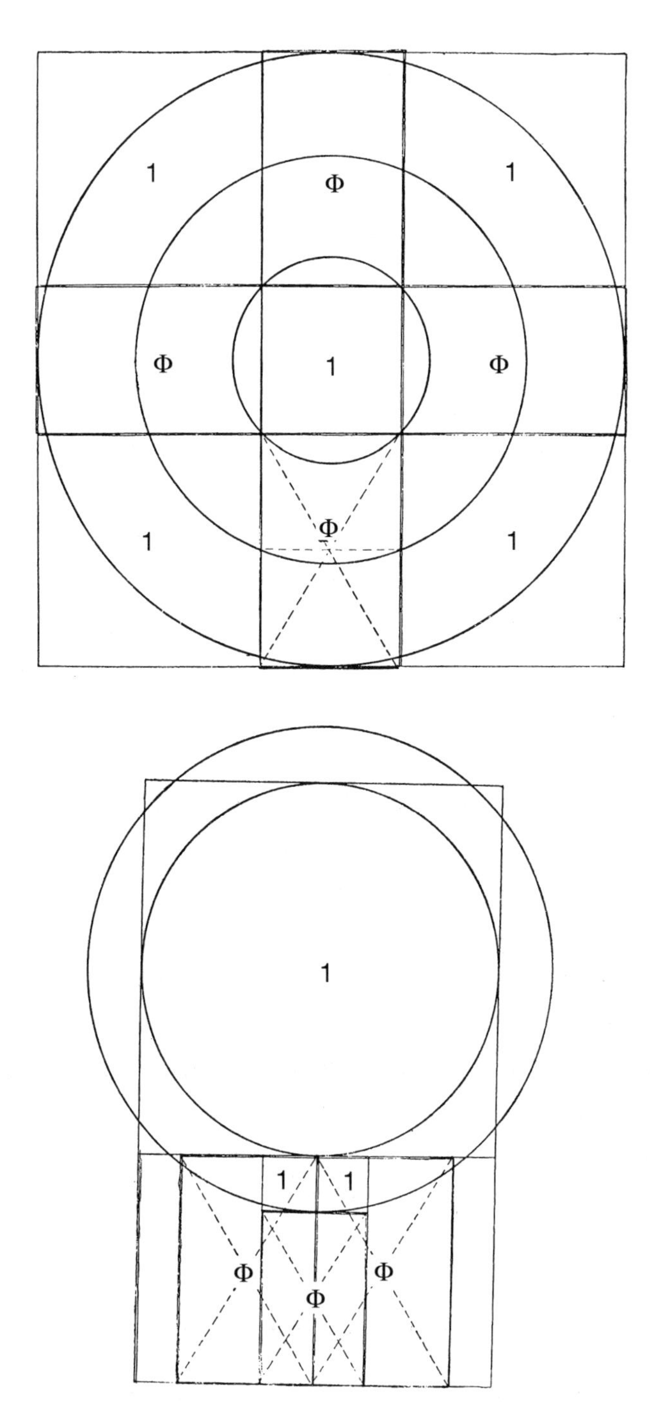

Plate 43. *Top*, explanatory diagram for the San Stefano plan; *bottom*, explanatory diagram for the Roman Pantheon plan

circle, could be considered as projections or sections of solid figures in space, these figures being polyhedrons inscribed inside the sphere. This is especially true for Mössel's diagrams with their directing circle, and most especially for the complex diagrams with a double directing circle and blended themes in which the projection could present different polyhedrons that were still harmoniously connected by virtue of their inscription within the same sphere.

Mössel also examines the volumes formed by the principal envelopes of Egyptian, Greek, or Gothic monuments and often finds true parallelepipeds with rectangular bases, what in my *Esthétique des proportions* I called "Egyptian volumes" after the most remarkable among them: the King's Chamber in the Pyramid of Cheops (edges with the proportions 1, 2, $\frac{\sqrt{5}}{2}$, large diagonal of $\frac{5}{2}$; or edges with the proportions 2, 4, $\sqrt{5}$, large diagonal of 5; one of the "diagonal rectangles" is formed by two "sacred" 3-4-5 triangles, another vertical one is a double square like the base rectangle). Other examples include, in tomb no. 105 at Giza and the Hathor temple at Deir-el-Medina the proportions 1, Φ, Φ^2; the *prosekos* of the great temple of Philae displays the same proportions but with the Φ rectangle flat. For Ulm cathedral (large nave) we have the proportions 1, $\sqrt{5}$, Φ^3, and so forth.

Once again the positions and proportions resulting from the three systems (Hambidge, Lund, Mössel) of designs are generally identical, and it is even worth the time to compare the same plan among them.[23] Lund's transverse sections (for example, the stellar gemmation of the seven concentric pentagrams that allows him to move from the transverse diameter of the nave of Cologne cathedral to that of the pillars, see plate 35) often coincide with Mössel's diagrams. His directing circles can be placed over the rectangular harmonic designs of Hambidge and vice versa (precisely for the reason that the constructions of Hambidge and Mössel are only two different ways—orthogonal projection and

23. I have reproduced (plates 40, 41, and 42) the harmonic analyses of several ancient buildings made by D. Wiener; they can be transposed equally into Hambidge's or Mössel's system.

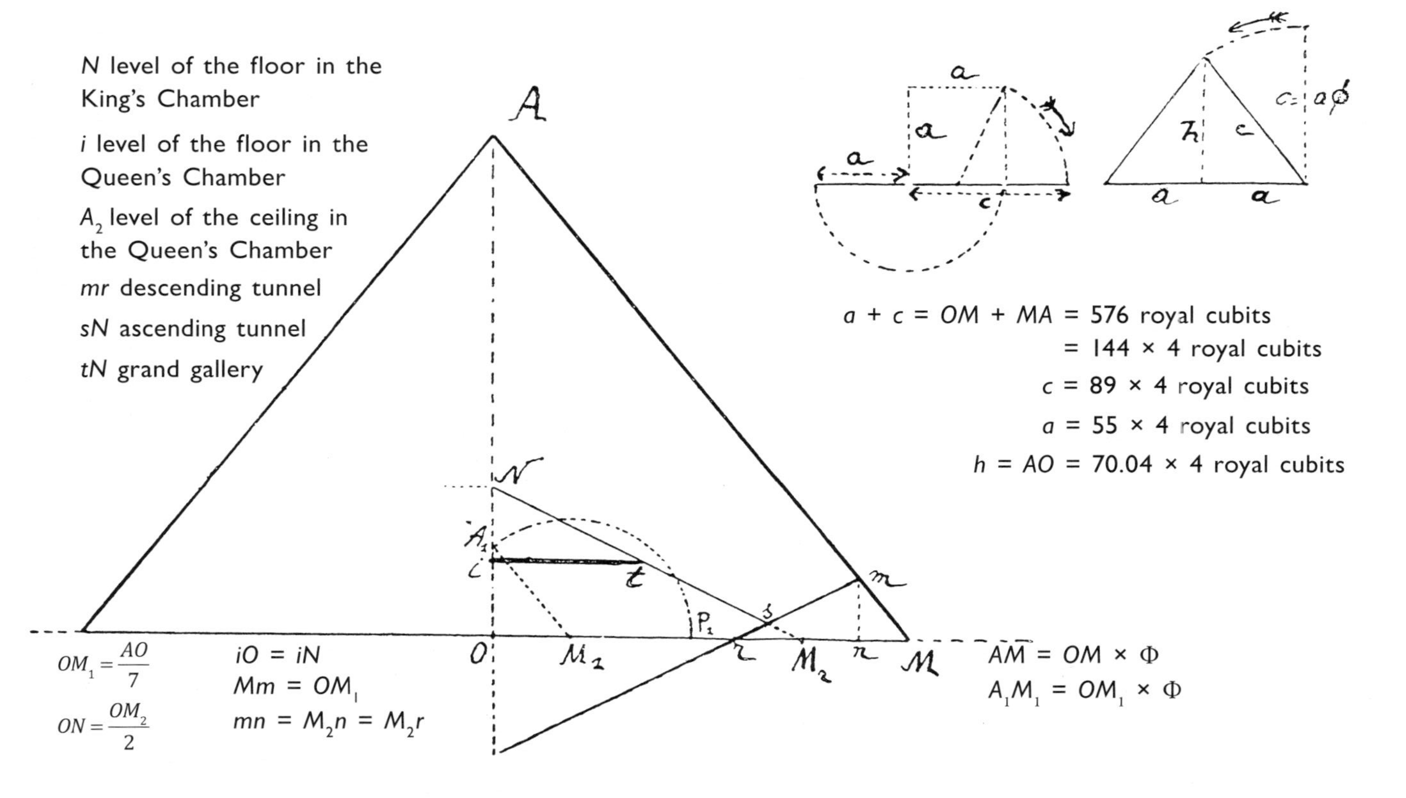

Plate 44. The Great Pyramid, meridian section

central projection—of projecting on a surface plane the same "event" in three dimensions).

It is impossible to not be impressed by Mössel's results in this or that design as well as by the remarkable synthesis in which his research has culminated. His theory is so logical that it is tempting at first glance to adopt it exclusively, but this decision may not be justified. To the contrary, after studying these three systems concurrently, in the light of Greek mathematical texts on proportion and the text of Vitruvius, it now appears to me that each contains one part of the secret of the Egyptian, Greek, and Gothic architectural symphonies, and even that the three together may contain the whole truth, and deliver the whole secret.[24]

The fact that the compositions that arise from this appear terribly subtle (when in reality, once one knows the geometric and arithmetic use of the golden section, they are quite simple) is no argument against them—quite the contrary.[25] The Greeks, in mathematics as much as in aesthetics and metaphysics (and, in the composition of the proportions

---

24. The rectangular triangle whose sides are proportional to 1, 2, $\sqrt{5}$ and that allowed Mössel to reconstruct with a rule and compass all the dimensions, in whole and in detail, of the typical Doric temple with six columns (plate 31) is nothing other than the (oblique) half of the double-square rectangle that plays such a major role in Lund's theories. Lund's double squares and Mössel's directing circles can be found as well in the Como version of the text of Vitruvius.

As a reminder, here is how Mössel connects the specifications of the typical Doric temple to the triangle 1, 2 $\sqrt{5}$. If the small side is equal to the interior width of the cella, the hypotenuse $2a\sqrt{5}$ is equal to the width of the stylobate, the other side $2a$ to the distance apart from axis to axis of the outer columns; then the total height h will be equal to $a\Phi$, the height from the ground of the capitals (complete columns with pediment) to a; the inferior diameter of the columns to $\frac{h}{10}$, and so on and so forth.

25. It is even likely that the continuous geometric proportion, Plato and Vitruvius's *analogia,* which (especially in its most fruitful form, the golden section) makes it possible to realize in an infinite number of ways the analogical recurrence of forms, of unity in variety, and which is at the base of all the systems of symmetry explained in this book, was not the only one to be used in the outlines of Greek temples, and that the search for musical correlations sometimes led to the use, among others, of harmonic proportion, if not to say, as Georgiades does, of the actual transcription in the proportions of some temples of musical rhythms as such.

of a temple, all three of these disciplines come into play), had no fear of subtlety or difficulty. It could even be said that, paradoxically, they revered both clarity and mystery, on condition that this clarity, the final unity toward which their philosophical and religious thinking tended, was of no value unless it was attained by an arduous journey through a labyrinth of symbols and analogies to the center where, in their true perspective, ideas and forms, Truth, Beauty, and Harmony, were illuminated in the revelation of Unity.

And through the esoteric Pythagorean geometry and the ever-parallel tradition of architects and artisans of stone, the great harmonies, one and many, of the music of temple and life, macrocosm and microcosm, were handed down from Egypt to the Gothic cathedrals.

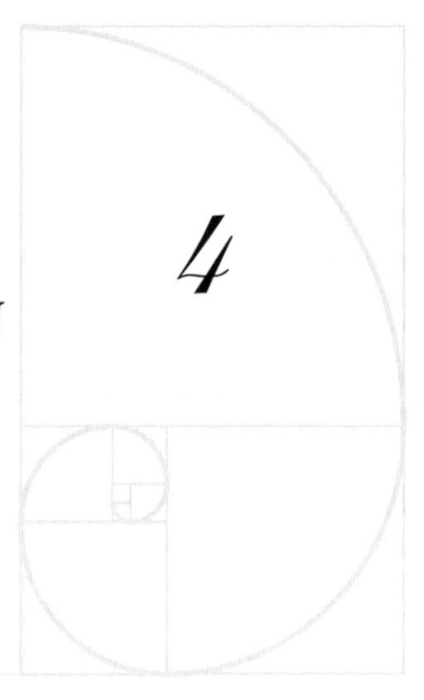

The Orchestration of Volumes and Architectural Harmony

This luminous rupture
sets the soul I once had to dreaming
of its secret architecture.

Paul Valéry, "The Pomegranates"

The present chapter is a gloss for verifying from the perspective of their concrete application the graphic procedures and theories presented in the preceding chapters. While it may offer some value to architects, I advise other readers not to linger too long over it.

The diagrams and "harmonic canons" mentioned or examined earlier are two-dimensional drawings, in other words "plans," each of which represents a construction or assemblage of imaginary geometric constructions situated on a single surface plane. Now, in architecture it is undeniably a matter of creating volumes, therefore of "conceiving," of thinking in three dimensions. The appealing manipulation of linear proportions drawn from the Φ series or other related schemas, or even those of surfaces that have been harmoniously decomposed or connected by Hambidge's "dynamic symmetry," should obviously not allow the creator of volumes to forget this. In my *Esthétique des proportions,* I especially emphasized the importance of the "science of space," particularly that of the study of the five regular solids (Platonic) and their inscription inside the sphere, and analysis of what I called right-angled parallelepipeds, orthogonal elements of volume whose proportions can be compared.[1]

The most interesting are the Egyptian volumes related to the golden section. I again must highlight the volume of the King's Chamber in the Great Pyramid. It has a double square for its base, and its height is half the diagonal of this double square, which is to say, its dimensions are proportional to 2, $\frac{\sqrt{5}}{2}$, and 1, or 4, $\sqrt{5}$, and 2. If we take 5 Egyptian royal cubits as the unit (5 × 0.524m), its dimensions are exactly 4, $\sqrt{5}$,

1. The shape of a rectangle, the orthogonal element of surface, is completely characterized by two numbers *a, b,* proportional to the lengths of its sides, or even, if you like, by the ratio $\frac{a}{b}$ between these two numbers.

This is what Hambidge called the $\frac{4}{3}$ rectangle, the $\sqrt{2}$ rectangle, and what I call the Φ rectangle, and so forth. This refers to rectangles whose long and short sides are expressed, respectively, in the ratios $\frac{4}{3}$, $\Phi = \frac{\sqrt{5}+1}{2} = 1.618...$, $\frac{\sqrt{2}}{1} = 1.414...$, and so on.

In similar fashion, one can characterize the shape of a right-angled parallelepiped, the orthogonal element of volume, by three numbers *a, b, c,* proportional to its dimensions, or even, strictly speaking, by two numbers representing the ratios $\frac{a}{b}$ and $\frac{b}{c}$ (this also results in the ratio $\frac{a}{c}$ as $\frac{a}{b} \times \frac{b}{c} = \frac{a}{c}$). The Greeks (Plato in particular) were greatly occupied with how the "solid numbers" of the form $a \times b \times c$ were able to accurately represent the volumes of right-angled parallelepipeds with the specifications (dimensions) proportional to *a, b,* and *c,* and to whom a series of three quantities appeared interesting especially if they formed a continuous proportion (geometric, for example, $\frac{a}{b} = \frac{b}{c}$) and gave their close attention to the orthogonal volumes characterized by numbers in geometric, arithmetic, or harmonic proportion. Plato and his successors bestowed special names on the different shapes of orthogonal solid numbers or volumes. This thereby gives us, in addition to the cubes $a \times a \times a$, the altars $a \times b \times c$ (the three dimensions are different), the bricks or plinths $a \times a \times b$ (with $b < a$), and the beams $a \times a \times b$ (with $b > a$).

It appears that the Egyptians displayed a marked preference earlier for the orthogonal volumes of the proportions 1, 1, Φ; 1, Φ, Φ; 1, Φ, Φ^2 (S. Colman's "golden volume"), and 1, Φ^2, Φ^3, governed by the golden section, as well as for the "Fibonaccian" approximations 6, 6, 10; 6, 10, 10; and 10, 16, 26, and so on. These volumes in which, separately or combined, the principal types of proportion figured, especially the tenth (see chapter 1, note 19), $\frac{c-a}{c-b} = \frac{b}{a}$, which applies equally to three consecutive terms from the Fibonacci series, 1, 1, 2, 3, 5, 8, 13, 21, and to three consecutive terms, Φ, Φ^{n-1}, Φ^{n-2}, of the Φ series, were handed down from Greece to the Middle Ages (see chapter 3).

and 2 ($\frac{4}{2}$ being the base double square). The "great diagonal" of the solid is equal to 5, and the two principal "diagonal rectangles" are characterized by $2\sqrt{5}$, $\sqrt{5}$ and 4, 3; the first (vertical) is therefore a double square, the second is formed by two "sacred" or Pythagorean 3-4-5 triangles, side by side.

It was most assuredly necessary to think and create architectural volumes "from the inside out." The ancients pursued this course, based on the secret revealed by the Pythagorean initiate Hippasus.[2]

The importance Plato granted to meditation on geometry in space is illustrated by the odd passage from the *Republic* in which he states that the state whose rulers were wise enough to impose the extensive study in school of the geometry of the solids would acquire a marked preeminence over all others. Plato personally reflected (more perhaps than anyone since) on the application of the notion of proportion to the regular solids. We have proof of this, in addition to the passages on this subject from the *Theaetetus,*[3] *Critias,*[4] and so on, in the theorem on the proportions between solids that he briefly puts forth in two lines from the *Timaeus,*[5] and the solution that he found for the famous problem of doubling the cube on which his friend and mentor, Archytas of Tarentum, had worked, and which also comes down to a question of

2. Proclus (commentary on Euclid) even attributes to Pythagoras the construction of the cosmic figures (the five regular polyhedrons). The first scholium of Euclid's thirteenth book only attributes to the Pythagoreans as such the rigorous construction of the cube, the tetrahedron, and the dodecahedron (this latter was divulged by Hippasus of Metapontum, who was excommunicated for it, bringing about a schism in the sect) and to Plato's friend Theaetetus (who fell in battle near Corinth in 360 BCE) the construction of the octahedron and the icosahedron and the synthetic and comparative studies of the five "Platonic" solids that Plato used in the *Timaeus.*

3. "And for the (proportions between) solids as well, we have found analogous laws."

4. This concerns the height of the great temple of Poseidonos, the capital of Atlantis, whose base was a double square and whose height was "proportioned in measure" ("συμμέτρον"), an expression we find later in Vitruvius.

5. It is always the "solid" (στερέοι) numbers or the three-dimensional figurate numbers, of the form $a \times b \times c$, that Plato uses indiscriminately for arithmetic or geometric problems concerning space.

the proportions between solids, specifically related to "Plato's theorem."

Here is what the problem of doubling the cube consists of:

An oracle, having enjoined the priests of Delos to replace their cubic altar with a similar one but double in volume, turned to the greatest mathematicians of antiquity for a solution. It involved constructing (or calculating) the length x of the edge of a cube so that its volume would be equal to double the volume of a given cube with edge a. The problem, despite its apparent simplicity, cannot be solved using euclidean methods, that is, by means of a strict construction by rule and compass, because the equation of the third degree $x^3 = 2a^3$ has no constructible root (one that can be reduced to a root or a combination of second-degree roots). The best mathematical minds of antiquity grappled with this problem: the Pythagorean initiate Hippocrates of Chios (excommunicated, as we saw earlier, like Hippasus, around 450 BCE for having revealed a secret about irrational numbers, which were—to use Plato's expression that was then borrowed by Hambidge—"potentially commensurable") broke it down in Pythagorean fashion to a question of proportions to demonstrate that it involved the placement of two continuous "medieties" y and z between two lengths, the second of which was twice as long as the first (from a letter to Eratosthenes preserved in a commentary on Archimedes by Eutocius in the sixth century CE, phrased algebraically $\frac{a}{y} = \frac{y}{z} = \frac{z}{2a}$ or $y^3 = 2a^3$; in it we again see the third-degree equation taken to the highest power).

Plato, Menechme, Nicomedes, and Eratosthenes found "mechanical" solutions to the problem, with the most elegant one provided by Plato. It is likely that it was the studies related to this "Delian" problem

The elliptical formulation of the theorem in the *Timaeus* says: "While a single mediety is enough to bind (in one proportion) two plane numbers (of the form $a \times b$), two medieties are necessary to connect two solid numbers."

The commentaries of Nicomachus (who explicitly states that this theorem is due to Plato himself) show: (1) that the envisioned solid numbers are two cubes *a*3, *b*3, *a* and *b* being prime numbers between them; (2) that the two medieties sought for should also be whole or rational "solid" numbers or volumes. The two sought-for medieties are then a^2b and ab^2, because $\frac{a^3}{a^2b} = \frac{ab^2}{b^3}$. R. D. Archer-Hind.

that led to Plato's discovery of his famous theorem mentioned earlier (see footnote 5).[6]

Luca Pacioli, who grasped Plato's ideas on this subject perfectly, recommended that architects take as their models and as objects for profitable harmonic meditations not only the five regular polyhedrons ("for . . . they serve as suggestion and an object of meditation for scholars and philosophers because of the 'divine proportion' that connects them," *De Divina Proportione,* bk. 2, chap. 18) but the semi-regular Archimedean solids (among others) as well.[7] To obtain full understanding, he advocated the study of small models of the solids. He recalled that he had

---

6. The riddle of the nuptial or marriage number in the *Republic,* which connects to a lunar cycle the conjunctions favorable for generation, is also solved by a construction of proportions between solids. It produces the interesting relationship among 4 cubes $6^3 = 3^3 + 4^3 + 5^3$. It should be noted that $6^3 = 216 = 3 \times 72$, and 216 in the legend of Pythagoras is the number of years that elapse between two successive incarnations of the master; 72, meanwhile, is one of the numbers that recurs most often in astrology and mythological cosmogony. It is the 360th part of 25,920, the duration of the precessional Great Year (the Metacosmesis of the Pythagoreans), with 72 years thus corresponding to a movement of 1 degree of the vernal point on the elliptic. In the Egyptian calendar, Thoth-Hermes carried off $\frac{1}{72}$ of each day from the moon, and from these 360 seventy-seconds he drew $\frac{360}{72} = 5$, the five epagomenal days during which the Egyptians celebrated the birth of the gods. In pure mythology, we have the 72 accomplices of Typhon (against Osiris), the 72 spirits corresponding to 72 portions of the sphere (Pacioli attributed great importance to a polyhedron with 72 faces used as a model in antiquity for certain spherical vaults, that of the Pantheon among others), the 72 arrows that pierced the prophet Hossein, and we can also cite the 72 articles of the Constitutions of the Templars.

As an angle, 72° represents one-fifth of the circumference, $\frac{360}{5} = 72$, which is to say, the angle in the center that subtends the side of the regular pentagon. This would therefore be one of the important central angles in the dodecahedron and the icosahedron. Von Laban noted it as the maximum angle of torsion and flexion of the body in his treatises on rhythmic dance (cf. chapter 2).

7. These are polyhedrons that can be inscribed inside a sphere, each having all their edges equal, all their solid angles superimposable as faces of regular polygons of two or three different kinds. All told, there are thirteen. Pacioli cites the use of the icosahedron as an abstract model for Ceres's temple at Cercio near Rome.

The small circular temple of Minerva Medica (also called the temple of Vesta) in Rome, whose plan I provided in plate 40 in chapter 3, could have been composed this way by taking an icosahedron or a dodecahedron as its ideal armature.

personally carved and painted three series, with sixty solids each; he still owned the first, the second was the property of Galeazzo Sanserverino, and the third was in the palace of the gonfalonier P. Soderino in Florence.

By doing this (which is to say, the plan of a building or architectonic whole once established on the basis of an abstract "Platonic" composition, a creation "from the inside out"[8]), limited, controlled, and dominated sometimes by the material conditions of its realization (I might even say the "incarnation" of the work—utility, duration, cost, price, and the whole of this composition in three dimensions), after an initial inspection of scale models, once accepted, the plans are all, at least, schematics,[9] revealing wholly or partially the sequence of the proportions between the lengths and surfaces, which will suffice (they will be both "necessary and sufficient") for the complete study of the whole and its details, in particular for establishing numerically and graphically this play of proportions. I mean to say that the principal aspects and projections, which are enough for the architect to be able *to see* his building in advance, because of the planes and axes of symmetry that both monuments and living beings possess, to establish it in surfaces, in two-dimensional schemas, and thereby its decomposition into harmonic rectangles based on Hambidge's ideas, can suffice for studying the proportions and "commodulations" of each schema in whole or in part. This is because Hambidge's modules of Φ, $\sqrt{5}$, and so forth, are not linear subdivisions but ratios characterizing the surface proportions. He used them to analyze and compare the modulations of surfaces and their harmonic decompositions by means of "potentially commensurable" irrational modules that produced systems of surfaces connected by rational ratios, in which similar figures but different magnitudes were

8. The creation could be the result of an inspiration, a spontaneous notion, or a conscious choice. These two sources, passion and meditation, are in fact blended in variable proportions in all works of art; the rhythm of passion and that of idea can alternately wed and separate from each other.

9. The very word *plan* used for the concept of the general directing composition reminds us that, practically speaking, two-dimensional diagrams are sufficient to execute an architectural project.

grouped rhythmically while reflecting the fundamental shape on various scales. This is clearly how Vitruvius understood the *commodulatio* or play of proportions in *symmetria*. Euclid, whose theory of proportions was borrowed in whole by Eudoxus of Cnidus (408–353 BCE), direct heir of the system of Theaetetus and Plato, understood it no differently when he made the distinction between rational proportions expressible by numbers and the others that were depicted by lines, surfaces, and solids.

And this Pythagorean-Platonic notion that Hambidge rediscovered and rebaptized with the name of "dynamic symmetry" can be clearly seen in Pacioli and Alberti,[10] to such an extent that with them it does not appear to be a question of rediscovery but of an unbroken transmission.

The very ratio of the golden section, $\Phi = \frac{\sqrt{5}+1}{2}$, although obtained by a priori "linear" reasoning, is in reality the pulse of an optimum rate of growth (homothetic through successive accretions) in two times, in two dimensions.[11]

---

10. The passage from Pacioli is quite important, especially as Hambidge appears not to have known it: "When it is not a case of employing simple symmetries ($\frac{1}{2}$, $\frac{1}{3}$, $\frac{3}{4}$, $\frac{2}{3}$, and so on), and you have fallen into the domain of irrational proportions—for example, one determined by the diagonal and side of the square—you will use a level and compass for placing the important points in your drawing. In fact, even if a proportion cannot be expressed in numbers, this does not prevent it from being set by means of lines and surfaces, because proportion can extend much farther in continuous than in discontinuous magnitudes (*che la proportione sia molto piu ampla in la quantita continua che in la discreta*)." *De Divina Proportione*, bk. 2, chap. 20.

This is a condensed version of Hambidge's entire doctrine on rectangles with "dynamic" modules ($\sqrt{2}$, $\sqrt{5}$, Φ, etc.) or "static" ones ($\frac{1}{3}$, $\frac{3}{4}$, etc.)!

11. I showed in chapters 5 and 7 of *Esthétique des proportions* that the attempt to achieve a pulsation of growth in three dimensions, in three times, an ideal theoretical proportion for the growth of volumes, does not end up with a practical result because we run into, as with the problem of doubling the cube, a third-degree equation that cannot be solved by euclidean methods ($x^3 = x^2 + x + 1$).

This is just one more indication for contenting ourselves with graphic plans and "commodulations" of surfaces for the practical treatment of proportional questions with regard to volumes.

OPTICAL DISTORTIONS

Let us suppose that in an architectural or plastic creation, the compositional side of the volumes had been set in abstracto, with plans and a scale model, and the controlled, calculating side of the resulting proportions treated by drawn plans of the Hambidge or Mössel type. This inevitably brings us, when it comes time for executing these plans, to the problem that I scarcely touched on in my previous book under the general name of "optical corrections." I will quickly draw up a sketch of the two principal aspects.

First, the execution of the project could produce, from certain vantage points, an accidental deformation, a truncation of the desired "commodulation." This distortion could simply be a consequence of the respective positions of the monument and the eye.

Such is the case in an assemblage, for example, when a superimposition of the architectonic volumes, calibrated according to a certain proportion or commodulation[12]—let's take the very simple instance of two superimposed cubes whose sides are in a Φ ratio—causes a portion of one of these volumes, here the little upper cube, to be concealed by the horizontal cornice side of the base, making it so that the observer's eye cannot truly see the Φ ratio between the heights, or the Φ^2 ratio between the cubes' front surfaces. The same situation occurs if, instead of a simple ratio between two sides, it is a more complex building with a sequence of ratios and proportions that can arise from the same truncations or are masked by the same sides, projecting cornices, or so forth. The commodulation whose consonance (a harmony creating a *consensus*) the eye can perceive would be distorted, and no eurhythmy would flow from it.

In fact, the consequences are not that serious. When any living organism (animal, plant, human being), or its representation, or an

12. I prefer using this term instead of *symmetry* so as to avoid having to continuously remind the reader that it is a matter of the symmetry of Plato and Vitruvius, and not the modern meaning of this word.

organic, harmonic, man-made creation[13]—a building, for example—realizes in the reciprocal proportions of its elements the eurhythmy that is supposed to provide the observer with aesthetic pleasure (whether pleasure arising from a liberating, "cathartic" rhythm, from a perception of recurring analogies, from unity in variety, or so forth), this eurhythmy continues to be perceived subconsciously even if some parts are concealed or removed. A lost female profile, the end of a shoulder, the curve of a hip, a fragment of a faraway silhouette are enough for the subconscious to reconstruct or surmise the harmony of the whole. A mutilated Greek statue from the classical era, reduced to a fragment that must be "formless" (the marbles of the Parthenon, and so forth), offers intact the melody expressed during its creation; because architectural, tonal, and plastic rhythm, the rhythm of a living being, are perceived wholly. The fact that some level of proportion is missing or concealed does not generally affect the rhythmic unity of the whole (if it existed!) or the perception of the same; its reconstitution in the perceiving mind is, so to speak, automatic. However, if the observer is, by virtue of the position of the object or the building and its surroundings, placed in an abnormal way, in a position (one that is too close, for example) that reverses the scale of the plans, we find ourselves in a situation in which "optical corrections" (that is, intentional distortions in the execution of the plan intended to compensate for the optical distortions) are necessary.[14] This issue will be discussed more fully in the next section.

---

13. "A work of art is anatomically, although not physiologically, an organism. It is a harmony, a unity." Professor J. Macmurray, "The Unity of Modern Problems," *Journal of Philosophical Studies,* April 1929.

"The work of art is in the first place an individual concrete being, imposing the impression of possessing a unique individuality, like a person . . . its criterion is precisely the unity that it shows in the variety of its parts, its organic character, the fact of manifesting a design." S. Alexander, "Philosophy and Art."

14. For example, the famous case of the extremely complex, intentional distortions that Michelangelo found in the horses of the Dioscuri (Montecavallo) when he sought to measure them, an event that converted him to the "law of number."

Second, the chain of ratios, the rhythm, could be influenced not by the truncation or concealment of an element, but by true "projective" distortion, that is, a façade seen obliquely might not present the rhythm that it is intended to present when seen from the front, and so on.

Here again, we can state at first glance that in reality the projection, even when oblique or repeated, of a whole that has been given rhythm on a plane or on several successive planes still provides a rhythmic image. A rhythmic progression of lines, surfaces, and volumes projected over any plane retains a rhythmic nature, and even generally keeps the same theme, the same *commodulatio,* the same play of proportions between the parts and the whole; in particular the play of the analogy is reflected—and reverberates—in the image if this rhythm existed in the real model.

This applies most especially to the baroque musical cascades, for example, those of a building like Melk Abbey. The magisterial orchestration of these volumes "sings" in some way as one nears this immense symphony of stone, and this applies as well to many other "baroque" abbeys of Austria or Germany, just as it does to their "archetype," the apse of Saint Peter's in Rome by Michelangelo.

A perspectival distortion does not therefore distort the play of proportions—so long as they are of isogonic projections or central projections of flat surfaces as in classical perspective.

OPTICAL CORRECTIONS AS SUCH

But in reality, the images seen by the viewer are not the result of central projections on a vertical plane like the projection-images of classical perspective, which has been that of painters and draftsmen for some four hundred years.

The truth that the ancients knew, and that was rediscovered by Viollet-le-Duc, is that unless the observer is fairly far from the monument or object under consideration, the perceived image does not correspond (as it does in classical perspective) to a projection on a vertical plane perpendicular to the line joining the eye to the center of the figure or the symmetry of the object in question. There are, to borrow the expression of Miloutine Borissavlievitch,[15] "optico-physiological" distortions stemming from the fact that the eye is neither a camera obscura nor even a camera lens "snapping" in a single image the object that presents itself to the retina.[16]

Vision is not instantaneous; it is an operation consisting of the fitting together of successive images, and as the eye rises, for example, to examine the vertical façade of a building, there is no vertical plane of projection but a series of planes that are no longer vertical but perpendicular to the fleeting axes of vision (thus increasingly slanted), which produces with this "optico-physiological" vision (which takes place as much in time, or rather "duration," as it does in space) an image made up of the joining of fragmentary projections over these revolving planes whose envelope is a curved cylindrical, or rather spherical, surface. If we take this same phenomenon with regard to horizontal vision: the eye must turn from left to right, for example, to embrace all the elements of the width of the façade, and the segments on the extreme left or right will seem foreshortened in comparison to those of the center, just as when the eye sought to encompass, by raising its gaze from the bottom to the top, the vertical elements of the façade, those at the top appear similarly foreshortened when compared to the equivalent segments at eye level (plate 45).

Guided by their spiritual father—the Jesuit scholar who painted the fabled ceiling of Saint Ignatius—the scenery designers of the baroque-era theater perceived this, as did the painters of the architectural panoramas and backdrops for the music-hall revues that followed in their footsteps. By the same token, the painters who sought to encompass a vast field of vision on a single canvas were not unaware of the need to

15. *Les Théories de l'architecture* (Paris: Payot, 1926).

16. The image is not printed on the retina as if on a photographic plate, which is to say simultaneously, but by a series of stimulations, therefore successively by virtue of the "fovea centralis," the point of clear vision. Borissavlievitch, *Les Théories de l'architecture.*

use several planes of projection: Veronese in his *Wedding at Cana* used seven "points of view" and five different horizons.

The apparent foreshortening effects of the vertical or horizontal lines located at the extreme ends of the sectors of the eye's rotation are not the only customary distortions of perspective;[17] the ancients had also noticed that in a building with columns, the columns at the very end appear to move apart (and by the same effect, a truly cylindrical column appears to splay toward the top), the lines of the cornice and the entablature appear to point toward the center, and so on. Hence the "optical corrections" prescribed by Vitruvius (he explicitly called the part of architectural science concerned with determining these corrections once the theoretical plan had been established *scenographia*): the axes of the outer columns would not be vertical but slanted slightly inward, the lines of the top would bulge upward, the columns themselves would "bulge" (entasis) at about a third of their height (closely corresponding to the height of a person), and the cornices would be pushed slightly forward, and so on.

These corrections were not noted on ancient monuments until the beginning of the nineteenth century. It was next observed that the Byzantine and Gothic architects had used similar corrections tailored to the forms of the buildings they built (their corrections appear more empirical; the corrected profiles in Vitruvius, to the contrary, verified by Choisy, strictly correspond to mathematical solutions: parabolic arches, hyperbolic arches, and so on).

The law of the apparent foreshortening of vertical segments (which does not respect their linear ratios) was, as we have seen, rediscovered by Viollet-le-Duc. He attributed this distortion to the fact that in the eye the projection of images occurred on a spherical portion of the retina.

---

17. Customary distortions: those produced in the observer examining a monument at a normal height and distance, in other words, not too far away or too close, as the eye is neither at ground level nor at the level of the central figure of the façade, but rather at the height of the eye above the ground. It could thus be interesting to make the projective symmetries of the façade automatically coincide with the eye's projection. This idea of Borissavlievitch will be explored later.

His explanation was not quite correct; Borissavlievitch has shown that in reality the phenomenon results from the successive movements of the eye when seeing; its curved surface, in fact, replaces the vertical plane of projection of classical perspective, being, as I noted earlier, the enveloping of different shifting planes, each of which is perpendicular to the instantaneous line of vision. This is what Borissavlievitch calls the "optico-physiological" perspective. He, moreover, admits that Viollet-le-Duc's practical conclusions are accurate: "When it comes to putting a building into proportion, there is a great need to take into account the point or points from where it will be possible to see it, and the reductions produced by the heights, the offsets, and the overhangs."

Borissavlievitch, after having established the premise of this "optico-physiological perspective," and using the probable or optimum position of the observer, established an entire system not only for optical corrections but also for determining the proportions of the interior elements (doors, windows, columns, the height of the entablature, and so on), as well as for a façade or an entire monument, and the contour or frame being chosen beforehand or provided by the other conditions of the problem. Once the perspective "pole," the eye's horizontal projection, is placed (on paper) in the frame, the positions and dimensions of these interior elements are obtained or directed by demi-diagonals going from the pole to the corners and to other important points of the frame (the diagonals are broken because the eye's projection does not coincide with the center of symmetry of the frame but is generally far below it),

> The little surfaces are the perspectives of the large ones, the small column has the perspective of the large one, the base and capital of the first column are equal in perspective to those of the last . . . the abacus has only the perspectives of the entablature. . . . In a word, all the framed architecture has the perspective of the frame . . . the interior frames share the perspectives of the exterior frames . . . we immediately see why the entablature is smaller when the columns are placed farther apart, and vice versa [plate 46].

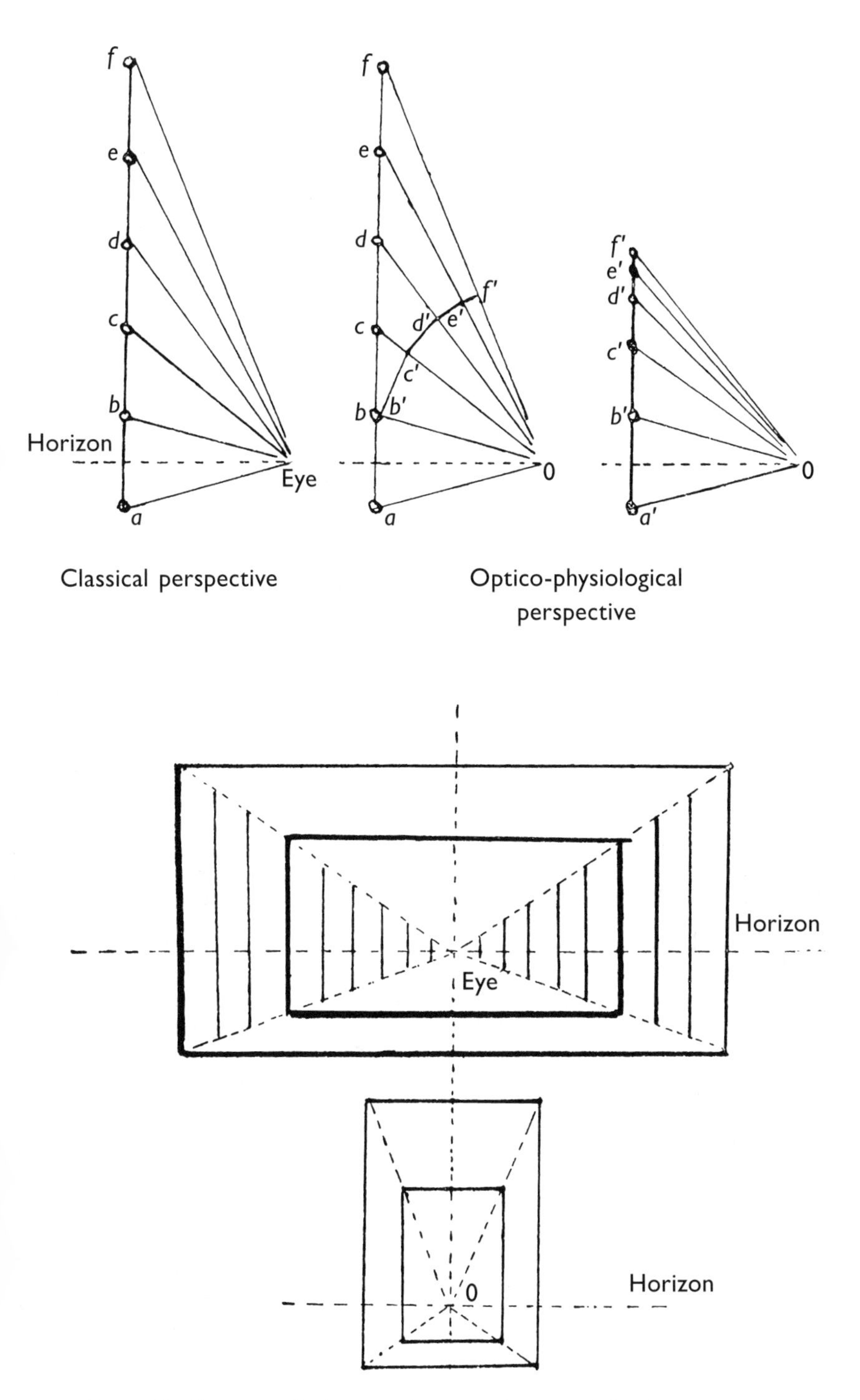

Plate 45. Miloutine Borissavlievitch's optico-physiological perspectivism

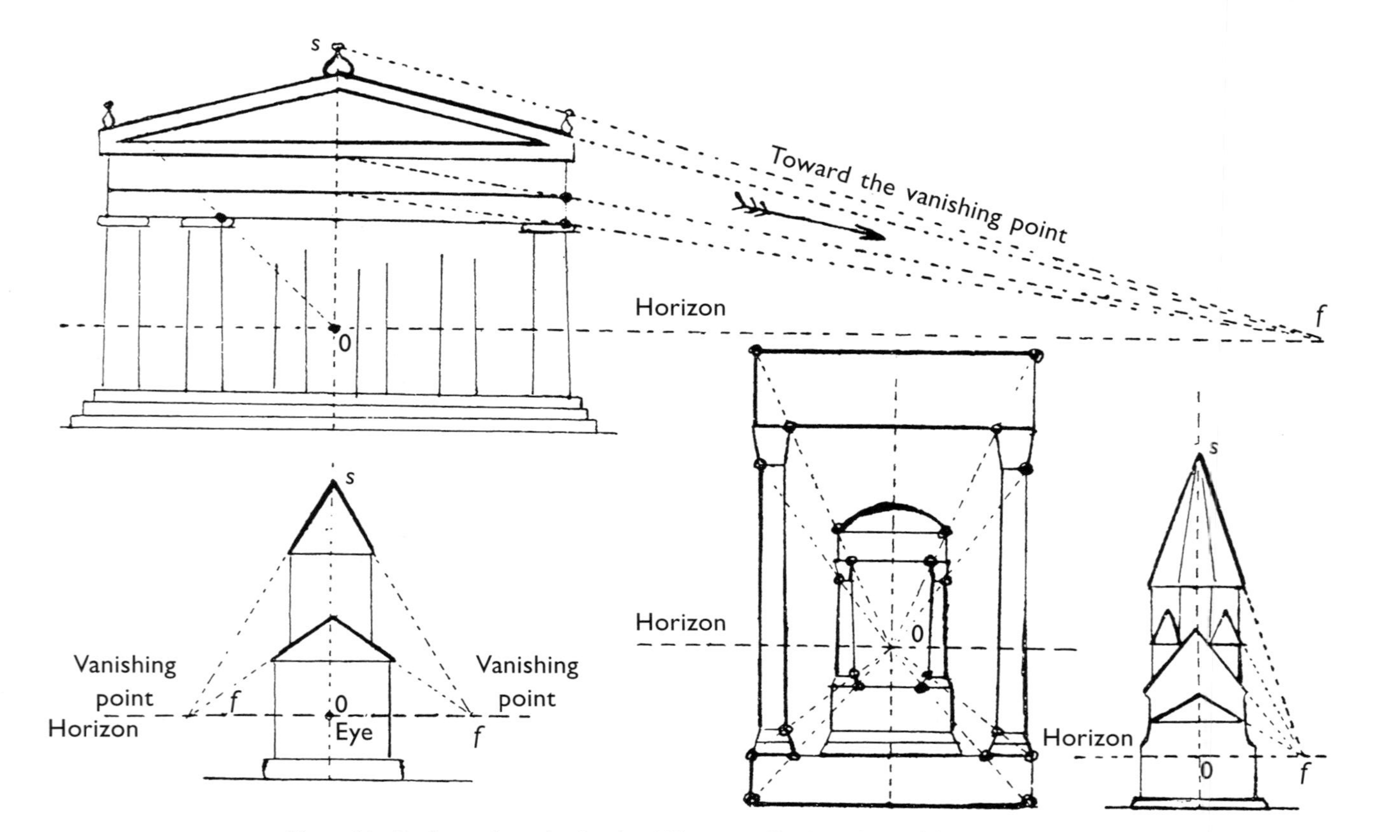

Plate 46. Projected analogies by Miloutine Borissavlievitch's perspectivism

Furthermore, the horizon line, which is the horizontal one passing through the pole, carried over this same diagram, will contain, on all parts of the vertical axis crossing through this same pole, lateral vanishing points and the intersection points of the converging extensions of the roof or tympanum lines.[18] This grouping, on two perpendiculars passing through the pole, of the points of convergence of all the diagonals, not only provides additional reference lines, but a dual "centering" that is certainly quite pleasing to the eye.

The "perspectivism" used in this kind of composition thereby produces, says Borissavlievitch, shapes that "accommodate" or conform to our vision. "Everything shrinks like the example of an avenue whose perspective is depicted on a painting."

We see, in short, that this is in fact a matter—by using the probable distance and height of the observer's eye—of composing the same plan (vertical) of the building according to the image that the classical perspective (of the painters) would produce for the first theoretical diagram if one were to be created. This procedure incidentally provides in advance the automatic corrections for a portion of the optical distortions.

This method of an optico-physiologically based "perspectivism" is, at least in theory, quite ingenious. Like Hambidge, its author relies on the principle of analogy as defined by Thiersch.

> We have found by observing the most successful works of all time that a *fundamental form repeats itself* in each of these works, and that their component parts form, through their composition and arrangement, similar figures. . . . Harmony only results from the repetition of the principle figure of the work in its subdivisions.[19]

---

18. Which is to say that with these vanishing lateral points, once produced by the choice of one of the ridge slopes, the other slopes will be obtained by joining them to the vertical vanishing points or vertices of the different stages of roofs or tympanums carried over the vertical axis (plate 46).

19. The principle of analogy occurs in music: "It is no exaggeration to say that systematic repetition under one form or another is the most important principle of musical structure." W. R. Spaulding, *Manual d'analyse musicale.*

> Because of the simple relationships of the parts to each other (analogy) and the more frequently these relationships repeat, the interior vision of the *psychical intuition* is more easily produced . . . it is in the faculty of organic action that pleasure is found.[20]

This is a declaration of the role of the "hedonistic" principle of least action in aesthetics. Borissavlievitch develops it further as follows:

> Pleasure is found in everything that *conforms* to our nature, everything that *corresponds* to it. The "perspectivism" of the forms of a composition produces the phenomenon of harmony, because the laws of perspective are those of our vision, and it is only through imitating its nature that we create forms that accommodate it, in other words, beautiful forms. . . . The beauty of a composition due to the "law of the same" is explained by . . . this "short-term memory" or "background sensation" that makes the act of perspective easy.

And also: "A tree appears beautiful to us because of the 'perspectivism' of its forms. Because it concerns the 'same' and its perspectival deductions, which we call 'the perspective of the same.'"

We find again in this original theory the "law of the same and the other," or unity in variety, since the "law of the perspective of the same" is established on arguments that are no longer metaphysical but purely

---

(cont. from p. 127) Pius Servien, in his remarkable sketch of the fundamentals of a philosophy of music, *Introduction à une connaissance scientifique des faits musicaux* (Paris: Albert Blanchard, 1929), studies the various transformations, variations, or themes that invariably produce one same "leitmotif." "We see elements everywhere in the domain of music that respond like images. . . . The music that soothes therefore appears cyclically. It opens cycles: creates dissymmetry; and closes then, restoring rest. . . . There are transformations that a musical figure can undergo without ceasing to be recognized as itself."

Servien ingeniously applies the "theory of groups" to the study of these musical transformations, and mentions among others the "homothetic," or "projective," transformations (through the increase or diminution within a same ratio of the duration of notes), which exactly correspond, temporally, to the transformations in geometry that respect "similarity," the "analogies," and the proportions of figures.

20. Thiersch, "Die Proportion in der Architektur," cited by Borissavlievitch.

pragmatic, psychophysiological ones based on the relationships between the object perceived and the subject that feels the "aesthetic" impression.[21]

And although this way of creating groupings of similar figures through radiating homothetic triangles starting from the pole and vanishing points is completely different from the procedures of Hambidge and Mössel, Borissavlievitch also notes that among the various continuous proportions that can allow for this play of analogies through harmonic propagation, the continuous proportion par excellence—which through its dual geometric and additive essence condenses in two terms an entire series of infinitely reflected proportions and directly introduces, to use Keyserling's expression, the infinite into a limited form—is the golden section. He also says: "Among all the proportions that create analogies, it is also the most beautiful individually."

I have dwelt on this issue of optical corrections because, except for a building designed to be seen from a great distance (the case in which the retinal image can be considered as projecting itself on a single vertical plane, as in classical perspective, and the distortions are imperceptible) it can, for a carefully designed plan like harmonic rhythm, give rise to a complementary geometry that is almost as complex as the plan itself. The "law of number" plays as strict a role in this case as it does in the plan. Erecting models and especially taking photos of models from various angles is naturally highly recommended.[22]

---

21. In conformance with Lalo's observation: "The laws of beauty, quite far from residing either in the objects thought or the subject that thinks them, consist in certain ratios between the two; they are one of the forms of their multiple mutual reactions." C. Lalo, *Esthétique* (Paris: F. Alcan, 1925).

22. Though their very symmetry (in the modern and not Vitruvian sense of the word) the large, rigorously geometric volumes like Chicago-style skyscrapers and so forth, divided into equidistant elements by countless floors and windows, show that the problem of optical corrections can be pushed aside by the architect's own intention; the natural "perspectivism" of the series of windows and the verity of the immense orthogonal volumes are evident.

The same considerations can be applied to certain modern structures with a plan and modulations of a starkly obvious geometry. According to the quite ingenious idea of Le Corbusier, the eye (or visual awareness, rather) perceives the desired schema and automatically supplies the necessary optical corrections (just as it automatically straightens out all the retinal images).

These corrections complete the finalizing of the plan. Once the general "symmetries" (consonances and rhythms of proportions among the parts and between the parts and the whole) have been adopted, they complete the eurhythmy, which is the harmony subjectively perceived. They therefore form part of what, to extend the musical analogies of Plato and Vitruvius, I call "the orchestration of volumes."

It is therefore necessary in architecture to compose with science, in accordance with the precept *Ars sine scientia nihil* of the Parisian master architect Jean Mignot, who was summoned in 1398 to consult on the building of the Milan cathedral. But this composition must be *creation:* any copying, clever and learned as it might be, would not remedy the absence of enthusiasm and inner passion (for passion, pulsation, and rhythm are in ideas just as much as in feelings).

It is this enthusiasm and tension that produces the great living rhythm, the one that the "lesser rhythm," the cadence, can prepare and bring forth as in poetry. There is always the distinction between the inner rhythm, the gust of "wind" of which Foch speaks with respect to leading a battle, and the simple cyclical return, periodicity, the quiet accompaniment from which eurhythmy emerges. And even in this preparatory rhythm-springboard (and this applies just as much to architecture and structural design as it does to poetry) it is, to use Paul Claudel's expression, the number that counts and not the numeral.

An architectural composition cannot help but be geometric, but this geometry includes the geometry of life and growth, and should be a conscious creation, not a simple grid of lines. The fact that the points of a drawing are selected among the intersections of the lines on a diagram are not enough to make it a geometrical design: the diagram and the selection must have a meaning.

A real creator (architect), equipped with "inspiration" but who does not compose geometrically, who has not contemplated proportions (who incidentally does not possess a "layout" and a controlling method like Hambidge's or Mössel's to handle the surfaces or volumes arising from the golden section, for this proportion and its related themes are imperative as fundamental accords[23]) could create magnificent things, but he or she could also through botching a detail fall just short of the perfect work.[24]

By the same token, between two architects with no inspiration, it will be the one who composes geometrically, and through use of a harmonic technique, whose work will be more pleasing, because his technique will automatically control itself. The other will create a veneer, a copy; his works will bring to mind constructions from a box of blocks, a builder's kit for a giant child (everything that was built in the Parisian suburbs from 1880 to 1914!). This has its importance because the mediocre is the majority, and was so even in the great eras; but there we do not notice because either intentionally, or by means of scales handed down, architects applied the "Platonic" procedures for establishing proportion. They were also their own "calculators" (from the perspective of strength of materials, and so on). Today, it is often an engineer, a specialist architect-calculator, who computes the strength calculations for the architect-draftsman, the one who draws the blueprints, which steals away another chance for the result to have organic unity.

When, moreover, as often occurs in today's architecture, the sole criterion adopted is the adaptation to purpose (fitness), including solidity and economy, with the idea of the engineer winning out, we can find unity anew. Even in these cases, we can sometimes choose among several diagrams of proportions. And that of the golden section is not bad; Le Corbusier will back me up here.

---

23. Two different drawings (one with "dynamic" rectangles of the Hambidge type, the other with directing circle and pole of the Lund and Mössel type) can be applied to the same system of points, because they are the aspects, the projections in two dimensions of the same *idea.* Again, I am not advocating for the golden section to the exclusion of all other proportions; I rallied to Hambidge's ideas, whose various dynamic themes allow for infinite choice (see his analyses of Greek vases and human skeletons in *Esthétique des proportions*).

I am perfectly happy to accept that an architect can *deliberately* use static themes.

24. Like Soufflot, who did not leave sufficient space between the pillars supporting the dome of the Pantheon (see Borissavlievitch).

For flat decoration, bas-reliefs, the creation of profiles, surfaces, or volumes in the decorative arts, the harmonic systems like those of Hambidge, Mössel, and Borissavlievitch also appear equally useful to me, if not of obvious necessity.[25]

Hambidge's analyses of Greek vases and objects, and Mössel's of Greek bas-reliefs, appear to confirm the use in their composition of rigorous directing diagrams.

In Egyptian art, the proportioning of volumes and shapes (including hieroglyphs) displays a world of harmonious perfection that was sought for and achieved. I will not attempt any analysis here but simply observe that Hambidge's method produces results of surprising precision when applied, and that the very rectangle of the golden section and its direct modulations appear to play a predominant role as well.

This is for proportion.

As for the effects the Egyptians drew from rhythm as such,[26] from its ordered, intentional periodicities, we can also say that the reliability of their compositions was such that their rites, transformed into rhythmic images, have attained eternal life. When we contemplate the motionless theories on their friezes of gods, men, beasts, and signs, the symbols speak, and *incantation* is produced.

This is because, from the point of view of pure form, the work of art can work on its beholder in two different ways in order to produce pleasure or even ecstasy (Ruskin said "all art is adoration"): either through the least harmonic or hedonistic effort, the organizing of the chaos of sensations (with sympathy, *Einfühlung*), comprehension, bringing the observer's mind into accord with the proportion or symphony that the artist has realized; or through incantation, where rhythm can be sovereign, by putting the emotional flow of the observer or listener in tune

25. For example, Hambidge's procedure automatically imposes the important law of the "non-blending of themes" stated earlier by Alberti.

26. Which is to say, the "discontinuous" rhythm with countable elements, like the tonic rhythms examined in the following chapter; proportion, to the contrary, and timbre form part of the continuous domain.

with the inner rhythm of the artist, who himself reflects the pulsation of the Great Rhythm, of the Grand Harmony. It so happens that there is a "law of number," in this case the number that organizes this action of incantation as much as other forms of artistic expression. This is what we are going to try to discern by ceasing to concern ourselves with the "numbers" of volumes or other visual shapes, so as to move into the category of the pure rhythm that acts through incantation as such.

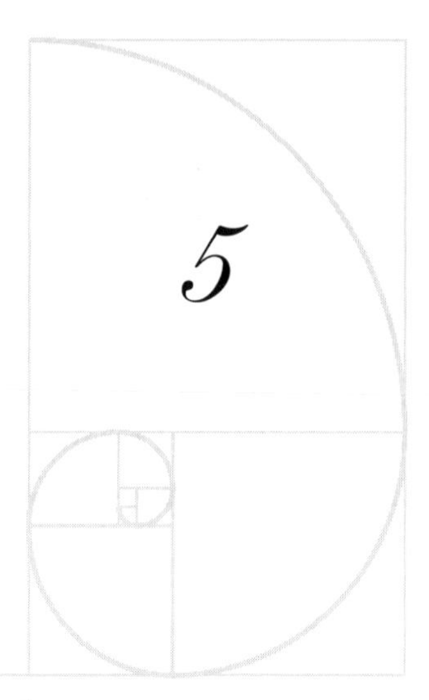

5

From Rhythm to Incantation

You whose bow is silver, god of Claros, listen!

André Chénier

In the preceding chapters we have used the word *rhythm,* summarily defined as "perceived periodicity,"[1] when speaking of the recurrence of elements or groupings that are identical or similar in a spatial artistic composition (in architectonics, for example).

Theoretically, the word *rhythm* should be reserved for something that characterizes the periodicity of events in time, with *symmetry* (a sequence of commensurabilities among the different parts and between these parts and the whole) being applied to the mutual relationship of elements and the whole in a "spatial" series.

But the Greeks themselves, who accepted no confusion of ideas or definitions when it came to aesthetic matters, here intentionally blended the terms belonging to architecture and music. What is more, architectural concepts and aesthetic morphology were consciously discussed and perceived in musical analogies. If in music the notions of harmonious chords and chord progressions are established according to numerical or geometric proportions and ratios, at the same time the Greeks give the name *symphony* to the harmonious sequence of proportions in

1. Pius Servien, *Essai sur les rythmes toniques du français.*

an architectural whole, and that of *eurhythmy* to its perceived effect.[2]

In the preceding chapters, I have also often interchanged terms related to the succession of elements in duration and to the juxtaposition of spatial elements (correspondences justified by the fact that visual sensations are neither overall nor instantaneous, but are often sorted and connected over a time span like auditory impressions), as Pius Servien has explained in a major work[3] in which he successfully established a general theory of rhythm (from which I will permit myself to draw generously over the course of this chapter). The concepts of periodicity and proportion, and their sequences—which can be used equally well for temporal progressions and spatial juxtapositions—are important notions to differentiate from the start since rhythmic groupings can be continuous or discontinuous, reversible or irreversible, symmetrical or asymmetrical.

Having said a great deal about rhythm and with a great deal left to say, I am now going to try to approach it in a purer, more "native" state. Leaving the arts with "visual" effects in which time, duration, and evolution are "solidified" and incorporated into "reversible" lines, surfaces, and volumes, we shall now examine the art in which rhythm and number (the Greeks made indiscriminate use of ῥυθμος and ἀριθμός in this sense)[4] are perceived, or even "experienced" directly and "irreversibly," by

2. Equivalents:

Music	Architecture
interval (consonant or dissonant chord of two notes)	ratio (of 2 lengths, surfaces, and so forth)
chord (combination of three or more notes)	proportion
harmony	symmetry
melodic eurhythmy	architectural eurhythmy

3. *Les Rhythmes comme introduction physique à l'esthétique,* completed by *Lyrisme et structures sonores* (Paris: Boivin, 1930).

4. Both derive from *ῥέω* ("I flow"). However, *rythmos* corresponds more specifically to the *symmetria* of a series of numbers, *arithmos* more specifically to the "measurement" aspect, therefore condensed for "λόγος ἀριθμῶς."

Examples: οἱ ἀριθμοι τοῦ σὼματος (the proportions of the body), Plato, *Laws,* and ὁ τοῦ σχήματος τῆς λὲξεως (the measure of the diagram of the discourse), Aristotle, *Rhetoric.*

addressing the rhythmic incantation of which pure music is a modality.

It is common knowledge that music is greatly involved with ratios, measurements, and measure—in a word, numbers. For the Greeks, music formed part of mathematical philosophy (which for the Pythagoreans and Plato, the precursors of Russell, Einstein, and Eddington, was the whole of philosophy); or, if one would prefer, the mathematical theory of harmony formed part of a general theory of the harmony of the cosmos.

His direct disciples as well as non-Pythagorean authors are in agreement here in attributing to Pythagoras himself the discovery of the numerical laws of harmony, and all are in agreement as well on the importance of this discovery.

It is in large part the correlation established between musical chords and intervals, and the arithmetic ratios that arose from the trinity of tetractys-pentad-decad,[5] to which is due the mystical importance in the

(cont. from p. 135) The word *ἄρρυθμος* conversely means "devoid of rhythm" (the opposite of *ἔνρυθμος* "rhythmic," "endowed with rhythm"), as in the passage in which the comic poet Alexis says that walking in the street without rhythm (*ἀρρυθμως*) is the mark of a coarse man; cited by E. A. Sonnenschein, *What Is Rhythm?* (Oxford: R. Blackwell, 1925).

5. The superimposition of two sounds is called an "interval." The intervals used in music are such that the number of vibrations of the two sounds makes a simple ratio. The superimposition of two (or even several) sounds having the same number of vibrations is called a unison, the ratio $\frac{1}{1}$; when the ratio is $\frac{2}{1}$, we have the octave (a note and its corresponding note an octave lower have their number of vibrations in the ratio of 2 to 1).

The other usual intervals (producing "pleasing" sounds) are:

$\frac{3}{2}$ the fifth	$\frac{6}{5}$ the minor third
$\frac{4}{3}$ the fourth	$\frac{5}{3}$ the minor sixth
$\frac{5}{4}$ the major third	$\frac{8}{5}$ the major sixth

When more than two sounds are heard at the same time, the resulting chord is all the more pleasing to the ear the simpler the different ratios (of vibrations) of the notes are between them.

Among the chords formed of three notes, what we call "consonant chords," are *the perfect major chord,* formed from a major third and a fifth: do–mi–sol (also do–mi–sol–do, because the notes of the octave and the repeated notes do not change the nature of the chord); and *the perfect minor chord,* formed from a minor third and a fifth: mi–sol–si (also mi_1–sol_1–si_1–mi_2).

The other chords of 3 notes and the chords of 4 and 5 notes are called "dissonant chords," and their use requires what is called preparation or resolution to "save" the doctrine and even the Pythagorean ritual of the tetractys, which as we have seen is nothing other that the "figurate" form of the decad as the fourth triangular number (1 + 2 + 3 + 4 = 10).

We also saw that the decad was the very symbol of universal life, or macroscosm; the pentad, its "condensed," analogical reflection, the number of harmony, health, beauty, and love, was—in a time when different disciplines of Pythagorean origin or inspiration (Gnosticism, alchemy, Kabbalah) standardized their symbols—directly assigned to

---

dissonance (by linking a note to a note heard in the preceding chord, therefore by introducing a "mediety" or mean in the Platonic sense).

The ancients made no attempt to count the vibrations corresponding to the different sounds, but worked directly with the law of simple ratios by working on a vibrating string whose length they varied by means of a cursor; these lengths are inversely proportional to the number of vibrations.

If the octave is obtained acoustically by comparing the sound produced by a taut string with that produced by half this string, the fourth will be obtained by the comparison of this string with $\frac{3}{4}$ of its length, the fifth by that with $\frac{2}{3}$ of its length. The sounds perceptible to the ear run from 32 (the lowest organ pipe) to 73,700 vibrations per second (the screeching of cicadas). Sounds have a musical character in the range of 60 to 33,000 vibrations. If we take the musical note C_1 at 258.6 vibrations per second as the fundamental tone we will have as its first harmonic C_2, at 2 × 258.6 = 517.2 vibrations. This last note, an octave above the first, will form with it the ratio 1 to 2, and these two will form with the six intermediary notes a sequence of eight notes that form the diatonic scale, the major or minor mode following the distribution of tones and half tones on either side of the fourth note (two tones and a half above the fundamental) and the fifth note (three tones and a half above the fundamental).

The art of stringing successive notes or chords together in a melodic phrase or line is harmony. It is easy to see the analogy with the sequencing of proportions in eurhythmic lines, surfaces, or volumes in the "visual" arts. We should note the very important fact that a chord does not change in nature if the notes are heard at the same time (block chord) or successively in any order (broken or arpeggiated). It can be seen that reversibility is not incompatible with a sonorous progression. A harmony can have several components (melodic series of tones), with counterpoint being the art of the simultaneous combination of these parts. Here the following observation by Servien concerning the "homothetic" transformations of musical themes can be compared, as we have seen (chapter 4, note 19), to the homothetic transformations (respecting proportions) in geometry: "Projective considerations are at the base of harmony, if we assume harmony as only meaning a melody heard simultaneously; though that has no effect on the considerations a chord might inspire, whether heard as is or as an arpeggio."

man, or microcosm, the sound box in harmony with the world soul.[6]

This "harmonic" conception of life was effortlessly applied to the operations of intelligence: perception and combining of "just" ratios in series in which syllogisms linking concepts like Plato's "medieties" filled the interval between two numbers, and from which easily sprung analogies like the essential "logical proportions."[7]

Musical analogies can be found in a being's very nature, aside and sometimes apart from its intelligence as such. A female individual may be harmonious without being intelligent. In such a woman, for example, the intuitive or acquired harmony composed of tact, on the one hand, and of a perfect adaptation of her "charm" to her physique, on the other hand, could perhaps be the supreme intelligence.

This inner harmony, which in no way excludes potential, organizes and eliminates chaos. The great psychological problem for a complex individual is often to "find" him- or herself, like Kipling's ship, to symphonically harmonize his or her component personalities,[8] in order, here again, to achieve unity in diversity.

---

6. Similarly, geometric symbols were generally inscribed inside a circle (projection of the cosmic sphere), especially the corresponding star polygons: a pentagram for man, the star decagon for the cosmos, or—when it is the created, material world, *natura naturata,* especially in the Kabbalah—the hexagram or seal of Solomon, a most appropriate symbol for the crystalline, static balance in inorganic matter, for the reasons examined in chapter 2.

7. "The secret—that of Leonardo as well as that of Bonaparte, as that which all the highest intelligences hold at one time—is, and can only be, in the relationship they can find—which they were forced to find—*between things whose laws of continuity escape us.*

"It is certain that at a decisive moment they had only to perform some definite acts. And the supreme achievement, the one that so impressed the world, was quite a simple matter—like comparing two lengths." Paul Valéry, *Introduction to the Method of Leonardo da Vinci.*

8. A similar problem in everyday life: "harmoniously" choosing the guests for a meal; a fourth person inserted as a "mediety," as a transition, in a trio of disparate personalities, is often enough to "resolve" the dissonances.

The metro passengers in Paris have been able to observe that despite the almost instantaneous start of the trains after brief stops in the stations, they feel practically no jolt. This is no accident but the result of an expert calculation that consists of calibrating the successive power rates in couples of force (therefore the intensities of the electrical

We shall now examine more closely the concept of rhythm itself, the transposition into duration of the Greek concept of symmetry[9] (or *commodulatio,* the proportioned extension of elements in space).

The old definition by Aristoxenus of Tarentum (the very same who knew and spent time with the last Pythagoreans affiliated with the group founded by Archytas) is not bad for all its pithiness: "Rhythm is a determined ordering of times."

Professor Sonnenschein, in his scholarly work on rhythm (see footnote 4), attempted to make a rigorous transposition of space into time in his definition: "Rhythm is that property of a sequence of events in time that produces in the mind of the observer the impression of proportion between the duration of the several events or groups of events of which the sequence is composed."

Eugene d'Eichthal says: "Rhythm, taken generally, is the division of time by phenomena that are perceptible to the human organs, in periods whose total durations are equal between them or that repeat in accordance with a simple law."

It is in fact the explicit highlighting of periodicity that especially distinguishes sound rhythm (as much in poetry as in music) from visual, plastic, or architectural rhythm. I would like to cite another passage on this topic taken from Servien's brilliant essay on tonic rhythms in the French language:

> Rhythm is periodicity perceived. It acts to the extent to which such periodicity alters in us the habitual flow of time. . . . This is how all periodic phenomena perceptible to our senses stand out from all

current received by the motors) in such a way to achieve these two conflicting effects: the rapid procurement of normal power (because every second lost or gained represents money lost or gained) through a gradual process that ensures the shocks due to inertia are practically impalpable; the numerical elements of this harmonic problem that would have enchanted Plato are the diminishing interpolated "resistances" in the motor circuit during the start up (proportional to the lengths of the wires that make up the rheostats).

9. "Rhythm is in time what symmetry is in space." E. d'Eichthal, *Du Rythme dans la versification française* (Paris: Lemerre, 1892).

> irregular phenomena . . . in order to act singly upon our senses, and impress them in a way that is entirely disproportionate to the weakness of each active element.

The "habitual flow of time" is the inner rhythm that for each individual accompanies the perception of psychological "duration," distinguished by Henri Bergson as being the model for the idea of time. His introduction of a psychological element capable of being influenced by an external rhythm is all the more accurate, as the Greek word ῥυθμός, as noted earlier, is derived from the verb ῥέω ("I flow"), and of the three Greek concepts of number, the most specifically Pythagorean notion considers number to be like a flowing wave (cf. chapter 1).

Servien was even more explicit in his last book, *Les Rythmes comme introduction physique à l'esthétique* (Rhythm as a Physical Introduction to Aesthetics) (Paris: Boivin, 1930), that, in addition to the analogous transposition into duration of the notions of proportion and symmetry (like musical chords and timbre) arising from the domain of the continuous, there are countable (discontinuous) periodicities that make up the analyzable framework of a sound rhythm.

> Let us suppose that this framework can be actually grasped as numbers. It would constitute, within a matter still so little accessible to scientific research, a domain in which science would immediately feel at home. We can therefore posit in principle that each time we speak of rhythms, we have perceived, more or less confusedly, numbers. . . .
>
> It seems that the sole numerical notion capable of following the notion of rhythm to its full extension would be this one: *A series of whole numbers in which one discovers a simple law.*

We shall later see the notations imagined by Servien for musical and prosodic rhythms, which, because the almost physiological nature of pulsation, of flow, of *cadence,*[10] of lyrical or musical effusion are in fact enough to set fast the essential characteristics.

Periodicity in the case of sound rhythm stands out from a frame, a trellis of consecutive countable elements: whether syllables and feet in prosody (the feet signify stressed elements, phonemes, or, as in Greek or Latin prosody groups of equal duration—in oratory—of two or more syllables, each syllable being long or short, and the law of alternation for these longs and shorts[11] inside the groups, as well as the law of succession and alternation, already forms two superimposed periodicities, and there are others),[12] or notes and measures (groups of notes corresponding

10. Let us not overlook this play of fairly subtle antinomies: the psychological stream is continuous, the living phenomenon is continuous and irreversible; the physical universe is discontinuous and, up to a certain point, reversible (time has no direction, it is the psychological duration [*durée*] of the observer that creates the illusion of an absolute time), but the pulse of the srream of life is punctuated by a discontinuous rhythm.

11. I will designate the longs and the shorts with the symbols — and ˘ ; the principal prosodic feet or meters of antiquity were the following:

trochee — ˘	dactyl — ˘ ˘	first paeon — ˘ ˘ ˘
iamb ˘ —	anapest ˘ ˘ —	fourth paeon ˘ ˘ ˘ —
spondee — —		

One long is equal to two shorts in duration.

12. For example, the grouping of the feet can produce lines of tetrameter, pentameter, hexameter, and so on (that is, of four, five, or six feet, and so on), and groups of lines can form stanzas, for which more extensive rhythms can be placed over the first. Similarly, in music, measures can be composed into "phrases" or "periods" (a phrase generally consists of four measures or multiples of four), and periods into musical stanzas.

But inside the foot or meter itself (sometimes in classical prosody a group of two feet, or a "dipody," is taken as a unit, in a dimeter, trimeter, or tetrameter line, corresponding, respectively, in this cases to four, six, and eight feet, strictly speaking), the presence of a secondary accent, or of the simple transition, inflection of a long syllable to a short syllable or vice versa, breaks the foot into two durations, generally uneven, thereby creating new interior ratios ($\frac{1}{1}, \frac{2}{1}, \frac{3}{2}, \frac{3}{1}, \frac{4}{3}$) and a new periodicity, a new rhythm parallel to the others, but like proportion, within the domain of the continuous.

This inner rhythm, created by the opposition of the two parts of the foot ("thesis" and "arsis"), plays an extremely important role in classical prosody, and its related numerical ratios were calculated quite strictly. The same inflection inside the measure (whether it was composed of two, three, or four notes, or more) can be found in music generally marked by a downbeat.

to equal durations) in music. We can almost always find in prosody, as in music, several superimposed periodicities, the isochronal recurrence (at intervals of equal time, like the tick-tock of a clock) of the feet or measures, of the syllables or notes being a basic framework, one that is, so to speak, static (like a mosaic, a tiling with identical, indefinitely repeating motifs), over which run, superimpose, and combine more complex periodicities that form the "dynamic" rhythms as such. It is the frames with countable elements that make it possible to notate discontinuous rhythms.

Isochronal, static periodicity, of which the most elementary arithmetic type would be the "monotonous" tick-tock of the watch, is already a rhythm. Other differentiated forms of isochronal periodicity would include identical drum rolls (identical in frequency, intensity, and duration) separated by pauses of equal length; the four-beat purring of a locomotive (groups of four beats of different intensity with the intervals of time between beats being equal); and the anapestic cadence of a forge (two short blows, one long, sharp blow). These series of groups or measures of three or four beats are isochronal by the repetition of identical elements, but inside each element the rhythm can be "dynamic."

From discussing stress with regard to the quantitative measure (based on the duration of the elements) we will now touch on another factor of rhythm—mentioned earlier (in note 12) in regard to inner rhythm—the inflection produced in the feet or measure by the accents or downbeats. These can also obviously be used to divide periodically the series of sound elements, to create proportioned and countable progressions in time. Here it is no longer a quantitative rhythm superimposed on another rhythm of the same kind, but a rhythm of a different nature, an accented or "tonic" rhythm, entwined with the other. The simple arithmetical rhythm mentioned earlier—the drum rolls separated by pauses without any longs or shorts, or downbeats or upbeats—represents a third type corresponding in prosody to the strict isosyllabic rhythms (verses composed of an even

---

(cont. from p. 141) We can see that the correlation between music and classical prosody was perfect, from both the perspective of the elements and that of the rhythmic phenomena.

number of syllables that are equal in duration and missing tonic accents or having a caesura accent at the end of each verse or hemistich). An instrument that makes it possible to achieve these three interlacing rhythms in practically their pure state is precisely the drum, hence Rousseau's line: "Without rhythm, melody is nothing, and through it, it is something, as can be felt in the effect of the drum" (or castanets).

I noted earlier that the prosodic or musical frame is the projection or the expression of another cadenced pulsation: that of life itself. Let me cite Paul Claudel here:[13]

> The sound expression spreads out in time and consequently is subject to control by a measuring instrument, a counter. This instrument is the inner metronome we all carry in our chest, the beat of our life pump, the heart that infinitely says:
>
> One. One. One. One. One. One.
> Pan (nothing) Pan (nothing) Pan (nothing).
> [All]
>
> The fundamental iambic, an upbeat and a downbeat.

It is not merely a question of the isochronal physiological cadence that is normally parallel to a monotonous series of iambs or trochees (a downbeat, an upbeat). This is only the accompaniment or prelude to a series of more complex impulses, for, as Claudel says, rhythm "consists of a measured impulse of the soul, responding to a *number* that is always the same and that obsesses us and draws us along. It is a kind of poetic dance[14] that involves the intertwining of a certain at least approximate numerical combination."

---

13. *Positions et propositions* (Paris: N.R.F., 1938).

14. This is an opportune moment to recall that dance rhythm was analyzed and put into number by the Greeks exactly like prosodic or musical rhythm (cf. Plato, the *Republic*); it was the gestures that were grouped together in measures of equal duration. Jaques-Dalcroze has reconstructed this rigorous transposition.

And lastly:

> The poet is set in motion, in accordance with a style that the studies of Father Jousse have shed some light on, by a certain kind of rhythmic stimulation, of repetition and verbal vacillation. . . . In a word, poetry cannot exist without emotion. . . . Similarly, before the voice there was breath, before expression there was the desire to express oneself.

This is just how Paul Valéry confided to E. Le Fèvre that the decasyllabic rhythm of "Le Cimetière marin" ("The Cemetery by the Sea") had haunted him before the subject and its verbal elements had taken shape in his mind.

It is the rhythm of stresses or tonic rhythm that in prosody appears to best reflect the inner psychophysiological rhythm of the poet and that acts through induction, through incantation on "the habitual flow of time" in the reader or listener. It was used together with the quantitative system in Latin and English scansion and with the arithmetical system (isosyllabic) in Latin religious prosody (hymns like the "Dies Irae," the "Stabat Mater," and so on). It is also, when combined with the isosyllabic (arithmetical) system in alexandrines of other classical isosyllabic verses, the dominant rhythm in practice, if not in theory, of French prosody.[15]

For if the first French alexandrine verses were purely isosyllabic (an arithmetical rhythm, the number of syllables in each verse or hemistich counted singly, with neither the accent nor the duration of the syllables taken into consideration, but a pause or a beat struck marking the end of each verse and generally the caesura of each hemistich, dividing the flow of syllables into groups of six), one[16] accent of secondary stress appeared during the classical era inside each hemistich, thereby

15. The geometer Huyghens, however, observed in a letter to Corneille (1663): "In my opinion, the maxim that dictates a French rhymed verse should only consider the number of syllables without considering the cadence of the feet is dangerous and far from true."

16. Sometimes, but more rarely, two.

transforming the alexandrine into verses of four "feet," each scanned by a final stressed accent, the second and fourth stresses always falling on the sixth and twelfth syllable. Here are some examples from Jean Racine's *Phaedra* (the stressed syllables are italicized):

D'un *secret* / que mon *coeur* // ne peut *plus* / renfer*mer* / . . .
Et mes *yeux* / malgré *moi,* // se rem*pli-* / ssent de *pleurs* / . . .
Et *Phè* / dre au laby*rin-* // the avec *vous* / descen*due* / . . .
J'ai- / me, je prise en *lui* // de plus *no-* / bles ri*ches*ses / . . .

(*Translation*) The secret that my heart can hold no more . . .
Against my will my eyes fill up with tears
and Phaedra to the labyrinth with you descending . . .
I love in him a richer, rarer prize . . .

We see that the number of syllables in each "foot" (we could say "meter" or "measure" to avoid confusion with the term *foot* applied generally to the syllable itself in French prosody) vary from one to five (even six in the very rare case where no secondary accent appears before the final stressed accent), and by reading the verses this way scanned naturally by the secondary accents, we can note that this tetrameter or quaternary division also has a quantitative rhythm (of durations), because despite the inequality in the number of syllables, it automatically establishes an equal approximate duration for each foot.

We have thus seen in the preceding examples that the secondary accent can fall in different places in the two hemistichs of the same line. We even come across two (three, in fact) accents before the downbeat of the sixth syllable. This latter case is quite rare, as is also that of the hemistichs with a secondary accent on the fifth syllable (which produces two consecutive stresses at the end of the hemistich). These additional secondary accents above all accompany the recitation of proper names. Here is a complete series of the different types of normal meters inside the first hemistich (the rhythm of the second hemistich is also marked):

1. *Monstre qu'a trop souvent épargné le tonnerre*[17]
2. *Ce fils qu'une Amazone a porté dans son flanc*
3. *Dans le fond des forêts votre image me suit*
4. *Si je reviens si craint et si peu desiré*
5. *Par notre roi David autrefois amassé*
6. *Je ne me souviens plus des leçons de Neptune*

1. Monster spared too long by the thunder
2. This son an Amazon carried in her womb
3. Your image follows me into the depths of the forest
4. If I return so feared and so little wanted
5. Long ago amassed by our King David
6. I no longer recall the lessons of Neptune

In these verses (from Racine's *Phaedra* and *Iphigenia*), the secondary accent of the first hemistich moves successively from the first to the sixth syllable (identifying in this latter case with the downbeat of the caesura).

It is for this tonic rhythm, whose overriding importance he emphasizes ("there is only one truly independent rhythm that commands the others, the tonic rhythm"), although little known in French poetry, that Pius Servien has conceived a very simple notation that makes it possible to represent by a series of whole numbers the armature of the flow of lyrical or musical phonemes. His "representative number," *N,* is defined (this applies equally to a sentence in prose or a poem) this way:

17. It will noted that in this system of decomposing a line into tonic and atonic (or stressed and unstressed) syllables, I am leaving out the mute syllables at the end of the lines. There are fairly rare cases, however, in which the final mute syllable should be counted, just as, conversely, silent syllables are sometimes eliminated from inside the line when they are spoken and should not be counted. The explicit apocope (the actual loss of certain mute letters) that plays such a huge role in English prosody can be found in the work of Ronsard, but has not been adopted in France.

1. *N* will have as many numerals as there are tonic accents in the phrase (each numeral therefore represents a phoneme).
2. Each numeral indicates the number of syllables per phoneme, from one tonic syllable to the next (inclusive).
3. The silences will be noted by either adding a punctuation mark to the numeral or, when there is no mark, leaving a blank.
4. The atonic (unstressed) syllables after the final tonic accent of a group, which is to say, before the silence separating this group from the next, do not count in reality . . . these (silent) syllables that don't exist, will be noted as necessary, by adding the mark (') to the last numeral of the group. For example, this passage from René Chateaubriand's *Atala*:

La lune brillait au milieu d'un azur sans tache (23332')

et sa lumière gris de perle descendait (444)

sur la cime indeterminée des forêts. (353)

The moon illuminated heaven's azure vault, unspotted with clouds
and her pearl-colored rays descended
to float among the vague crests of the forests.

"All the rhythmic properties of the text (from the arithmetic and tonic standpoint) have moved into number, everything foreign to these properties has been removed."

Applying this form of notation to the six lines of verse by Racine quoted earlier gives us the following numbers:

1533 2433 3333 4233 5133 0633

The reader can see that I have used the signs customarily employed for shorts and longs for the atonic and tonic (stressed) syllables. This transposition of symbols has its uses. It first corresponds to an approximate reality (the accented syllables generally are or appear to be long in comparison to the others), and it moreover makes it possible to immediately "see," then hear, these tonic rhythms.

The idea is not new insofar as Quicherat (*Traité de versification française*) noted earlier that by calling the long syllables stressed and the short ones unstressed, we find in French prosody the principal "feet" of the ancients, and cited, as an example of anapestic rhythm:

> ⏑ ⏑ — ⏑ ⏑ — ⏑ ⏑— ⏑ ⏑ —
> *Le moment où je parle est déjà loin de moi.*
> The time I spoke is already distant from me.

The cadence of the alexandrine in four anapests ($N = 3333$) is quite frequent in Racine's work.

> *J'ai langui, j'ai séché dans les feux, dans les larmes . . .*
> *Si ta haine m'envie un supplice si doux . . .*
> *Je ne crains que le nom que je laisse après moi . . .*

> I have been drowned in tears, and scorched with fire . . .
> Or if your hatred envy me a blow of such sweet torture . . .
> I only fear the name I leave behind . . .

This parallel is all the more accurate as, let me repeat, there is not only similarity but often concordance, and the rhythm in fact is often generally both tonic and countable, with the tonic syllables being or appearing to be long and the others short.[18]

---

18. The relative durations do not produce, moreover, the exact identities or ratios that were de rigueur in Greek meters. However, it is interesting to note this concordance of the two rhythms in the lines for which the duration of the syllables have been measured with recording devices during their recitation by a random reader. Here is an example provided by L. Estève for a line from Paul Valéry's poem "Narcissus" (from a deleted passage):

> *Voici dans l'eau ma chair de lune et de rosée . . .*
> 33 70 37 60 33 74 32 61 24 24 31 45 (numbers are hundredths of a second)

(*Translation:* Here in the water my flesh of moonlight and dew . . .)

The classic alexandrines cited earlier all have the second hemistich made up of two anapests (the tonic anapest ˘ ˘ — and the iamb ˘ — are, because of the characteristic position of the accent in French words, the normal meters of French rhythm). I will next cite some more of Racine's tetrameters in which the second hemistich is no longer subject to this condition (being formed of two anapests).

Phèdre, dans ce palais, tremblante pour son fils (1524)

La fille de Minos et de Pasiphaé . . . (2406)

L'éclat de mon nom même augmente mon supplice (2424)

Elle meurt dans mes bras d'un mal qu'elle me cache (3324)

Par un chemin plus lent descendre chez les morts (4224)

Si je n'offense point les charmes que j'adore (4224)

Ne verrez-vous point Phèdre avant que de partir? (0624)

The tonic (iambic) rhythm is:

˘ — ˘ — ˘ — // ˘ — ˘ ˘ ˘ —

If we add the respective durations of the two pairs of iambs we find exactly 200 for each, which is fairly remarkable (it amounts to hundredths of seconds). The concluding paeon "*et de rosée*" only provides 124. It seems to me, by ear, that the final syllable was cut a bit short, and that its duration could at least be equal to the caesura "*chair.*"

Here is another example of phonetic measures provided by M. Gramont, from *Le Vers français* (Paris: Champion, 1923):

— ˘ ˘ ˘ ˘ — ˘ ˘ — ˘ ˘ —

Li	*bre*	*du*	*joug*	*su*	*perbe*	*où*	*je*	*suis*	*at*	*ta*	*ché*	
100	29	18	19	15	68	19	19	59	19	18	73	(durations in hundredths of a second)
41	4	3	5	8	16	3	3	11	3	6	13	(relative stresses)

(*Translation:* Free of this superb yoke to which I am attached)

The average duration per measure (except for the one with five syllables) is one second.

Furthermore, for each reader there is a personal quantitative cadence. There is also an acoustic perception and instinctive compensations. As Sonnenschein so rightly puts it: "Rhythm produces an 'impression of proportion.' The following sequence cannot in

Phaedra, in this palace, trembling for her son
daughter of Minos and Pasiphäe . . .
The renown of my very name adds to my torment
She dies in my arms of a disease she will not name
By some slower path go down among the dead
If I have offended her whose charms I adore
Shall you not see Phaedra before leaving?

All these classical meters, 3333, 3324, 3342, 2433, 4233, 2424, 2442, 4242, 4224, and the less frequent or intervening ones including the hemistichs 15 and 06[19] (the cadence 51 is, itself, exceedingly rare) can be found in Romantic, Parnassian, and Symbolist alexandrines, in the work of Valéry and Mallarmé as well as in that of Baudelaire. Here are some examples:

Les souffles de la nuit flottaient sur Galgala . . . (2424) Victor Hugo

Cheveux bleus, pavillon de ténèbres tendues . . . (3333) Charles Baudelaire

La Floride apparut sous un ciel enchanté . . . (3333) Maria de José Heredia

Le printemps vient briser les fontaines scellées (3333) Paul Valéry

Il colore une vierge à soi même enlacée.[20] (3333) Paul Valéry

Le vierge, le vivace et le bel aujourd'hui (2433) Mallarmé

(cont. from p. 149) reality have strict metronomic proportions, but it is necessary for it to create this impression to be called 'rhythmic.'"

Edith Sitwell, in her *Pleasures of Poetry* (London: Duckworth, 1932), reveals (in English prosody based, like the meters of antiquity, on the convention of the equal duration of the feet) a purely cinematic element that accompanies the cadence: the differences in speed and inflection that it produces for its declamation to be actual or cerebral.

19. I have added this 0 to 06 (60 would be more logical) to underscore the fact that it is a hemistich missing the usual secondary accent.

20. The last four of these lines by Valéry are from "La Jeune Parque" ("The Young Fate").

◡ — ◡ ◡ ◡ — ◡ ◡ — ◡ ◡ —
O Mort, vieux capitaine, il est temps! levons l'ancre! (2433) Charles Baudelaire

◡ — ◡ ◡ ◡ — ◡ ◡ — ◡ ◡ —
Le gel cède à regret ses derniers diamants (2433) Paul Valéry

◡ ◡ ◡ — ◡ — ◡ ◡ ◡ — ◡ —
Tel qu'en lui même enfin l'éternité le change (4242) Mallarmé

◡ ◡ ◡ — ◡ — ◡ — ◡ ◡ ◡ —
Tout l'univers chancelle et tremble sur ma tige. (4224) Paul Valéry

— ◡ ◡ — ◡ — ◡ — ◡ ◡ ◡ —
L'arc de mon brusque corps s'accuse et me prononce. (additional accent)[21] (13224) Paul Valéry

The night's breath floated over Galgala . . . (Victor Hugo)
Blue hair, pavilion hung with shadows . . . (Charles Baudelaire)
Florida appeared beneath an enchanted sky . . . (Maria de José Heredia)
Spring has just broken the frozen fountains. (Paul Valéry)
It pictures a virgin wound upon herself. (Paul Valéry)
Will the virginal, strong, and handsome today (Mallarmé)
O Death, old captain, it is time! Weigh anchor! (Charles Baudelaire)

---

21. This hemistich, due to the fact that the verse in which it belongs must be spoken aloud, can require a change or an addition of the accent. For example:

◡ — — ◡ ◡ — ◡ — ◡ ◡ ◡ —
Le ciel, tout l'universe, est plein de mes aïeux (213)
(The heavens, the whole universe, is full with my ancestors)

◡ — ◡ ◡ — — ◡ ◡ ◡ — ◡ —
Me dit que sans périr je ne me puis connaître (231)
(Telling me that without dying I cannot know myself)

— ◡ ◡ — — — ◡ ◡ ◡ — ◡ —
J'aime à vous voir frémir à ce funeste nom (1311)
(I like to see you shudder at this dire name)

◡ ◡ ◡ — — — ◡ ◡ — ◡ ◡ —
Tu frémiras d'horreur si je romps le silence (411)
(You will shiver in horror if I break the silence)

◡ ◡ ◡ — — — ◡ ◡ — ◡ ◡ —
Je reconnus Vénus et ses feux redoutables (51)
(I recognized Venus and her dreadful fires)

But where lines not meant to be declaimed are concerned, these exceptional hemistichs (213, 231, 1311, 411, 51) are not encountered, so to speak.

Ruefully frost surrenders its last diamonds (Paul Valéry)
Such as eternity at last transforms into himself (Mallarmé)
The whole universe quivers and shakes on my stem. (Paul Valéry)
The arc of my sudden body reveals me pronounced. (Paul Valéry)

This notation by Servien is especially valuable for explaining the repetitions, inversions, responses, symmetries, asymmetries, and so forth, as well as the tonic elements in the groups of two or more lines, as the tonic rhythm generally changes from one line to the next.

— ⏑ ⏑ ⏑ ⏑ — ⏑⏑ ⏑— ⏑ —
Ah! que le monde est grand à la clarté des lampes! (1542)

⏑ — ⏑ ⏑ ⏑— ⏑ ⏑ ⏑ ⏑ ⏑ —
Aux yeux du souvenir que le monde est petit! (2433)

Ah! How large the world is in the light cast by lamps!
In the eyes of memory, how small and slight! (Charles Baudelaire)

⏑ ⏑ — ⏑ ⏑ — ⏑ — ⏑ ⏑ ⏑ —
S'accomplir dans la nuit l'hymen des amazones, (3324)

— ⏑ ⏑ ⏑ ⏑ — ⏑⏑ — ⏑ ⏑ —
Fier, et semblable au choc souverain des combats. (1533)

Fulfilling in the night the hymen of the amazons,
Proud, and similar to the sovereign shock of battles.

⏑ —⏑ ⏑ ⏑ — ⏑ ⏑ ⏑⏑⏑ —
Vos lèvres et vos yeux ne profaneront pas (2406)

⏑ ⏑ — ⏑ ⏑— ⏑⏑—⏑ ⏑ —
L'immortel souvenir d'héroïques étreintes . . .[22] (3333)

Your lips and your eyes shall not profane
the immortal memory of heroic embraces . . .

22. These two lines and the two preceding them were taken from Renée Vivien's beautiful poem "Amazones."

Je t'adore à l'égal de la voûte nocturne, (3333)

Ô vase de tristesse, ô grande taciturne! (2424)

I love you as I love the vault of night,
Vessel of sorrow, deeply silent one! (Charles Baudelaire)

Tes ancêtres dompteurs des vagues atlantiques (3324)

À l'abîme ignoré des océans antiques (3342)

Ont ravi le trésor des pays merveilleux (3333)

Your ancestors, tamers of the Atlantic waves
From the unknown gulf of ancient oceans
Have stolen the treasure of wonderful lands.
(Sébastien C. Leconte)

C'est notre heure éternelle, éternellement grande, (3333)

L'heure qui va survivre à l'éphémère amour (1542)

Comme un voile embaumé de rose et de lavande (3324)

Conserve après cent ans la jeunesse d'un jour. (2433)

It is our eternally great hour,
Hour that shall survive ephemeral love
Like a veil scented with rose and lavender
Preserves after one hundred years the youth of one day.
(Pierre Louÿs)

À la molle clarté de la voûte sereine (3333)

Nous chanterons ensemble assis sous le jasmin, (0624)

Jusqu'à l'heure où la lune en glissant vers Misène, (3333)

Se perd en pâlissant dans les feux du matin. (2333)

By the soft light of the serene vault
We shall sing together seated beneath the jasmine,
Until the hour when the moon, gliding toward Miseno,
Turns pale and vanishes in the fires of morning.
(Alphonse de Lamartine)

This stanza was cited by Pierre Guégnen as being "the most sonorous of all the verses written by Lamartine."

And a stanza from Valéry's "La Jeune Parque" ("The Young Fate") in which, between the cadence of the anapests in the beginning and end verses, the rhythm bursts, climbs, and falls again in waves parallel to the heroine's bursts of enthusiasm.

Mon cœur bat! mon cœur bat! . . . Mon sein brûle et m'entraîne

Ah! qu'il s'enfle se gonfle et se tende, ce dur

Très doux témoin captif de mes réseaux d'azur . . .

Dur en moi . . . mais si doux à la bouche infinie . . .

My heart beats! It beats! My burning breast impels me!
Ah, let it swell, dilate and stretch, that hard
Too soft witness prisoned in my nets of azure . . .
Hard in me . . . yet so soft to infinity's mouth! (trans. David Paul)

The "soft" of the third line, the "hard" of the fourth can be "accented" or not in accordance with personal taste, but they are certainly longs. We can see in general that the tonic rhythm and the counted rhythm are precisely superimposed, able to find completion when one gives way to the other. Here, "acoustic perspective" comes into play, which restores the proportions.

In music the intentional movement of the regular downbeats makes it possible to obtain the breaking, choking, or extending of the given rhythm, called "syncopation."

The corresponding effect in prosody is also obtained by the end positioning or enjambment, shifting, or sliding of the regular tonic accents (those of the caesura and the end of the verse).

— ⏑ ⏑ —⏑ ⏑ ⏑ —⏑ ⏑ ⏑ —
Ah! . . . que de choses, qui sont mortes, qui sont nées . . .[23]

Ah . . . so many things, those that are dead, those that have been born

Another example taken from "La Jeune Parque" is quite interesting.

⏑— — ⏑ ⏑ — ⏑— ⏑ ⏑ ⏑ ⏑
L'étonnant printemps rit, viole . . . On ne sait d'où

⏑— ⏑ ⏑ ⏑ — ⏑ — ⏑ ⏑ ⏑ —
Venu? Mais la candeur ruisselle à mots si doux

⏑ ⏑ ⏑ —⏑ ⏑ ⏑ — ⏑⏑ ⏑ —
Qu'une tendresse prend la terre à ses entrailles . . .

Astounding spring, laughing, raping . . .
Where can it come from? Its frankness brims with speech
So soft, earth's entrails are seized with tenderness . . .

The classic cadence of the alexandrine is completely broken; it could just as easily be written as:

⏑— — ⏑ ⏑ — ⏑—
L'étonnant printemps rit, viole . . .

⏑ ⏑ ⏑ ⏑ ⏑— ⏑ ⏑ ⏑ ⏑ ⏑ ⏑ —
On ne sait d'où venu? Mais la candeur ruisselle

⏑ ⏑ ⏑ — ⏑ ⏑ ⏑ —⏑ ⏑ ⏑ — ⏑⏑ ⏑ —
À mots si doux qu'une tendresse prend la terre à ses entrailles . . .

23. The first syllable is an expletive that by completing the alexandrine almost succeeds in masking the tonic beauty of the trimeter: ⏑ ⏑ —⏑ : ⏑ ⏑ —⏑ : ⏑ ⏑ —⏑; the silent *nées,* —⏑, counts here.

which is an octosyllabic alexandrine, a series of four fourth paeons, or even:

◡— — ◡ ◡ — ◡— ◡ ◡ ◡ ◡ ◡—
L'étonnant printemps rit, viole . . . on ne sait d'où venu

◡ ◡ ◡ ◡ ◡ — ◡ ◡ ◡ —
Mais la candeur ruisselle à mots si doux

◡ ◡ ◡ —◡ ◡ ◡ — ◡ ◡ ◡ —
Qu'une tendresse prend la terre à ses entrailles . . .

These shifts of the caesura lead us, like the last line of this stanza, to the ternary or *romantic* line division (trimeter) of the alexandrine (444):

◡ ◡◡ —◡ ◡ ◡ — ◡ ◡ ◡ —
Et l'étamine lance au loin le pollen d'or (Heredia)
And far the pollen darts the pollen's gold.

— ◡ ◡ — ◡ ◡ ◡ — ◡ ◡ ◡ —
L'or des cheveux, l'azur des yeux, la fleur des chairs (1344) (Verlaine)
Gold hair, blue eyes, and blossoms of the flesh.

◡ ◡ ◡ — ◡◡ ◡ — ◡ — ◡—
De monde en monde, allant plus haut, plus haut encore
(Sully Prudhomme)
Going from world to world yet higher and higher.

◡ ◡ ◡— ◡ ◡ ◡— ◡ ◡ ◡ —
J'ai déchaîné les sangliers parmi les fleurs (Pierre Quillard)
I have unleashed the wild boars among the flowers.

Here the caesura is no longer even respected for the eye, with the wild boar riding astride the hemistich.

Using an iambic cadence the trimeter can also be articulated in hexameters (222222).

◡ — ◡ — ◡ — ◡ — ◡ — ◡ —
Il faut qu'il marche! Il faut qu'il roule! Il faut qu'il aille!

◡ — ◡ — ◡ — ◡ — ◡ — ◡ —
Le sceptre est vain, le trône est noir, la pourpre est vile

◡ — ◡ — ◡ — ◡ — ◡ — ◡ —
Marcher à jeun, marcher vaincu, marcher malade (Victor Hugo)

He must march! He must roll! He must go!
The scepter is vain, the throne black, the purple vile
Marching hungry, defeated, and sick, . . .

Lastly, among other phonetic "partitions" or variations, it can furnish the very dynamic line division 354, as in Hélène Vacaresco's beautiful line:

˘ ˘ —˘ ˘ ˘ ˘ — ˘ ˘ ˘ —
Les mains pleines des jours légers que nous portons . . .
Our hands full of the insubstantial days we are carrying . . .

This tonic analysis naturally applies in the same way to verses of ten, eight, or six syllables. Here are some of Valéry's iambs:

˘—˘ — ˘—˘ — ˘ —
L'argile rouge a bu la blanche espèce ("Le Cimetière marin")

˘ — ˘ — ˘ — ˘ —
O Roi des ombres fait de flammes ("Serpent")

˘ — ˘ — ˘ —
(*La belle devant nous*) / *Se sent les jambes pures*[24]
("Cantique des colonnes")

For the red clay has swallowed the white kind
O King of shadows made of flames
Beauty before us / Feels her thighs blameless

Equally interesting is the application to prose of the notation and tonic analysis conceived by Pius Servien. The results gleaned from texts by Rousseau and Chateaubriand are extremely suggestive. From them Servien concludes that among the great artists of prose those who have the innate gift of harmony express themselves spontaneously in tonic rhythms in the passages of their books that were written with pleasure or in the grip of emotion—in a state of auto-incantation.

24. It is obvious that for some syllables this tonic accentuation depends on the reader's taste or pneuma. There are personal tonic accents, but there are also statistical averages.

Generally in Rousseau's work, lyricism and tonic regularity are two faces of the same phenomenon; conversely, tonic rhythm disappears or is diminished in the "thesis-like" passages (Rousseau) or during the correcting and pruning of the first draft to lighten the phrases and disengage the timbres (Chateaubriand).

Tonic notation makes it possible to immediately distinguish in all prose, from one side the amorphous areas "that do not tend to form numerical strictures in which some simple law can be seen," which translated into numbers show numerals following one another at random ("this absence of any law in the sound structures coincides with a total absence of lyricism"), and on the other side, "organized" islets where the tonic rhythms and numbers that characterize poetry as such can be found. "A precise phrase is, among other things, a phrase that has defined tonic accents."

Readers of the previously mentioned books by Pius Servien found in them detailed tonic analyses of extracts from Rabelais, Rousseau, and Chateaubriand. This last author was the "master of number" (and incidentally of timbre; it could rightly be said that Chateaubriand introduced music into the French language).

> *et sa lumière gris de perle descendait . . .* (444)
>
> *et l'on respirait la faible odeur d'ambre*[25] (55)
>
> and its pearl-gray light came down . . .
> and one inhaled the faint odor of amber.

---

25. And the end of a renowned period:

> *Mais au loin, par intervalles, on entendait les roulements solennels*
> *de la cataracte du Niagara*
> *qui dans la calme de la nuit se prolongeaient de désert en désert*
> *et expiraient à travers les fôrets solitaires* (34433 55 44433 4333)
>
> (But faraway, at intervals, one heard the solemn roaring / of Niagara Falls / that, in the night's tranquility, passes from wilderness to wilderness / and ceases among the lonely forests.)

As recalled by d'Eichthal, the supporting syllables that divide both verse and harmonious prose into natural members and phrases always have (in French) final or penultimate accents (when the word ends in a silent *e*), or monosyllables: "Consequently, they are those on which the accent derived from the Latin is carried, and in which the vitality of the word is concentrated (*Fémina, femme* [woman]; *ámat, aime,* [verb love]; *amórem, amour* [love], and so forth)."[26]

But these natural tonic syllables (derived from the Latin accent) do not always or exclusively carry the sound accent of intensity or tonality. The word's position in the phrase or part of the phrase and the punctuation play an important role. Paul Claudel brilliantly established these characteristics of French cadence in the fallowing passage from *Positions et propositions,* which I would like to cite in its entirety:

Servien also points out that by replacing *les roulements solennels* with *les sourds mugissements* (the solemn roaring, the muffled thundering), Chateaubriand lessened the tonic periodicity for the benefit of the timbre. He also notes in this regard the interesting finale on which the last (seven) stresses are echoes of *é.*

And the anapests of Napoléon on Saint Helena:

◡ ◡ — ◡ ◡ — ◡ ◡ — ◡ ◡ —
Je désire reposer sur les bords de la Seine
◡ ◡ — ◡ ◡ — ◡ ◡ —
Au milieu de ce peuple français
◡ ◡ — ◡ —
Que j'ai tant aimé.

(I wish to rest on the banks of the Seine / among these French people / I have loved so much.)

Non-articulated silent syllables do not appear in the notation.

26. It can be seen that in Latin sometimes the first syllable is stressed, and that it is often the change of the final syllable into a silent one that causes the apparent shift of the tonic syllable in French.

In English and German, the tonic is quite often on the first syllable, as it is in Latin. The prosodic rhythm will therefore resemble the trochaic rhythms of Greek or Latin prosody, and the tonic syllable can coincide with the downbeats of the beginnings of musical measures without any artifice, as in Wagner's

— ◡ — ◡ — ◡ ◡ — ◡ —
Winterstürme weichen dem Wonnemond.
(Winter storms have waned in the moon of delight.)

> The French phrase is composed of a series of phonetic members or short vocal waves with a more or less long accentuation and emphasis of the voice on the last syllable ("dix sé *sous*" and "un franc dix *sett*," "tous les *enfants*" and "ils y sont *louss*"). By themselves in French the syllables are therefore neither short nor long, and the phenomenon consists of a long that is always the last syllable of the phoneme and a varying and almost indiscriminate number of silent syllables that are always short in comparison to it. . . . We could say that French is made up of a series of iambs whose long element is the last syllable of the phoneme and the short element an indeterminate number, that can go as high as five or six, of the indiscriminate syllables preceding it.[27] It depends, moreover, on the speaker, guided by intelligence or emotion, to vary the phoneme on a given measure by placing the point of emphasis here or there.

Claudel is harsh on alexandrines ("mechanical Malherberisms"*), which he accuses of distorting "the essential principle of French phonetics by attributing an equal value to each syllable," and not only mechanizing the delicate rhythm of the timbre (by enslaving it to the coarse punch of rhyme) but also manipulating the *white spaces,* even the visual suspensions of one line to the next, which are the true punctuation in poetry of rhythms interwoven with sounds, inspirations, and ideas. It is certain that, through the properties of internal "static" symmetry, through the tendency to homogenous, isotropic equipartition in the balance exhibited by the numbers 12 and 6 and the rhythms and shapes related to them (cf. chapter 2 on the crystalline, inorganic quality of hexagonal systems), the alexandrine can

27. We have seen in the earlier tonic analyses that the dominant notes are iambs in the broad sense, consisting of iambs as such (˘ —), anapests (˘ ˘ —), and fourth paeons (˘ ˘ ˘ —). Conversely, German and English are languages of a more trochaic rhythm (trochees, — ˘, dactyls, — ˘ ˘, and first paeons — ˘ ˘ ˘) like Latin.

[*François de Malherbe was a precursor of the classical style in French poetry.—*Trans.*]

in fact, when it is not handled by true poets, degenerate into an arithmetical and even "mechanical" rhythm that attracts clichés and monotony. But, on the other hand, by virtue of the fact that 12 is the multiple of 3 and 4 (as are 2 and 6), and that the majority of the usual French phonemes number 3 or 4 syllables, the French alexandrine, compared to all the other regular groupings of syllables, provides, despite its symmetry, the greatest possible number of rhythmic combinations, of natural "divisions"—both symmetrical and asymmetrical. Servien's notation has allowed us to detect this in the examples possessing a large number of these partitions and corresponding cadences; I have also borrowed from him this "Bergsonian" demonstration of the alexandrine's value for allowing, despite the "satisfying" appearance of its regular structure, the greatest possible choice of cadenced combinations of the usual phonemes.

I do not know if the introduction or revelation of a tonic rhythm in French prosody will cause harm to isosyllabic rhymed versification. In any case, it does bring us an extremely interesting analytical instrument,[28] and I share the opinion of Pius Servien: "The most seductive arithmetical verses for the ear are also tonic verses."[29]

---

28. For example, these thirteen-syllable verses of Banville:

> *Le chant de l'orgie avec des cris au loin proclame*
> *Le beau Lysios le dieu vermeil comme une flamme . . .*

> (The orgy's song with far-off cries proclaims / the handsome Lysios, the god vermillion as a flame . . .)

in which Servien finds the ascending tonic rhythm ˘ — ˘˘ — ˘˘˘ — ˘˘˘ — (˘). (*N* = 2344' 2344').

29. The importance of the tonic rhythm in prosody has its parallel in music. Servien, having studied all the transformations (variations) that leave the "nucleus" (the leitmotif) of a "musical figure" unchanged (which permits its recognition—these transformations can be produced by acting on the following five elements: the number of notes, volume, timbre, pitch, and duration), reached the following conclusions: "Volumes vary the least and are almost constants. The first accent of the theme is an integral part of its leitmotif. The accents remain in place for as much as the measure allows even during profound thematic alliterations. They therefore belong to the leitmotif. . . . In this way the object behind the music, this invariant of the themes called leitmotifs,

When it comes to the translation into French of Greek or Latin verses, his theory appears logical: the use of "arithmetic" (isosyllabic) rhymed verses for these translations, like the alexandrine, "rhymes with nothing." He proposes the translation of the prosodic stanzas of antiquity into French tonic stanzas based on this following law of transmutation: "For each long syllable of the metrical ictus we find a corresponding tonic syllable, for all other syllables, an atonic one."

He illustrates his theory with sample translations made in accordance with this principle. Here, with the author's permission, I cite three stanzas paraphrasing an ode to Aphrodite:

Dame aux yeux baignés de nuit inquiète, vertige
fauve, aux épaules lunaires qu'un lourd frisson enveloppe!
mords, Aphrodite, le fruit d'où tout notre sang s'échappe,
mords la grenade et souris: nous souffrons, ta beauté s'en éclaire.

Temples profonds où l'amour se blesse à l'amour, où l'étreinte
rêve d'étranges néants, d'infinis supplices de joie:
faites rouler sur les cimes des pins ces volutes bleuâtres
d'humble encens suppliant vers la fille des vagues changeantes.

Nul ne baisa ta cheville d'argent plus pieux et plus pâle,
nul ne comprit plus tôt l'infini de tes tresses profonds;

laisse tes doigts si frais passer sur mon front comme un rêve;
laisse ton sein crispé peser sur mon âme, ô Caresse
lourd comme toutes nos peines, brûlant comme Éros l'avide
laisse tes boucles d'étoiles flotter sur mes yeux qui s'aveuglent . . .

(cont. from p. 161) possesses for the necessary skeletal structure the distribution of tonic accents. The primordial, elementary leitmotifs are thus the tonic rhythms themselves: the combinations of stronger and weaker accents. Incidentally, we have seen the same thing in language."

(Tonic notation of the last 4 lines: 132233 132233 133332 133333)

Lady whose eyes are bathed in troubling night, savage
vertigo, with the lunar shoulders clasped by a strong
shudder!
bite, Aphrodite, the fruits from which seeps our blood,
bite the pomegranate and smile: we are suffering and beauty
takes its light from it.

Deep temple where love wounds itself with love, in which the
embrace
dreams of strange nothings, of infinite torments of joy:
send rolling over the tops of the pines those bluish wreaths of
smoke
from the humble pleading incense toward the daughter of the
changing waves.

None has kissed your most pious and most pale silver ankle,
none grasped sooner the infinity of your deep tresses;
let your cool fingers pass over my brow as if in a dream;
let your taut breast weigh upon my soul, o Caress,
heavy as all our pains, shining like greedy Eros,
let the ringlets of stars float over my eyes going blind.

This adaptation successfully reproduces the spirit of the ancient rhythm by reversing at the beginning of each line the natural iambic rise to the French "phoneme" (˘ —, ˘ ˘ —) by using a trochaic attack (— ˘, — ˘ ˘).

An interesting example of this reversal in which the first syllable of the verse triumphantly gathers tonic intensity in order to let it fall in cascades can be found in the next-to-last trochaic stanza of Valéry's "Le Cimetière marin" ("The Cemetery by the Sea").

— ˘ ˘ — ˘˘—˘ ˘—

Oui! Grande Mer de délires douée, (1333)

— ˘ ˘ — ˘ ˘—˘ ˘—

Peau de panthère et chlamyde trouée (1333)

˘ — ˘ — ˘—˘ ˘ ˘ —

De mille et mille idoles du soleil, (2224)

— ˘˘——˘ ˘ ˘ ˘ —

Hydre absolue, ivre de ta chair bleue . . .[30] (1315)

Yes, great sea endowed with deliriums,
Panther skin and chlamys-riddled
By thousands and thousands idols of the sun,
Absolute hydra, drunk on your blue flesh.

This very "colorful" stanza offers us an example of the way in which a tonic rhythm can weave rhythm and timbre together, the sole one that Rousseau's drum is powerless to reproduce, melodic coloring that the vowels toss on the black line drawing of the consonants.[31] A rhythm is also involved in the sense that the successions, oppositions, and harmonies of the timbres can certainly be grouped in prosody (as in music where the corresponding play of notes and chord progressions creates a melodic pattern) into "perceived periodicities."[32] But except for the very simple and very basic case of the rhyme, the play of the inner harmonies of the sounds in prosody, as in prose, falls more into the domain of an intuitive harmonic gift (exactly as the sense

30. We could also count as tonic the "*du*" of the third line, for one is literally pulled by the rhythm of this line (which interrupts that of the other three and rises in rushed iambs toward the summit where sounds the "tuba" fanfare of "*Hydre absolue*") to mark this normally unstressed syllable with an accent.

31. Let us say straight off that the rhythmics of the timbre is similar, if you like, to the rhythmics of the pitch, a measurable element (in number of vibrations) for the fundamental sound of a syllable, as for a note. In reality, the tops of the diagram of the pitches follow the tonic rhythm, the high syllables (the *i* especially as in the stanza of "Le Cimetière marin" quoted above) always coincide with the downbeats. Some authors even presume that the downbeats were obtained in Greek prosody by acting on the pitch of the corresponding syllables.

32. Cf. Gramont, *Le Vers français.*

of color is differentiated from the science of drawing) than that of a "law of number" analogous to the one that allows us to analyze, if not more precisely correct, the counted and tonic rhythms; or at least, as noted by Servien, the timbres (in the same way as chords in music and proportions in space) act as a continuous material, and do not lend themselves to notation in simple whole series like the ones that concern us here.

They differ especially in this way from musical "values" (tones provided with timbre and pitch), which have an absolute meaning that makes music a subject as international as any branch of mathematics or physics,[33] in that the orchestration of the timbres, the very elements of the scale of timbres and their number, differ in each language in beat as in sounds perceived. As an example, consider the wealth of timbres in French due to the stranglehold of vowels[34] over the neighboring consonants, which produces the faint differences among *am, om, un, in,* and so forth. In German, thanks to their independence with regard to vowels, consonants have retained their full value as constructive

33. Music offers us an example of "number" appearing there where it was not expected to be explicitly found—in what we earlier called the domain of the continuous—by the presence of proportion in the very arrangement of the periods inside the "movements" of a musical suite (corresponding more or less to a psychological development). In Beethoven's sonatas (the Seventh, for example), the ratio between the duration of the exposition and the rest (development and recapitulation) of the piece is often that of the golden section (or rather the opposite $\frac{1}{\Phi} = 0.618...$, approximately $\frac{5}{8}$). The fairly rare exceptions (in Beethoven's sonatas) produce the ratios $\frac{1}{1}$ or $\frac{1}{2}$. See also Gustave Ernst's communication to the Musical Association on January 20, 1903.

In the majority of these examples the repetitions of the exposition or other parts of the movement are not counted in the durations.

The same observation has been made about some of Haydn's symphonies (for example, the first movement of Symphony no. 13 in G Major), some of Mozart's (second and fourth movements of the Symphony in G Minor), and so forth. We should note that according to Zeysing, the Fibonacci intervals 2, 3, 5, 8, 13, appear in the most important chords of the diatonic scale.

34. "Vowels are timbres composed of any fundamental note, harmonic notes, and an almost invariable note called its characteristic as it alone characterizes the vowel; the characteristic of *i* is the most high-pitched of all with 3,698 double vibrations." L. Estève.

connections, as condensers or valves for tension and movement, and it is the plasticity, the heightened relief of words that emerges rather then their music. In Spanish, the two factors of relief and harmony are pushed to their maximum and bring us to an architecture of sound.[35]

Before bringing these considerations on timbre and tonic rhythm to a close, I would like to mention a collection of prose stories (*La Canne de jaspe* [The Jasper Cane]) in which Henri de Régnier gave to the French language its maximum harmonious resonance. The oppositions of timbres, the assonances and alliterations, are meticulously conceived, with an extraordinary musical result, and it is interesting to analyze in tandem the rhythm of certain passages using the tonic method.

˘ — ˘ — ˘ —
des ponts bombés sonnèrent . . . (222 or 21111)

˘ ˘ ˘ — ˘ ˘ ˘ — ˘ ˘˘ —
J'ai fait la guerre; les clairons d'or m'ont précédé . . .[36] (444)

˘ ˘ — ˘˘ — ˘ ˘ —˘ ˘ —
par des cordes de soie ou des chaînes d'argent . . . (3333)

cambered bridges sounded . . .
I waged war; golden bugles went before me . . .
by silken ropes or silver chains . . .

I have arbitrarily carved up into verse the prose lines from the end of the tale "Le Chevalier qui dormit dans la neige" (The Knight Who Slept in the Snow):

---

35. Let's continue the comparison between prosodic rhythm and musical rhythm. Servien, after having shown that in music as in prosody the themes become unrecognizable if the tonic accents are shifted or altered (this even applies to words), notes the following difference: if we change the timbres, nothing essential is changed in a musical theme, but in words timbre plays a role of capital importance. "In sum, the closest connection between words and melodies is in the major role played by tonic accents in both domains. The greatest divergence is the diametrically opposing role played by the timbres."

36. Intentionally, as in some preceding examples, I have not notated the silent syllable in "*guerre.*"

Aussi, quand vient le soir (24)

au delà des vitres gelées en arborescences de forêts (32354)

et en arabesques de grèves imaginaires . . . (535)

je regarde, (3)

en maniant délicatement les verreries fatidiques et vides (45433)

où s'amusent mes songes de soif et de philtres, (3333)

je regarde, au-dessus des fleurs des consoles, (3323)

sur le mur, dans son cadre d'écaille et d'ébène (3333)

debout en ses armes glacées (233)

l'antique portrait taciturne, (233)

avec sa face pâle et son épée (64)

du chevalier qui a dormi dans la neige! (443)

So once evening fell / beyond the windows frozen into the arborescence of forests / and in arabesques of imaginary strands / I gaze, / while delicately handling the empty and fateful glassware / in which my dreams of thirst and potions amuse themselves / I gaze above the flowers on the console tables, / on the wall, in its frame of shell and ebony / standing in his frozen armor, / the taciturn antique portrait, / with the pale face and sword, / of the knight who slept in the snow!

We can therefore say, while recognizing with Claudel that harmony is an innate gift, that "number" plays almost as important a role in poetry as in music—that inner number of expression into which the poet, in a kind of auto-incantation, fits the gushing forth of images and ideas. The delays and backups due to the composition of the first rhythm, that of the ordering of the syllables into a dual alternation of tonic beats and timbres, does not prevent the emergence, when present, of a deeper, sometimes parallel, cadence of the movement of emotions. I have often seen the measured gallop of the anapests, even when governed by the symmetrical framework of alexandrines, perfectly fulfill this role of incantatory springboard for the sudden soaring of passion, whether from fervor, pain, desire, or pride.

But for this auto-incantation to succeed, the passion must all the same precede it, at least in potential. It is often true that when the poet is *inspired* from the beginning, the "number" is not the stimulant but the preliminary expression, the very sign of the entrance of feelings or ideas into vibration (for passion moves just as easily through the domain of abstract thought as it does through that of the affections).

In Plato (who, as Paul Souday once wrote, may have been the greatest poet ever known to humanity) we can see number, in his notion of poetic creation, appear as the master of harmony, but connected precisely to passion, which takes on the nature of divine possession, as he remarks in his *Ion*.

> For all good poets, epic as well as lyric, compose their beautiful poems not by art, but because they are inspired and possessed. And as the Corybantian revellers when they dance are not in their right mind, so the lyric poets are not in their right mind when they are composing their beautiful strains: but when falling under the power of music and metre they are inspired and possessed; like Bacchic maidens who draw milk and honey from the rivers when they are under the influence of Dionysus but not when they are in their right mind. And the soul of the lyric poet does the same, as they themselves say; . . . For the poet is a light and winged and holy thing, and there is no invention in him until he has been inspired and is out of his senses, and the mind is no longer in him: when he has

> not attained to this state, he is powerless and is unable to utter his oracles. (trans. B. Jowett)

It is always the "gust of wind"—here the brush of a wing—(following the essential preparation, the organization of all the details) in which Foch saw the sign of genius in action.

A rhythm that is not constructed from passion and inner tension is nothing more than a symmetrical assemblage, in the modern sense of the word, of sound motifs, which can hold the static charm of a tapestry of regularly juxtaposed designs, of a crystalline configuration, but absolute symmetry, here as in physics, is indicative of a lack of life, of a fall to a uniform leveling of the least amount of effort.

This is an opportune moment to touch on the famous antinomy between Plato's opinion of poetic inspiration and that of Paul Valéry. The antinomy is a fruitful one because it compels reflection, but it is only an apparent difference as Plato had passion and inner tension in mind as essential for the creative act, and Valéry, by scorning inspiration as an efficacious element, or rather by eliminating present and direct cerebral intoxication *for himself* from among the factors useful in poetic composition, had in mind the facile exhilaration of the "first draft," whether verbal or sentimental. For, when passion and tension are there, the important, the difficult, and the salutary thing is in fact to tame them. Their repressed presence will suffice, if the art of technical composition, the potency, and the will to achieve are equal to the desire to endow the work with life.

The research, trial and error, and elimination are necessary in this "symphonic" notion of the creative work, as demanding in and of itself as was the work of the architect geometers whose designs we discussed in the previous chapters.

A more direct term of comparison will once again be given to us from the domain of musical creation. I am going to cite some more passages from Servien's work, because the profound idea that emerges from it appears to deliver, better than I could, the true key to this pseudo-antinomy of inspiration.

> Often a theme arises from an emotion playing the role of a stimulant. The theme arrives therefore already altered by the emotion, and more expressive . . . but not in the form that best exhibits musical structure. . . . Therefore fiddling with a theme to find a more characteristic form, work that seems almost unnatural from the perspective of expression and psychological verity, is in fact just as unnatural from this point of view. But, this is because these things are of no importance here just as in the analogous work of the mathematician: the latter has no concern either for expressing his idea in the same emotionally charged shape that it certainly had on the day of its revelation; he allows it to become clarified at its own pace in his mind; he suspects it will lose its mist of emotion or alter it; but he has no way to cure that. An emotional theme would therefore often be an overly idiosyncratic case, and although this emotion could be an element of beauty, it must sometimes be sacrificed, because this beauty will reroute some of the specifically mathematical or musical beauty one is seeking to attain.

Later, regarding the operation of extracting the leitmotif, the invariant of a set of musical variations, he notes: "Bach has already performed research in this direction. A work like *Art of the Fugue* has this one intent among its principal objectives: to systematically study the principal types of transformations that can be enforced on a theme without altering the invariant."

This is the impression produced on a French music critic by the execution (for the first time in Paris) of this work (a series of forty-eight fugues and preludes) that Bach composed solely for himself:

> On analysis, everything is calculated and reasoned, and submitted to rigid formal rules, with nothing left, it seems, to what is called

> inspiration. And yet inspiration rings out with a fullness and freedom that find few examples in music.
>
> Is the emotion gripping us here from the contemplation of a certain formal perfection, a purely classical beauty? But why then isn't Stravinsky's *Capriccio,* which is also a perfect piece in its genre, so deeply moving? The theme on which Bach worked, and which he broke down and drew out in every way by himself is equally disinterested. We could almost say that if the material of *Capriccio* is literally run of the mill, in the *Art of the Fugue* there is no material at all: it is uniquely a set of "operations" in the mathematical sense of the word. However, these abstract formulas prove to be charged with a human meaning, and these pages in which Bach invested all his knowledge, convey the sound of personal confession. (B. de Schlœzer, *Nouvelle Revue Française,* February 1, 1930)

We know that the comparison from the perspective of rhythm between music, poetry, and architecture has stimulated the critical faculties of aestheticians since the time of Plato to the present. As we have seen, Pius Servien has succeeded in extracting the parallel importance of the tonic rhythm (the rhythm of stresses) as an invariant in music and in prosody; he has even managed through the Cartesian rigor of his method, through going back to the old notation by "neumes," and by replacing absolute stresses with the ratios of stresses between them, to obtain a veritable assimilation of the melody to an abstract curve. If we keep the ratios of the stresses, the number of the notes, and the rising or falling nature specific to each interval, we can vary at will the timbres, durations, stresses, and length of the intervals. We will always obtain the same leitmotif. "From this invariant, we note with surprise that it seems to belong less to the specific domain of the senses that to that of movements, of the dynamic."[37]

37. To figure these musical leitmotifs or invariants by series of whole numbers ("the musical essence stripped of the emotional charge that might encumber it"), Servien finds that two numbers are sufficient: The first marks the intervals between the successive

Similarly, Paul Valéry explains (*Nouvelle Revue Française,* February 1, 1930) the proper value, both architectural and symbolic, of the composition, of the form in poetry, in which the "optimum" adjustment of words, accents, and timbres acquires a value independent of the exact meaning that the artist had in mind while working. The parallel is complete with the work of an architect, and with the music that is both abstract and charged with the potential of a Bach (see above): "While unique content is required from prose, here it is the unique form that establishes order and survives. It is the sound, it is the rhythm that are the physical approximations of words, their effects of induction or their mutual influences that dominate, at the expense of their ability to consume themselves in a certain, definite sense. . . . A beautiful verse is born endlessly anew out of its own ashes, it becomes anew—as the effect of its effect—the harmonic cause of itself."

(cont. from p. 171) notes (the numerals in italics for falling intervals, regular numerals for rising intervals; these intervals in tempered semitones therefore give us the relative pitch); the second number marks the duration of each note ($\frac{6}{8}$ measure); the numbers in bold mark the notes struck with an accent of principal intensity, the italicized numerals denote an accent of secondary intensity.

For example: a leitmotif from *Tristan,* the "deliverance through death," is notated this way:

{ *344*1
{ *23*1**2**1

Servien also applied the theory of transformation groups to the overall question of examining and classifying all the possible tempered scales. He first set down that among the 12 scales imaginable within the octave, the class of the scale of seven notes is predominant because it allows the richest play of cadenced combinations (a property analogous to that of the alexandrine in French prosody); it is followed by the pentatonic scale (five notes). For the heptatonic scale, we find 66 different unlimited scales (if one holds that between two piano keys or notes in an octave there is always a whole number of tempered semitones). These 66 scales, moreover, produce 7 × 66 = 462 possible *modes* depending on the way these 66 types have been divided into octaves (following the note selected as the developmental center). And among the 66 possible heptatonic scales, Servien demonstrates that all rule themselves out as "amorphous" except 4. Out of these 4, the richest type (because it is the most asymmetrical), the one that allows the greatest variety and freedom of movement, is the diatonic scale (of which our major scale is one of the modes). Following this optimum type are two "crystalline" types: the scale of equal (whole) tones (atonal) and the scale of equal semitones (chromatic).

Claudel envisions rhythm as both auto-incantation (see above) and as manifestation of inspiration. His definition here is concrete and pithy: "Poetic inspiration is distinguished by the gifts of image and of Number."[38]

We have now performed a sufficient examination of rhythm as number; we shall now look at the image. This is not at all the simple, evocative, visual image, but the image in its capacity of containing an association of ideas, a comparison, a *metaphor*.

Leaving aside the dead or dormant metaphors that make up three-fourths of our language,[39] we are now going to examine "poetic" metaphors that are still alive in written works, or are in the nascent state.

A metaphor cannot hold just any visual image, but it will always contain, even in the most condensed and masked kind of allusion, a comparison, and the "transfer" (translation of the metaphor) of the resulting ideas.

I nurture an unfair antipathy for Aristotle,[40] but after a thorough search, I have found no better definition for the metaphor than his: "The greatest thing by far is to be a master of metaphor. It is the one thing that cannot be learned from others; it is also a sign of genius, since a good metaphor implies an eye for resemblance" (*De poetica*).

We see again the idea of the same and the other, of unity in variety, of the great principle of analogy that in Plato and Vitruvius governed plastic and architectonic compositions having here become a process of mental integration, and this instantaneous synthesis that reveals the unity or enchantment of a set of concepts or feelings that until then had been distinct in consciousness acts on the intellect by means of

38. Paul Claudel, *Positions et propositions.*

39. "Attempting a fundamental examination of the metaphor would be nothing less than an investigation of the genesis of thought itself." *Times Literary Supplement,* October 14, 1926.

40. This is all the more unfair as I have just realized this aversion is especially due to the antipathy that the very name of the Great Professor inspires in me for purely euphonic reasons. If his name were Diophantus or Alcibiades, I would at least feel respect for him. This absurd observation is appropriate here as we are going to soon examine the power of suggestion held by words and names.

the hedonistic principle of the least effort, of harmonic simplification, which we have also seen at work in visual and auditory perceptions.

At the risk of seeming pedantic, I would like to remind the reader that the implicit comparison found in every metaphor is the very essence not only of analogy in its broad sense, but also of mathematical analogy or proportion that introduces into a geometrical composition the recurring play of similarities. Proportion is merely the mathematical aspect of a comparison that, because it is concerned with measureable quantities, can be reduced to algebraic symbols or numbers.

These explicit comparisons

Ce que le titan chauve est à l'archange imberbe
Don Jayme l'est à don Ascagne . . . (Victor Hugo)
Tu n'as jamais été, même aux jours les plus rares,
Qu'un banal instrument sous mon archet vainqueur . . .
(Louis Bouilhet)

What the bald titan is to the beardless archangel
Don Jayme is to Don Ascagne . . .
You were never, even on the rarest of days,
but a banal instrument beneath my victorious bow . . .

have the same logical structure[41] as the statement

A is to B as C is to D,

an equivalence of two relations that, in geometry or algebra, assume the more precise aspect of the equality of two ratios, a *proportion,* which would be written in algebraic notation:

---

41. These comparisons are so precise and articulate, moreover, they would be equally at home in prose. What does strike a false note: the "poetic" image should be elliptical and unexpected enough to introduce the lightning flash of creation, the revelation, the "miraculous transmutation."

$$A : B = C : D, \text{ or } \frac{A}{B} = \frac{C}{D}.$$

In the following examples from Victor Hugo the comparison is less explicit, giving us semi-metaphors:

Dans le vaste palais catholique romain
Dont chaque ogive semble au soleil une mitre . . .
Charles fut le vautour, Philippe est le hibou . . .
On distingue des tours sur l'épine dorsale
D'un mont lointain qui semble une ourse colossale . . .

In the vast Roman Catholic palace
whose every ogive is like a miter in the sun . . .
Charles was the vulture, Philip is the owl . . .
Towers can be made out on the dorsal spine
of a faraway mountain that looks like a colossal she-bear . . .

But these comparisons are so ingenious that they, too, would not be out of place in prose. It is not yet the "gust of wind."

"The essential thing," as said in the study on metaphor mentioned in footnote 39,

> is simply for there to be that intuitive perception of a similarity between two different concepts that Aristotle describes. What we require in the first place is that the analogy be real and that it had been rarely seen up to the present, or rarely glimpsed, so that it strikes us with the force of a revelation. Something unknown has suddenly been brought to light. From this perspective, the image is truly creative; it denotes a step forward for the writer who perceives it and the reader who receives it, in the conquest of a reality.

The anonymous author of this remarkable article, which I wish I could cite in its entirety, justly notes that Shakespeare was, and will probably remain, the greatest master in the use of metaphors; tight series of them emerge in a rhythm that overlaps that of the verse and whose intoxication bestows the triple incantation of the cadences of timbre, images, and ideas.

Two examples (from a host of others) are given, taken from *Antony and Cleopatra;*[42] they are an excellent choice as the poet here allows his imagination to run wild, heedless of all the risks that would topple any other writer into the hackneyed and the absurd.[43]

Shakespeare almost always takes metaphor further than the simple image; the analogy dives into the depths of the subconscious and stirs up feelings and ideas whose harmonics begin resonating on the surface.

In addition to the "Shakespeare case" with its torrents of living metaphors, we have the numerous virtuosos of the image as such, of the

---

42. In the *Times Literary Supplement,* October 14, 1926.

43. Portrait of Antony:

> His legs bestrid the ocean, his rear'd arm
> Crested the world: his voice was propertied
> As all the tunèd spheres, and that to friends;
> But when he meant to quail and shake the orb,
> He was as rattling thunder. For his bounty,
> There was no winter in't; an autumn 'twas
> That grew the more by reaping: his delights
> Were dolphin-like: they show'd his back above
> The element they lived in. In his livery
> Walk'd crowns and crownets; realms and islands were
> As plates dropp'd from his pocket.

And the arrival of Cleopatra:

> The barge she sat in, like a burnish'd throne,
> Burn'd on the water: the poop was beaten gold;
> Purple the sails, and so perfumed that
> The winds were love-sick with them.

Also, from *Romeo and Juliet:*

> For here lies Juliet, and her beauty makes
> This vault a feasting presence full of light.

comparison that is evocative or allegorical. For example, in France, after the conventional mythological allegories of the two classical centuries, we see the spontaneous emergence of a Hugo in whose work the flood of comparisons or evocations is—even when its intoxication is purely verbal—imposed by the supreme control of the steadiness of his accents and timbres.[44] Then, following the perfect harmony of Baudelaire's work between poetic tension, verbal incantation, and image, we get the hermetic condensation of the metaphor transformed into symbol in the work of Mallarmé.

There are, finally, the modern prose writers who through their handling of illuminating, elliptical, terse images (the condensation of the subtext essential to true metaphor) are magnetizing their prose with the life unique to poetic creation. I am thinking of the best pages of Paul Morand in which, beneath the show of dilettantism and the intentional dryness of the pretense, metaphor and evocative analogy let fly incessant golden arrows with the passionate sting of thought: "the Slavs . . . those men with comet eyes looking at us through Baltic forests"; or the New York winter sky at night above Central Park: "All is dry and precise; the stars are sparkling in the sky like the nickel of the trapezes at

44. And Ruth asked,

> *Quel dieu, quel moissonneur de l'éternel été*
> *Avait, en s'en allant, négligemment jeté*
> *Cette faucille d'or dans le champ des étoiles . . .*
>
> (What god, what harvester of eternal summer / had, when departing, casually cast / this golden sickle into the field of stars . . .)

And the beautiful allegory in the *Tristesse d'Olympio* (Sorrow of Olympio):

> *Toutes les passions s'éloignent avec l'âge*
> *L'une emportant son masque et l'autre son couteau*
> *Comme un essaim chantant des histrions en voyage*
> *Dont le groupe décroît derrière le coteau.*
>
> (With age, all human passions pass away / One carries off its dagger, one its mask, / As bands of traveling players disappear / Over the hills, still singing at their task.) [trans. E. H. and A. M. Blackmore]

the top of the circus tent." And sometimes a simple, cadenced evocation worthy of the Great Viscount: "*c'est l'immense et salubre estuaire de l'Hudson, où les Hollandais chassaient la baleine et dans lequel souffle un vent gemissant et glacé*" (3333 55 4333) ["It is the immense and salubrious Hudson estuary, where the Dutch hunted whales and in which a frozen wind is moaning"].

The play of images can finally—while losing none of its power of visual evocation, its participation in the resonant rhythms, and its lyricism—express the most abstract of speculations in the realm of pure ideas. Each metaphor is then, so to speak, raised to the second power, because a less accessible symbol overlays the first image, and the visual and lyrical harmonic perfection of the evident, apparent series forms an organic entity that can perfectly do without the superimposed psychological or metaphysical poem. Especially as the fairly abstract symbolic meanings of the "second power" do not need to be easy—quite the contrary—but their linkage must be consistent, and the key, once discovered or transmitted, makes it possible to decipher all the symbols, to see the "mystery in full light."[45] For there is mystery here in the ancient sense of the word: the reader is actor and spectator of the symbolic drama, and subject and object of the incantation. The mystes of Eleusis saw with the eyes of the body the young goddess succumbing to curiosity and touching the forbidden narcissus, saw her seized by the Master of Hades and dragged down to his dark caverns, heard the modulation of the desperate calls of the *hierophantides* (priestesses).

Then, turned actor, clad in buckskin, he wended his way along the interminable black tunnel to rediscover she who was no longer anything but Queen of the Underworld; visions, incantations, ordeals, followed on each other's heels until the moment when, now become a "seer," the mystes understands that he has seen and experienced the

45. "Basically, next to the material images that build a bridge between two points of the world, there are transcendent images that, by a very fine ladder, lead from one world to the next. This ladder is ceaselessly swaying over the Valérian style, inviting our spirit to climb." P. Guéguen, *Paul Valéry*.

adventure of the soul for which Persephone was the likeness.

It is in this sense of an epitome that a poem with dual symbolism can be exactly the analogical condensation of antique mystery, lyrical vision, human drama, and incantation for everyone else, and "revelation" and divine drama for the initiate.

This condensing operation, in which the skeletal structure of the ideas should not break the lyrical waves, or the rigor of the metaphysical schema should not smother the passion, is very difficult to realize. When it does succeed, it is the "Great Work" of poetry, the both simple and indefinable criterion being the one that Abbé Bremond applied to all poetry that should be "magical" on penalty of not being poetry at all (prose, meanwhile, *can* be magical, but this is not essential or even always helpful).

An example of a successful poem is "La Jeune Parque" ("The Young Fate") by Paul Valéry. And from an earlier time, there is the *Divine Comedy*. I am thinking mainly of *Paradiso* in which is overlain, in the almost algebraic armature governing the proportions of the work, the architectural composition of the nine circles surrounding the "eternal rose," the purely metaphysical development that accompanies the progressive enlightenment of the soul on the road of truth, and finally, the tête-à-tête, against a stellar backdrop, of the fierce exile with this Beatrice to whom, when she was still of this world, he had never dared speak a word, witnessing the emergence of the burning devotion that, for his whole life, was the bittersweet flame of the imperishable memory of the two smiles he had received. Modulated by the inexorable rigor of the frame, parallel to the increasingly abstract dissertations that illustrate their solitary ascent toward the empyrean, unfolds the most unlikely and grandiose dream that a human being has ever dreamt to make amends for and even vanquish death and fate. The Florentine idyll fully resumes the rapture from the beginning of the *Vita Nuova;*[46]

---

46. "And I with my eyes fixed upon her, turned away from the spectacle above, and fed on that sight . . . like Glaucus of old when he ate of the fruit that gradually transformed

and the unprecedented dialogue begins (their first conversation!) and continues—in metaphysical syllogisms of the arduous revelation interwoven with the growing bedazzlement of love refound, love acknowledged, love accepted[47]—until the moment when, having reached the end of their sidereal ascent, his companion, after explaining how the Supreme Unity is realized through the great law of love, vanishes to retake her place in the tiered petals of the immense rose, so high, so remote, "that it cannot be imagined." But as the poet, in one last prayer is thanking her for all she has given him, he sees her turn back toward him to bestow on him one last time, despite the infinite distance, the beloved gaze and smile, recalling that first look that upended him when he was younger.[48]

* * *

We have noted the natural transition of the metaphor to symbol; the two concepts are often taken to be identical, the latter being more general as domain (the metaphor is specifically verbal) and more precise (but not always) as intention.

But like the metaphor, the symbol can be condensed into one word. In written language, moreover, the word, not the syllable is the definitive element, the true monad of expression. In a word we can rediscover the elements and the resultants of the rhythm and the proportion, and see therein the triple harmony in its initial state:

(cont. from p. 179) him into a god. . . . Rapture! words, this time, could not restore you. . . . These sounds, these unknown chords, the great light, awoke within me the immense desire to join with the essence of her being, a desire I never felt so violently. *Paradiso,* canto 1.

47. "O Beatrice, my guide so dear and beloved." And Beatrice: "You can now open your eyes—look at me—as I truly am. . . . For you have seen such things that you have become strong enough to support the flame of my smile." (And Dante looks, and, stupefied, sways under the glow of the "sacred smile"); then Beatrice: "Why does my countenance enrapture you so that you cannot pull your eyes away even to contemplate these heavenly gardens?" (canto 23).

48. *Cosi orai; e quella si lontana*
Come parea, sorrise e riguardommi;
Por si torno all' eternal Fontana. (canto 31)

a) Form and rhythm (anatomy or architecture, and tonic proportion[49]);
b) Timbre-color; and
c) "Metaphorical" quality (power of suggestion, evocation, liberation, incantation).

Without extending ourselves on the question of metaphors or associations of images, emotions, and ideas imprisoned in the words of every language, recall that Plato and Plutarch indulged in this exciting game that puts philology at the service of logic by seeking in the word the original metaphor or symbol.

Perhaps more interesting than this philosophical aspect of the question is the apparently irrational power of suggestion or charm that words can exert as the action of an affective or even logical harmonic symbolism that is entirely buried in the subconscious.[50] We have all been lulled by the almost magical sound of certain words, especially those of proper nouns (people or places) whose effect appears at the same time more direct and more obscure by virtue of the very fact that the word has no visible logical past, and because the subtle dynamism of crystallization, of liberation, or of awakening of complexes through association of sounds, rhythms, and images takes place in this veiled sanctuary of the subconscious.

The harsh flavor of biblical names, the melodious sweetness of the Gospels, still exercise their magnetic charm on a hundred million English

49. We should note with respect to the word considered as a series of several sound syllables creating harmonic ratios or gradations, the way in which children spontaneously arrange the words they find unpronounceable or dissonant in their normal or "adult" form. For example, a little girl I know quite well was taken to the zoo at the age of eighteen months and was greatly amused by the apes that were presented to her as "monkeys." She immediately harmonized the word "monkey," which was too abrupt for her taste, into "minkamalah." In the same way she transformed "Merry Christmas" into "memolly Kimmy." She was obviously inserting medieties (means), in the same way as Plato.

50. An interesting category of words created artificially or accidentally that became permanent because of their intrinsic qualities of sonorous or dynamic suggestion are some of those words derived from proper nouns, geographical or personal (objects designated after the name of the inventor, and so on), like:

speakers; the Romantic era noisily spouted Spanish sounds. I confess for me that even names of Spanish or Indian origin have, since earliest childhood and my reading of the works of Fenimore Cooper and Gustave Aymard, resonated like nostalgic calls. I have already spoken of the truly architectural, three-dimensional harmony (we think of Plato's two mathematical medieties, the two interpolated chords!) of Spanish words; and how Spanish architecture, with the crystalline purity of its Arabic inflections, clear-cut as scimitar strokes, models and distributes flawlessly the ample Latin harmonies.[51] This magic call was one to which I could never resist succumbing, and during my stays in the lands whose names are charms or thoroughbred gallops—California, Arizona, Oregon—I noted that

(cont. from p. 181)

parchment	sophism	algorithm (from al-Khwarizmi,
faience	masochist	author of the first treatise on
mayonnaise	sadist	algebra)
bayonet	theorbo	bombastic
pistol	*poubelle* (trash can)	macadam
phare (beacon)	galvanize	onanism
lesbian	voltage (from Volta,	*barême* (grade or pay scale)
meander	inventor of the electric	mansard
sodomy	battery; a curious	nicotine
laconic	convergence occurs	*marivaudage* (lighthearted banter)
Hermetic	here as the Etruscan	Machiavellian
daedal	god of thunder was	draconian
mausoleum	called Volta)	platonic
shrapnel	guillotine	Cartesian
academy	silhouette	batiste
praline		boycott

51. Spanish music owes its "shapeliness" and its incantatory power to these same components. The "harmonic" pride and gravity of the autochthonous races (Incas, Aztecs), which in America were in contact with the Spanish conquistadores, admirably wed with the corresponding notes in these latter.

With respect to the Arabic component, we should note here the luminously austere and crystalline nature of the words of Arab and Hebrew origin: *azure, sapphire, Altair, Aldebaran, Algol, Seraphim;* there is also an odd correlation with the crystalline equilibrium for which the Semitic hexagram is the geometrical symbol.

the incantation acted not only on me but on the very inhabitants without becoming blunted. The heroic or cooing names, the names of cities or rivers—Los Angeles, Monterrey, Colorado—wove as if from Aeolian harps a grid of magical arpeggios above the domain of the handsome race that in conformance to Walt Whitman's prophecy[52] had surged to the very shores of the Pacific and had ranged their capital around the triumphant Golden Gate, the "Ciudad Real de la Santa Fé de San Francisco."

I have mentioned Walt Whitman's name in regard to this subject; his is the first interesting case of a great poet in whose works the metaphorical image or symbol did not play, so to speak, any role. Passion and fervor, sometimes taut, sometimes overflowing, were, on their own, enough to nourish his rhythms. His images are direct evocations, memories,[53] or visions. But he, too, is also made lightheaded by the Indian or

---

52. A California song,
  A prophecy . . .
These virgin lands, lands of the Western shore,
To the new culminating man . . .
You promis'd long, we pledge, we dedicate. . .
I see in you, certain to come, the promise of thousands of years, till now deferr'd,
In man of you, more than your mountain peaks or stalwart trees imperial,
In woman more, far more, than all your gold or vines, or even vital air.
("Song of the Redwood-Tree")

Like Walt Whitman, like Keyserling after him, before the giant sequoias of "Mariposa Grove" ("It is in America where we will complete our development, if we are to do it anywhere.") I had the impression on the California banks of the Pacific that here the new Hellas would be built, that it would be (despite Hollywood) in two or three hundred years the center of gravity of Western civilization.

53. Once I passed through a populous city imprinting my brain for future use with its shows, architecture, customs, traditions,
Yet now of all that city I remember only a woman I casually met there who detain'd me for love of me.
Day by day and night by night we were together . . . all else has long been forgotten by me,
I remember I say only that woman who passionately clung to me.
Again we wander, we love, we separate again,
Again she holds me by the hand, I must not go,
I see her close beside me with silent lips sad and tremulous . . .

Spanish sonorities I spoke of earlier, and concerning the aboriginal name of the rock on which rises his proud native city, the infernal and divine New York of the foreseen skyscrapers and titanic bridges, he responds, at the beginning of the poem titled "Mannahatta" (which ends with the impassioned: "City nested in bays! my city!") to the "What's in a name?" of the other giant of his race, of the king of metaphors:

> I was asking for something specific and perfect for my city,
> Whereupon lo! upsprung the aboriginal name.
> Now I see what there is in a name, a word, liquid, sane, unruly,
> musical, self-sufficient . . .

And coloring almost all his poems is the resonance of all these barbaric or Mediterranean cadences that he found completely magical: Alabama, Oregon, California, Colorado, Nevada, Ontario, Savannah, Nebraska, Idaho.[54]

It is time to call a halt to this journey in the land of rhythm that has apparently led us far afield from the Mediterranean and the laws of number. But only apparently, for we have found that rhythm and harmony can be analyzed and studied through the concepts of periodicity and proportion, bringing us back to that mathematico-musical synthesis that is specifically Mediterranean and even Pythagorean.

The Pythagoreans would not have disapproved of the statement "A painting is a state of soul," in which Wang Wei summed up for the ages to come what some twelve centuries later would be adopted in Europe under the name of "expressionism," but they would have completed the phrase of the great painter-poet of the Tang Dynasty by adding that a

54. Hindu and Burmese names had the same effect on Rudyard Kipling ("On the road to Mandalay").

Concerning this semantics of proper names, these essences "deeply steeped and marinated in the black jam of the centuries . . . stars in the oil . . ." (Paul Claudel), we should not overlook Marcel Proust and "the golden brown sound of the name of Brabant" in *Swann's Way.*

state of soul is quite often a rhythm, and that a rhythm can be sometimes the cause and sometimes the effect of expression. And that the poetic or musical rhythm appears to precisely express or result in the "attunement" of the artist's rhythm (that of the "durations" weaving together the framework of his "ego," as Bergson would say), or of the one who perceives it through the work, with a greater rhythm.

In fact, following in their footsteps, we have found rhythm, proportion, and analogy everywhere, or, in the last place, the "analogy" and the "number."

About the American bard who led us to the beaches of the Pacific, it could be said that the feeling and expression of love in him, love of the individual, of his tribe, of humanity, finally of the pulsing universal life whose harmony he could see and hear, attained a cosmic range that connected him directly to Hellas. There are no boundaries between abstract Platonic intoxication and the more pagan variety of Walt Whitman. This pioneer of the Abraham Lincoln era, this ardent onlooker of Mannahatta, who wrote "I Hear America Singing" and launched the grandiloquent, savory "Hello World! Walt Whitman!" has heard Plato's own sirens and is deserving over all others of a seat at his symposium.

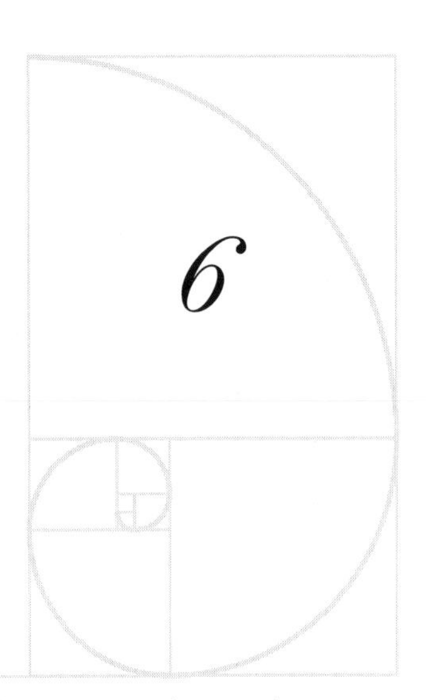

6

From Incantation to Love

I am Isis, the goddess of the words of power . . . the words whose voices are magic.

Several times over the course of the previous chapter I was led to observe that rhythmic harmony, whether musical or prosodic, had—in addition to its normal aesthetic or cerebral appeal, related to the action of the "hedonistic principle of least effort" (that which acts by harmonic organization of chaos or internal disharmony, by "sympathetic" perception, *Einfühlung,* of the perfect adaptation of the technical work or artwork to its function or reason to exist)—this incantatory effect that we are tempted, for want of a better term (Abbé Bremond also succumbed to the temptation), to call *magical.* All while admitting that the term *incantation* should be reserved in principle for the effect obtained by the repetition of a word, a formula, an assonance, a prosodic or musical periodicity, which is to say, the action of a rhythm, I have noted that the rhythm and its incantatory action are sometimes condensed into a word.

It is because of its immediate operation that we label this action magical or quasi-magical. Its tangible result is in fact the establishment of an ecstatic state in the person subjected to the effect of the "charm." This person has been rendered "receptive" and permeated by the rhythm of passion, of intoxication, or of simple euphoria that had vibrated in the composer at the time of the work's creation, with this creation having been anonymous or collective. Sometimes even the virtue, the "charm," appears to be inherent in the symbol or the word itself, equipped with a condensed dynamism that is one of the characteristics of magical action. In the physical world we have an analogous phenomenon in the instantaneous production of the current induced in the coil of a secondary coil, under the sole effect of the primary electrical current (alternating), without any contact or communication between the two circuits (this is the principle behind the transformer). It can be seen again in the magnetization of a soft iron core surrounded by a coil in which an electrical current is circulating (electromagnet), or in the instantaneous transformation of a heap of iron filings into an electricity-conduction system by a high-frequency discharge (the initial principle behind wireless communications). However, we do not call this class of phenomena[1] "magical," because although we may not understand their mechanics, we do know the energies they bring into play are purely physical. This distinction (which is not obligatory) gives us a second characteristic of the magical phenomenon: it involves condensation, followed by liberation, utilization, and application in a determined direction of essentially spiritual, psychical energies originating in living centers or reservoirs (transcendent in comparison to inorganic matter and even immaterial energies—like those of electromagnetic waves, gravity, and so forth—that are not mental or spiritual by nature[2]). These "transcendent" energy reservoirs,

1. The definition "in the broad sense" of magic will go so far as to say: everything that liberates a force, or harnesses it (even the sealing of a flask of alcohol). We can also speak of a "relativity of magic" and say that a dynamic operation, the controlled captation or emission of forces, is always magical—for those who cannot comprehend it.

2. These psychic energies are not simply parallel sensations ("epiphenomena") to the physiological metabolism, whose processes are purely physicochemical. Psychic reactions can certainly be influenced (stimulated or attenuated) by the functioning of the physiological circuit, but they can also influence or dominate it (suggestion, autosuggestion, or therapeutically by the Coué method, and so forth). No proof even exists that the milieu and the waves (let's use this word as it concerns an "action at a distance") that make communication possible between the minds of different persons through suggestion, thought transmission, or thought reading *are* the immaterial but also physical milieu (electromagnetic ether), and the waves function as the psychic-energy transmitters.

presuming they exist outside the mind and vital circuits of the "mage" and the recipients, can be formed from psyches communicating with each other without conscious awareness, and with all living beings at any given moment. They can also (this is the spiritist hypothesis) communicate with disincarnate psychic energies. It is only here the branch of magic begins that can be designated in bulk as "occultism."

However suspect this concept of magic may be—which suffers precisely from its kinship with that occultism and those occult sciences whose poor reputation is partially deserved by the misuse of superficial popularizations, by the fraudulent or puerile aspect associated with their practices, by the risk for those that devote themselves to them to fall into a dubious mysticism, and so forth, and so on—we cannot in a study of the forces that have influenced the evolution of human thought, specifically what I call Mediterranean thought, ignore the fact that magic existed as theory and technique, and played a much more important role in this cultural cycle than one might be tempted to think.

In any case, as we have seen, not only the idea of incantation, but the words *magic* and *magical* naturally come up in any attempt to analyze the effects of musical or prosodic rhythm.

Here is an explanation of the magical effect in which the word does not even appear:

> The Rites, which are only the *putting into play* of symbols having *natural* power over the astral world that contain in potential and in seed the full blossoming of the physical world . . . The word *symbol* means above all else "summary," "quintessence"; thus, by performing a symbolic ceremony, we draw the secondary cause into the orbit of our will, we set in motion . . . the productive dynamism of the phenomenon, our fingers exit the physical plane and communicate with the keyboard from which matter hears the harmonies but that itself remains perpetually hidden. But to be accomplished in a completely effective manner, every rite requires a state of mind and even a preparation of the body, a preliminary physical, psychic, and intellectual

foundation, without which it would be childish to dream of manipulating the keys of phenomena.[3]

This is the spiritualist point of view, which one can reject or push aside just as one can reject transformism, Sitter's five-dimensional world, or the subconscious of psychoanalysis, but there is nothing absurd or incoherent about it.

We cannot attribute to symbolism and to symbols the direct action granted them by the world system of a Yeats, for example, but we cannot discuss rhythm and poetry by removing the concepts of incantation, symbol, and rite, and if we wish to analyze them, it is logical and perhaps even fruitful to examine without bias the definitions and hypotheses in use among the thinkers who are specifically concerned with them.

These ideas of incantation, symbol, and rite appear closely linked together here. We can accept Larmandie's observation that the rite involves an "activation" of symbols; the rite is always founded on a rhythm (be it simply the periodic return of the ceremony itself), and often on an incantation. The religion of ancient Egypt was specifically based on magic ritual.

The Pythagoreans took special notice of the purifying and regulating value of music, and attributed to this effect (which they called "catharsis")[4] a very important role in the daily discipline of adepts. Moreover, rites that are unreservedly magical and even "occult," such as the evocation of the dead and so forth, figured in the practices of the

---

3. Count de Larmandie, *Magie et religion.* We can complete our definition of magic here: it is the rationalization and systematic study of the manipulation of forces (captation, condensation, application) and correspondences.

4. This musical catharsis also acts through a therapeutic effect comparable to that of Freudian psychoanalysis: the person revisits his complex, his conflict, as in a confession; the psychosis is lanced, like an abscess.

"This soothing music therefore appears to be cyclical. It opens cycles: creating asymmetries and closing them back up, restoring rest. . . . The listener is not a 'tabula rasa,' but a collection of highly varied agitations. Music offers asymmetries, to which they aggregate. When the music has captured everything in its asymmetries, it gradually closes them back up and restores everything to rest." Pius Servien

full initiates, as we can find traces of them in all the neo-Pythagorean literature.[5]

The musical "catharsis" can be seen again in the ritual of the Catholic church, as well as in the practice of incantatory spells and prayers[6] adopted directly from the Greek and Egyptian mystery religions, after the elimination (and this is one of the motives responsible for the battle between the Church of Rome and Alexandrian Gnosticism) less of the practices than of the "desire for magic." This elimination comes down to the fact that in Catholic ceremony and ritual, incantation, prayer, and incense are an homage to God and not the means of capturing or condensing, unbeknownst to the Deity,

5. See part 2 of the present book.

6. As noted by Count Hermann von Keyserling, the efficacy of the prayer-incantation can exist independently of reality or the good will of the deity. "Whether the gods are objective realities or simply subjective . . . in all cases a sincere prayer creates a channel, through which the representation can echo back on the person who prays."

And also, concerning the incantatory dynamism especially attached to words and phrases:

"In the ancient faith placed in magic spells, there is more truth than our era wishes to admit: the words and percepts possess virtues that communicate their message even to those whose minds can only grasp them literally." *Travel Journal of a Philosopher.*

This prayer-incantation is not a magical operation in the strict sense of the term, but a magical act that can also, according to some, induce vibrations in more or less vast zones of collective souls: "We can look at each community as having at its disposal, in addition to the purely material means of action in the ordinary sense of the word, a force formed by the contributions of all its members past and present, and that, consequently this force is even more considerable when the community has a longer history and a greater number of members. Each member can, when necessary, use a part of this force for his personal benefit; all it requires is for him to put his individuality in tune with the whole of the community of which he is a member, a result he can achieve by observing the rites, which is to say, the rules the group has established. . . . Sometimes the force about which we have been speaking can become concentrated on a determined symbol in a specific place and create tangible manifestations there (for example, the Ark of Covenant). Hence also certain 'miracles' that are not truly 'contrary' to natural laws." "Palingenesis," "La Prière et l'incantation," *Gnosis,* January 1911.

This is the theory of "Couéism" applied to a collective soul. [A self-help method founded on autosuggestion and positive thinking created by French psychotherapist Emile Coué. The slogan "Every day in every way I am getting better and better" was coined by Coué.—*Trans.*]

supernatural or extra-natural forces by means of the incantatory technique.

This elimination of the "desire for magic" by the Church, however, is not absolute. The technique of exorcism, excommunication, and the immediate effect attributed to the sacraments (like baptism) exhibit certain essential characteristics of magical activity, by virtue of the fact that the result should inevitably be produced through the sacerdotal virtue of the celebrant. Similarly, through the rite-incantation, the individual is desiring "to force" the magnetic, astral, or even natural or extra-natural psychic fluids to condense or operate in a specific direction; even astral specters, demons, and the dead are summoned against their will (*Höllenzwang*, "hell's compulsion"). While the technique has changed, there is no great difference between a sixteenth-century magician and a modern medium. With respect to Catholic ritual, it could also be said that transubstantiation by sacramental consecration in the Mass is a magical operation because it must always "succeed," absolution as well (it does succeed, moreover, as it makes an impression on those confessing).

In any case, the "cathartic" effect on the faithful is incontestable, as well as the production of psychical concentration that in tandem with asceticism[7] is necessary to emit or receive the magical "current" or draw from the hypothetical reservoirs of psychic energy mentioned above.

We could roughly say that the Mediterranean magic that we have been led to consider is Egyptian in origin: magic of the rite (rhythm to the second power), magic of the sign, and magic of the word.

The word, the Logos, the Word, can possess a condensed harmonic rhythm and a capacity for charm and suggestion due to this rhythm and its timbre, and the dormant metaphors it potentially represents. For Egyptian religion that, like the Egyptian social system, was based on magic, certain words had an actual magical incantatory value. These "words of power" or *hekau,* were mentioned as early as the

7. Éliphas Lévi called Saint Ignatius's exercises a "magical ritual."

sixth century BCE in a special chapter of the *Book of the Dead.*[8] They play a role here that is more important than the simply formal signs or talismans. The *Book of the Dead* shows that the soul of the departed should use these "words of power" and "words of free passage" throughout the entire journey in the otherworld.[9] We find these passwords again in the mysteries and the Pythagorean-Orphic funeral tablets, then again later in the guilds of stone carvers, masons, professional freemasons, and journeymen, and finally in "speculative" Freemasonry. Their line of transmission is linked to that of the graphic symbols, the most important of which is the Pythagorean pentagram, sign of the golden number.

In Egypt it was Isis who—like her counterpart Demeter-Ceres of the Eleusinian Mysteries, goddess of both fertility and death, and patron deity of initiations[10]—was the goddess who ruled over magic words. This comes from various passages in the *Book of the Dead* and numerous inscriptions (such as the one I chose for the epigraph of this chapter; the expression, "Isis, lady of the magic words," recurs frequently). Isis was most expressly the goddess of mystery and the mysteries, and this side of her nature is immortalized by the famous Saïs inscription passed on by Plutarch in *De Iside:* "I am all that was,

---

8. This is the chapter that deals with the use of the words of power needed to reach Osiris in the otherworld.

9. To the demon crocodile Sui, who guards one of the passageways to the beyond, the soul will say: "Because I live by virtue of the words of power I carry with me," and so on (*Book of the Dead*).

We shall see analogous phrases on the tablets of Petelia and Thurii (part 2). Here is another inscription of the same kind that was found in Thurii (this time it is a god or spirit that emphatically repeats a direction to the wavering soul): "Enter the path on the right! . . . take the right path if you wish to reach the holy meadows, the sacred grove of Persephone."

The Coptic Gnostic text called the Bruce Papyrus (at Oxford), in which gnosis and magic are intimately mingled, shows us Jesus providing in initiatory form "passwords" and incantations for crossing through the hostile layers of the astral world in the beyond. There is also a question of the secret, inexpressible "Great Name" that would play such a huge role in the Kabbalah.

10. "It is I who instituted initiations for men." Hymn to Isis discovered on the island of Ios.

all that is, and all that will be, and never has any mortal yet lifted my veil."[11]

As her associate in the domain of magic, Isis has the god Thoth-Hermes, "lord of the divine words," who at the creation condensed into words the will of the unknown and invisible Creative Power. Plato (*Phaedrus*) brought back the Egyptian tradition according to which Thoth was also the inventor of logic, arithmetic, geometry, chess, and writing. We can also say that Thoth, the father of Isis, is the god of reason, number, and the word. In Greece, as the god of eloquence, he held the cognomen Logios.

Thoth is also the chief of infernal protocol in the otherworld. As the psychopomp Hermes, he appears in the *Odyssey* performing his duty as herald of souls. We shall see the merger of these two deities—already noted by Herodotus—forming a new entity that was sometimes god, sometimes daimon or legendary superman, but a magician in any case: Hermes Trismegistus, who in the Ptolemaic era began his role as the patron deity of the seekers of secrets, magicians, alchemists, necromancers, and Kabbalists, in a word, all those concerned with the "Hermetic" sciences (one of these words that emerged from proper names due to their accidental power of suggestion).

Plutarch (*De Iside*) mentions the "Hermetic" literature or "Books of Hermes" attributed to this legendary figure. Iamblichus carries their number to twenty thousand. They were almost all destroyed during the burning of the Serpaeum of Alexandria by the Christians and at the time of the final pillaging of the famous library by Omar. The most important of those that survived into the present (aside from the *Book of the Dead*) is the *Pymander*. If Thoth-Hermes was the god of the word and of number, and more particularly the god of the science of numbers, mathematics, the female consort deity strictly corresponding with him is Maat, goddess of physical and moral law considered as harmonious order, and the regulator

11. "*Ego sum omne, quod exstitit, est, et erit, meumque peplum nemo adhuc mortalium detexit.*"

of rhythms, the "lady of the Hall of Supreme Judgment." This goddess of rule, harmony, and truth had as her symbol a vertical feather stuck in her diadem. The word *maat* itself originally meant a cut reed, then a ruler for measuring, then "that which is right," then "the rule," "the law," "the truth." The Greek word *κανών* (canon) passed through the exact same chain of metaphors. From the related Semitic root word *qanat,* which also means reed, then tube—the bellows (blowpipe) of the blacksmiths—also comes the tribal name of the Cainites (sons of Cain), the smith-miners of the Sinai, whose mysterious traces Robert Eisler has shed light on (we will revisit them in part 2).

Emerging from the incantatory magic by "words of power" is a fantastic flowering of Gnostic verbal "charms" in which Egyptian, Greek, and Hebrew syllables and words are combined together to form bizarre alliterations, assonances, and repetitions. The word is let loose and scatters into fragments, anagrams, palindromes,[12] steps, triangles, or "magic" squares, whose abracadabric[13] architecture has been handed

12. Words or phrases that can be read either backward or forward, like the famous SATOR AREPO TENET OPERA ROTAS.

13. The triangle:

A B R A C A D A B R A
A B R A C A D A B R
A B R A C A D A B
A B R A C A D A
A B R A C A D
A B R A C A
A B R A C
A B R A
A B R
A B
A

has a pentagramatical key: the A of the pentalpha is repeated $6 \times 5 = 30$ times.

The phrase ABRACADABRA has been found cited as early 250 CE (in a text by the doctor Quintus S. Sammonicus).

We find again in the Kabbalah and medieval magic the Hebrew and Chaldean names of angels and spirits who had invaded the Gnostic talismans and incantations, generally grouped in series of ten. Blessed spirits: Ophanium, Aralim, Seraphim (burning), Elohim, Cherubim, and so on. Spirits of evil: Samaël, Beelzebub, Lucifer, Astaroth, Asmodeus, Belphegor, Lilith (female demon, queen of the obscene spirits the Gamatiels), and so on. The fire spirits: Michaël, Anaël, and so forth.

down to us by the uninterrupted chain of magical treatises from the third to sixteenth centuries.

One category of magical practices that smacks of black magic (for criminal or immoral purposes), which is also of purely Egyptian origin, is that of enchantment through the application of incantations and magical rites to wax figures. These wax figures are mentioned as early as the Third Dynasty.[14] The last indigenous pharaoh, Nectanebo (360–350 BCE), was, according to Greek tradition, the best versed in the occult sciences and the most powerful magician of all the Egyptian rulers. Enchantment using wax figures and the creation of artificial beings played a huge role in the magical experiments of his inner circle. The use of the ushabti figurines that were doubles for the dead individual (but as beneficial magic) is related to these practices. The technique of the creation of homunculi by the insufflation of "pneuma" and the insertion of a written magical word into the clay doll passed directly into the Hebrew Kabbalah[15] through the *Sepher Yetzirah,* or *Book of Creation,* and was the source of the legends of the creation of homunculi

14. The "Judicial Papyrus" in the Turin Museum mentions the use of wax figures by the conspirator Hui and the ladies of the gynaeceum in their attempt to bewitch Ramses III (circa 1200 BCE) using a technique borrowed from the magical books of the royal library.

15. The Hebrew word *Qabalah* means "tradition." The Kabbalah was born in Alexandria at the same time as Gnosticism, Hermeticism, and alchemy, and represents the Hebrew version of the Pythagorean mysticism of number. The Babylonian sojourn of a portion of the Jewish intellectuals transported from Palestine by the Sassanids strongly colored the evolution of the Kabbalah between the fourth and eighth centuries with common magic and Chaldean astrology. The Arab conquest and the intellectually fertile influence of the early Islamic empires grafted back the high Neoplatonic speculation to this Hebraic neo-Pythagoreanism that was entering Europe though Italy, Spain, and the Midi region of France. Dating from the Babylonian phase is the *Shi'ur Koma,* or "Measure of the Stature of God," (precise shapes and measurements of the body and face of God), but the most remarkable book of this era is the *Sephir Yetzirah,* or *Book of Creation*, written in Hebrew (in Syria most likely) toward the sixth or seventh century (therefore after the Talmud was completed in 499), and already commented on everywhere by the tenth century.

The Gnostic and neo-Pythagorean influences are quite visible. God created the world through the intermediary of ten powers or words called *sefirot* [sing. *sefirah*] and the twenty-two letters of the Hebrew alphabet. The bible of the Kabbalah is the *Book of Splendor* (*Sepher ha Zohar*, generally called the *Zohar*), also written around the sixth

in the Middle Ages, especially the entire "golem" cycle in Prague.[16]

The Hebrew Kabbalah has two components, one of which is

(cont. from p. 195) century, in Syrio-Chaldean like the Talmud. It is a Kabalistic and Neoplatonic commentary on the Pentateuch (between God and the world is the decad of Mother-Ideas, or *sefirot,* forming the macrocosm or world of emanations, the intellectual type for the material world. The first *sefirah*, Kether or Crown, God's will, created the other nine. The tenth, Malkuth or Kingdom, is the harmony of the world. We can recognize the monad, the decad, and so forth.

A practical Kabbalah is combined with the mystical one: combinations of numerals, letters, magic names of God, angels, demons, geometrical "lesser keys," and finally methods of divination through the permutations of words, numerals, letters: gematria, *notarikon,* and *temurah* (isopsephy).

The "*sefirotic* triangle" replaces the "abracadabras" of the Gnostics in the Kabbalah just as the "inexpressible Tetragrammaton" YHVH replaces ABRAXAS or ABRASAX. The *sefirotic* triangle is a triangular decad or ascending scale composed of ten divine names. It is the incantatory decomposition of the "inexpressible Tetragrammaton" (or *Schem ha mephorasch,* or Shemhamphorasch, which literally means "the inexpressible divine name") YHVH, condensation of the divine occult force, which the profane pronounce YaHVeH or YeHoVaH. The initiates pronounce each letter: Yod, Heh, Vav, Heh. Its contraction Y H (or rather G H, because *G,* as the first letter in the name of the deity is the corresponding letter in our alphabet for the Hebraic Yod) is already an expression of the name of God (numerically 15 = 10 + 5, according to the numerical values of these two letters in the Hebrew alphabet).

16. *The Treatise on Sincerity* by Rabbi Moise Takko (thirteenth century) says: "And all the magicians of Egypt, who had created any kind of being, studied the order of the spheres by means of demons or a kind of magic . . . and created whatever they wished. Now, the rabbis, who created a man or a calf, knew this mystery. They took earth . . . spoke the 'Shem' over it, and the being was created."

We can find as early as the tenth century the legend of the creation of homunculi with the help of the *Sepher Yetzirah,* by "geometrical magnitudes" expressed in letters taken from the Shemhamphorasch (the divine names of the *sefirotic* triangle).

A sixteenth-century commentary on the *Sepher Yetzirah* (by Saadia Gaon) tells us that Ben Sira created a man (still with the help of this book) by writing on its brow the magic word *EMeT,* "truth." This detail is given in various stories about the creation of golems or artificial human beings (*golem:* "shapeless mass") by rabbis. To cause the death of the golem, it only requires the erasure of the first letter of the magic word, which then becomes *MeT,* "death." It can be seen that the Kabbalistic word for truth is strangely reminiscent of the name of the goddess Maat, the Egyptian word for truth.

But the most famous golem was the one created by the famous Rabbi Loew (1513–1609), the friend of Rudolph II and Kepler who inspired the fine novel by Gustave Meyrink.

precisely ancient Egyptian magic founded on "words of power" (the word, the incantation can be written on talismans), the other the Alexandrian neo-Pythagoreanism in which the mysticism of number—decad, pentad, tetractys—naturally plays the predominant role.

To the corresponding geometrical symbols—dodecahedron for ether-quintessence and for cosmic harmony, decagon and quaternary triangle

•
• •
• • •
• • • •

for the decad-tetractys and the living macrocosm (*natura naturanda*), pentagram for the microcosm (man) and for love—the Kabbalah and magic add the specifically Hebraic sign of perfection in balance, of the crystalline order in symmetry and homogeneity, the hard hexagram or seal of Solomon, which will become the symbol of the inorganic macrocosm (*natura naturata*).[17]

From these two tables of symbols and magical units—words and numbers—the Kabbalah will quite naturally draw from the fact that each Hebrew letter corresponds to a number (letters and numerals are interchangeable in this case—the yod or aspirated Hebrew *G* is thus the symbol for 10, the decad, hence the mysterious *G* of the blazing star or Masonic pentagram*), a combined system from which emerges the procedures of symbolic analysis and divination specific to this discipline, the best-known practice of which consists precisely of replacing each letter in a word by the corresponding number, or vice versa. The riddle of the "number of the Beast" in Revelation is a famous example of this "isopsephy"[18] [or gematria].

---

17. We have seen that number six already has this character in the work of Nicomachus and Vitruvius. But the hexagram became a specifically Semitic symbol; Arab legends say that it was carved on the diamond *shamir* on Solomon's ring.

[*The Hebrew *Y* (yod) is the first letter of the name of God and thus corresponds to the *G* in the Masonic pentagram.—*Trans.*]

18. The "Number of the Beast," 666, is the double transposition of Nero's name.

Several passages in Revelation and the Gospel of John reveal traces of Hermetic and

The decad therefore gives us the ten divine names (development of the Shemhamphorasch) and the ten *sefirot.* The tetractys gives us the Tetragrammaton or "inexpressible" name of the Supreme God. Names of demons, angels, saints, planets, and elements exchange or combine their graphic symbols alphabetically or numerically, the blazing Hebrew consonants alternating with the alphas and omegas, the IAO, the IO, and the ABRAXAS, sparkling at the points of the pentacles and in the case of magic squares, reinforcing with their dual ferment of breath and numbers the incantation of the Gnosis.[19]

We acknowledged previously that all *successful* verbal or musical incantation, in other words one that produces "enchantment" (the words *incantation* and *enchantment* are two shades of the same verbal entity, envisioned in the one case as action and in the other as the effect produced), ecstasy, liberation, or even bewitchment in the broad sense, is a phenomenon that we have good reason to call magical. We can distinguish three important variants of these "natural" magical states or actions, based on the nature of the effect produced.

One type is an indeterminate ecstasy, drunkenness, or general or

---

(cont. from p. 197) Kabbalistic influence. For example, the opening of the Gospel: "In the beginning was the Word, and the Word was with God, and the Word was God."

Professor Robert Eisler brought to my attention that the number of fish in the miraculous catch is only recorded in the Gospel of John: 153 fish. It so happens that 153 is equal to the sum of the 17 prime numbers ($s = n\frac{n+1}{2}$), number of the points of the triangular figurate number on a base of 17 = 7 + 10 (virgin septenary and decad). We shall see in part 2 that Saint John, the "beloved disciple," was claimed as a kind of patron saint not only by Gnostics but also by the stone-carver lodges of the Middle Ages, by the Templars, and by an entire branch of Freemasonry.

19. The IO, IAO, and related words that were already so frequent on Gnostic gems and talismans have retained since antiquity their nature of magical invocation. IOVAH, IAH, IAVE, IUWE, IOU, IOH, IOV, ION, IACCHOS, IANUS are modulations of the same root, or rather, of the same incantation-sound.

The *voivodes* [warlords] of Moldavia and Wallachia since the time of the first Bessarabia to the merger of the two principalities in 1858 have always prefixed their title with the mysterious syllable IO, about the etymology of which historians can still not agree. The role of Saint John the Baptist and Saint John the Evangelist in magic traditions, secret societies, Gnosticism, and so many Christian sects is obscurely connected to the incantatory value of this name.

"cosmic" euphoria: from the "charm" flows the sensation, conscious or not, of being in tune with the harmonious waves of the Great Whole. The two other types are more differentiated and specialized (but can melt into the first one): religious or mystical ecstasy (universal affection fixed on the deity), which I mentioned earlier with regard to the role of the prayer-incantation, the incantatory rhythms to the second power that are rites; and amorous ecstasy (in which the object of love, instead of God as in the preceding case, is a specific human being). When the ecstasy is strong or durable and is seen under its quasi-magical aspect, then we have the bewitchment of love.

If we choose to analyze the "incantatory" effect of one individual on another, we could say that shared amorous ecstasy also corresponds with entering into "resonance," with the "attunement" of two beings, or of the rhythms of two beings, or if we like, their accord (a chord of special "timbre") with the rhythm of life. There is also in the lover or faithful mystic as in the creative poet and those he touches, a lasting or temporary state of "trance." Plato, who postulates what he calls "divine inspiration" in poetic creation, also recognizes it in love. He says in the *Symposium:* "The smitten one is in fact more divine than the elect, because the smitten one is inspired by God."

The trance and the bewitchment of spontaneous love often have that important quality of the magical phenomenon of being immediate and having the power to act at a distance, or at least without any other contact but visual or auditory perception. I spoke at the beginning of this chapter about the analogies between certain psychological dynamics and electromagnetic phenomena by citing in particular the instantaneous magnetization of soft iron, the production of induced currents, and so forth. It could be amusing to raise an analogous phenomenon (like dynamism) to electromagnetic induction in what we could call the psychical circuits of two beings in "amorous resonance," which is to say, that when the waves of desire or fervor inspired in the smitten one by the presence, gaze, voice of the chosen individual (to use Plato's terminology) are active, there is an instantaneous magnetization (the field of iron filings becomes conductive, the primary circuit closes), and the secondary circuit—in other words the one "induced" (in the chosen one) by the primary current (of the love-struck one)—can be established in turn.[20]

The instantaneous effect of "love at first sight" forms part of this order of phenomena that music alone can translate (and sometimes produce); for example, the incantation during the first conversation between the fiancée of the old baron and the "Knight of the Rose" in the opera of the same name.[21]

The preliminary incantation can also be present in so called "profane" or physical love, generative love (which is also a part of the rhythm of the Great Whole), with its scale that ranges from the serious and gentle search for the "wild strawberry" of the Gothic lovers to the orgy, if not to

---

20. We can find numerous analogies between electromagnetic circuits and mental circuits: electromotive force, self-induction (inertia), resistance, capacity, hysteresis, residual magnetization have their correlations in the psychomotor systems or circuits. What we are looking at here are phenomenal analogies due to the similarity of the operations (as in the electromechanical correlations between a condenser and a spring, a choke coil and a flywheel, and so on), independent of the actual presence of tensions, currents, and electrical discharges in the physiological circuit. This presence is real, moreover; the electrical tension (potential) of certain parts of the human body and the electrolytic mechanism of the nervous currents offer biology a field of research that has been little explored at present.

21. The gaze and the voice can have an immediate "catalyzing" effect.

Here is how Dante describes his "love at first sight" for Beatrice in the introduction to the *Vita Nuova:* "Nine times already since my birth had the heaven of light returned to the selfsame point almost, as concerns its own revolution, when first the glorious Lady of my mind was made manifest to mine eyes; she appeared to me at the beginning of her ninth year almost, and I saw her almost at the end of my ninth year. Her dress, on that day, was of a most noble colour, a subdued and goodly crimson, girdled and adorned in such sort as best suited with her very tender age. At that moment, I say most truly that the spirit of life, which hath its dwelling in the secretest chamber of the heart, began to tremble so violently that the least pulses of my body shook therewith; and in trembling it said these words:

Ecce deus fortior me, que veniens dominabitur michi:

Here cometh a god stronger than I who will rule over me.

I say that, from that time forward, Love quite governed my soul; which was immediately espoused to him." *Vita Nuova,* trans. Dante Gabriel Rosetti.

say stupor, that in mystic love the perfect example remains Dante's fervor for Beatrice, with this new nuance, not found in antiquity, of the unselfish devotion and infinite tenderness for the beloved woman, supported by the sweet certainty of finding her, to never again lose her, among the angelic host that share eternal bliss at the feet of the divine throne.

This ineffable conviction is the specifically Christian part in Dante's love, but the fervent, adoring tenderness, which despite the allegorical and literary ornamentation of *Vita Nuova* provides such a passionate note, despite its infinite respect, to the "terrestrial" part of this love[22] (and which returns with an even more intense emphasis in the *Divine Comedy*) is rather the bringing in of what I sum up in the label of "Celtic-Nordic." Or to be more precise: it is not so much the respectful adoration of the Christian for the Virgin Mary that softened the attitude toward women

22. "When she appeared in any place, it seemed to me, by the hope of her excellent salutation, that there was no man mine enemy . . . and if one should then have questioned me concerning any matter, I could only have said unto him 'Love, [while at the gathering] I began to feel a faintness and a throbbing at my left side, which soon took possession of my whole body. Whereupon I remember that I covertly leaned my back unto a painting that ran round the walls of that house; and being fearful lest my trembling should be discerned of them, I lifted mine eyes to look on those ladies, and then first perceived among them the excellent Beatrice. . . .'" And to his friend who asked the reason for his emotion, he replied: "Of a surety I have now set my feet on that point of life, beyond which he must not pass who would return."

And then "at that spot [of the poem that the author is explaining] where it is a question of her *eyes,* which are the source and principle of love, and in order to turn aside all coarse thought of what I was saying, the reader should recall that it is written above that the salutation of this lady, expressed through her mouth, was the end, the objective of all my desires while she clearly wished to grant it to me."

Concerning the smile of Beatrice: "The look she hath when she a little smiles / Cannot be said, nor holden in the thought; / 'Tis such a new and gracious miracle."

This haunting presence of the eyes and smile is the specific note of Dante's complex, his erotic "fixation" for Beatrice. Their exchange of smiles and glances in Paradise from the time they first meet at the beginning of the first canto until the last smile (cf. chapter 5), rises in a crescendo of fervent adoration culminating in the moment when his companion, after temporarily forbidding him from gazing at her anymore for fear he would be struck as if by lightning by the glow of her smile (canto 21), spontaneously restores him, this right after the poet, strengthened by the new splendor of the spectacle parading before them, had become a "seer."

of the conquering "barbarians" of Roman Europe and increased their prestige in these clans of free warriors in the feudal structure; rather, in several centuries she transformed the relations between the sexes, what we could in a very general sense call the social erotic, including the Greco-Latin notion in which the woman, who could be given infinite respect as wife and mother and yet be feared as a sovereign, was from the erotic point of view only an instrument, or at best the priestess[23] of sensual pleasure (the Semitic variety can be seen in the Song of Songs).

Just as the Nordic graft (Visigoth, Frank, Norman, Celt, Saxon) on architecture based on Greek geometrical harmony and on the clarity and truth of Arab-Egyptian lines and volumes brought forth the forest

---

(cont. from p. 201) *Apri gli occhi, e riguarda qual son io;*
Tu hai vedute cose, ché possente,
Se fatto a sostener to riso mio (canto 23).

Here are other mentions of this dual leitmotif of eyes and smiles:

Beatrice me guardô con gli ochi pieni
De faville d'amor, con si divini
Che, vinta mi virtù, dredi le reni,
E quasi mi perdei, con gli ochi chini (canto 4).

E cominciô, raggiandomi d'un riso
Tal, che nel fuoco faria l'uom felice (canto 7).

Poscia rivolsi gli occhi agli occhi belli . . . (canto 22).

Ma ella, che vedeva il mio desire,
Incomincio, ridendo tnto lieta,
Che Dio pareal nel volto suo gioire . . . (canto 27).

Ch'io feci, riguardando me' begli occhi,
Onde a pigliarmi fece Amor la corda (canto 28).

Ché, como Sole, il viso che più trema,
Cosi lo rimembrar del dosi riso
La mente mia da sè medesma scema (canto 30).

23. From the aesthetic if not ethical perspective, we should give the society of antiquity credit for the "valorization" of physical love through the mystery and the rite (sometimes an "orgiastic" rite). This is, moreover, a result of the "magical" attitude toward life found in the social life of many so-called primitive races.

of Gothic stone that clothed in dream the most absolute geometrical and dynamic rigor, so did the heat and clarity of the Mediterranean desire for the bacchante, slave, or goddess become blended with the Celtic-Nordic "sentiment" of fervent tenderness for the woman-sister or fairy. Hence this new form of love that I would call "Gothic," for which the Madonna represents the heavenly type, Dante's love for Beatrice the mystical type, and that of Tristan for Iseult what the Middle Ages called the profane type, in which the desire for spiritual and carnal union with the beloved is an end in itself; here the one smitten is not seeking a means of surpassing himself in order to advance in the contemplation of pure beauty. This profane or carnal love that does not exclude what I have called Celtic tenderness, but requires reciprocal possession as an end in itself, was given a charming name in medieval symbology: "the pursuit of the wild strawberry," the exquisite taste of which is fleeting and leaves only melancholy behind.

Hieronymus Bosch immortalized this symbolic quest in the central panel of the magnificent triptych that Philip II installed in the Escorial's sacristy, and which the catalogs call *The Garden of Earthly Delights* (plate 47).

The preceding can be summarized by noting that at first glance the woman in love in antiquity never appeared to play the role of inspirer that she won and has held on to since the so-called age of chivalry, for which the most illustrious example is the lifelong fidelity to the vanished ideal that motivated the author of the *Divine Comedy* to compose that unique monument to love.

We could say, however, that the ideally loved individual who corresponds to Beatrice in the work of Plato is Dion of Syracuse[24] (see chapter 7), but all sexual components are absent from this love. What

24. Here is the translation of the final lines of the ode attributed to Plato and dedicated to his assassinated friend:

> These same gods
> Who led him to victory
> Had, o Dion, unraveled before you
> The most noble hopes . . .

Plate 47 Hieronymus Bosch, *The Garden of Earthly Delights*, detail of the central panel (Prado) (photo Alinari)

we are looking at here, if we carefully read the *Symposium,* which is the epilogue of Plato's divine comedy, is a fully sublimated love that leads directly to the vision of divine love.

Nevertheless, we have seen in chapter 1 that certain treatises by Nicomachus of Gerasa are dedicated to an "unknown noblewoman"; Plutarch dedicates his book on *Isis and Osiris* to someone named Clea, another highly educated noblewoman, to whom he also offered his treatise *On the Virtue of Women:* "That Osiris is the same as Dionysus, who would know better than you, o Clea, as you are first of the Thuiades[25] of Delphi, and your father and mother have consecrated you in the mysteries of Osiris?" he notes in *De Iside.*

Nor should we forget that the "revelation" on love passed on by Socrates to the guests of the *Symposium* is spoken by the Pythagorean woman initiate Diotima of Mantinea.

We are so accustomed to the idealization of woman as the supreme objective of human love that the ancient conception that chose the male adolescent in preference to the female adolescent as the embodiment of the charm of youth today requires a certain effort of transposition by the contemporary man. It is helpful in this regard to consider the social conditions of Greek life in which neither the woman nor the girl appear. The well-born male adolescent, to the contrary, was in the forefront in the city of antiquity; his mind cultivated by teachers like Socrates or Plato, his body cared for like that of a courtesan and toughened by exercises at the stadium; when this beardless youth with curly hair, his forehead encircled by a gold or cloth band, dressed much like a woman of today (or 1926 rather) in a tunic the color of lilies or purple, appeared in the midst of his elders, whether it was for a banquet or on

(cont. from p. 203) And yet, before your time was up,
You were, revered by all, sleeping in the grave,
O Dion, you for whom Love
Made my heart beat so passionately.

25. The Thuiades were female initiates who, on a certain date, celebrate the night orgy of Bacchus, on the high grounds. From ϑυέιν, "to leap."

the agora of Olympus, it was he who drew all eyes, inspired admiration, tenderness, and the whole gamut of these sentiments including feverish raptures and jealous torment.

We can explain one aspect of this phenomenon, which we find unsettling even in Shakespeare's sonnets: style as such, dress, and hairstyle play a major role in erotic fashion.

The reverse phenomenon can be seen today. Just as in the time of Pericles, Plato, and Alexander the male ephebe represented for the majority of educated individuals of both sexes the hermaphroditic (in the rich, not the negative sense of the word) and virginal charm of adolescence, in our time this crystallization targets the adolescent woman. As in the time of Plato, our physical and sentimental ideal is androgynous; as before, athletic fashions, and even hairstyle, are the cause (or effect?) of this oscillation of desire. It is undeniable that the Antinous in the Vatican, and such a saint as John the Baptist by Leonardo, have an "appeal" more feminine than some "boyish girl" (*garçonne*) of today, with her severe Eton haircut and sexless body.[26]

All of us have, in varying percentages, what could be called a "homosexual component." It is again a myth of Plato, who in no way gives us

---

26. I find the same idea in the following passage from an article by Émile Lucka in the *Neue Frei Presse* on the reign of what the Germans call the "girl-type": "Because the 'girl' is a very clean, clearly delineated type, who, under the progressive dictatorship of America and of the mind-set of those in Europe who believe in America, has become the erotic ideal of our era. How are we to comprehend this image of our desire? First and foremost, the girl must be quite young, just young enough for the feminine attributes to be already indicated but not fully developed. She has the features of the ephebe: her hair cut short is no whim but the transition to the boy-type, from the young teenager to an intermediary type who while remaining female, represses all that is specifically feminine in her appearance and in her habitual mental functioning.

"The secret erotic ideal of our time is the hermaphrodite, a being who is partly a budding young virgin and partly a young boy who is not yet manly, but who is still a woman, a girl with a tomboy nature, whereas the Greeks of the classical era were also drawn to a hermaphrodite ideal but saw this ideal in the youth who looked like a young girl. This ideal, which is valid for both sexes, can be summed up and seen in the formulation: type tending toward an average between both sexes avoiding both the virile and the extremely feminine nature."

a scientific demonstration but a symbolic shortcut: the beautiful legend of the androgynous individuals that Zeus cut in half in punishment for their pride. "Since this time, love is an innate quality of human beings; it takes us back to our primitive nature; it strives to make from two individuals one, and to repair the misfortune of human nature . . . the entire human race would most likely be happy if each achieved love and could find the beloved individual who could return him to his primeval state" (*Symposium*).

Should I apologize for talking at such length about incantation, charm, and enchantment with respect to love, or rather for having, with respect to rhythm and incantation, introduced this pedantic dissertation on love? It is because incantation, rhythm, and spontaneous magic can be found in all types of love: the vibrational call of the male spider, prelude to the bridal and funereal banquet where he will be both priest and burnt offering in a ritual very much of Eros, or the nuptial song of the birds, words and caresses turned into prayers and rites, or spiritualized into poems of human or heavenly love?

As for the incantations of conscious love and the "magic will" applied to the realization of an amorous desire, they form an important chapter of operative magic. On this subject, and legitimate as all the precautions that can be entertained may be, it is, let me repeat, silly to deny the role of magic and related disciplines in a text on the development of human thought in general, and of Mediterranean thought in particular. We will revisit the whole of Mediterranean operative magic in part 2, with its specifically numerical and geometrical nature. The pentagram, as emblem of Aphrodite "Gamelia" (goddess of generative love), naturally figures here (moreover, we shall find it to be the preeminent magic symbol used in all ritual incantations).

On a copper love talisman (copper is the metal of Aphrodite Kypria) reproduced by Paracelsus, we indeed see Venus with a large pentagram above her head. The famous magic square whose boxes hold the letters of the mysterious SATOR AREPO TENET OPERA ROTAS was previously published in a treatise on women (*De Secretis Mulierum*) attributed to Albertus Magnus as an effective talisman for seducing virgins.[27] It is pentadic (five squares on each side), and in the arsenal of magical weapons for erotic purposes it holds the place of a condensed incantation that can serve as a pocket weapon for quick use. The great love incantation of the "classic" rite is, on the contrary, a very long and complicated operation requiring, as do all the great incantations of this type, a moral and material preparation that is so difficult, requiring such concentration and implementation of nervous power, that few apprentice magicians will have the tenacity to prepare and execute it.[28]

---

27. This phrase is a perfect palindrome, which means it reads the same backward and forward. For those interested in this, here is the meaning that Henri de Guillebert (*Revue Internationale des Sociéties secrètes*) attributes to this riddle that has intrigued everyone who has devoted him- or herself to the exegesis of magic for the last six hundred years: "The sower, the ram in heat, accomplishes the great cyclical work (of fecundation)."

This palindrome can be found in a Latin manuscript from the Bibliothèque Nationale (no. 1505) dating from 822. It was carved on one of the outside stones of Saint-Laurent Church in Rochemaure-en-Vivarais, in the twelfth or thirteenth century by an Albigensian or Asian [Asian as in the Asian heresies of early Christianity—*Trans.*] journeyman (this is de Guillebert's hypothesis), who signs it with the strange name of QIROI. Under the name of the "Key of the Great Arcanum," this pentadic square is depicted with the same letters, all inscribed within a hexagram, in the chapter "De Magicis Amuletis" (On Magic Amulets) in Athanasius Kircher's *Arithmologia* (1665). He refers to its satanic nature. Finally, it can be seen in the pseudo-Faustian manuscript of the Ducal archives of Coburg. Here each letter additionally towers over (as an initial) two words written in minuscule letters. The whole design provides a proliferation of five word phrases, each reading horizontally and vertically. Below you can see the first and fifth lines:

S	**A**	**T**	**O**	**R**
ator / atanas	repo / negelus	enet / onans	pera / lympo	otas / ejectus

R	**O**	**T**	**A**	**S**
otas / ectifications	pera / rbem	enet / ribuetque	repo / mara	ator / inistris

The *rotas* are the cycles or wheels of generation that also feature in Eleusis.

28. I will do the favor of offering the reader the details for this love spell that should take place on a Friday, in which the caster is clad in sky-blue garments (with roses for flowers, musk and sandalwood for perfumes and the talisman: AVEEVA VADELILITH). He should wear or carry no metal (diadem, wand, etc.) but copper, which is sacred to Aphrodite-Kypria. We should note, however, the lineage of the tradition in the fact that the ring should be set with a turquoise. In the Sinai sanctuary (in Serabit in the "Malachite Mountains") where Egyptian engineers, Cainite miners and smiths, and Phoenician sailors worshiped Hathor-Astarte (goddess of love identical to the Greek Aphrodite), this goddess

Plate 48. Maat, goddess of truth
(photo Anderson)

As for the other incantations, we can admit that if there are sheets or waves of psychic energy that can be "captured" (constituted by either the emanations of living people connected consciously or unknowingly to the mind of the operator, or by agency of supernatural psychic fluid like that presumed to exist by spiritualist theory), a fairly formidable concentration of rites and symbols, founded on such an ancient tradition, can in fact achieve through an effect of condensation and implementation of these dynamic energies results that are just as effective as more recent techniques such as telepathy, suggestion, and so forth. We still have these three elements whose respective proportions are unknown: suggestion and autosuggestion due to the faith of the operator as well as that of the patient; the "natural" incantatory action of rites and symbols; and lastly, and here is the main question mark, the direct, intrinsic force that is actually concentrated in the symbols.

Highly interesting from this point of view is the account, which seems perfectly truthful and excessively prudent in its conclusions, of the famous incantation performed on June 24, 1854, in London by Éliphas Lévi (Abbé Constant) at the request of Lady Lytton.[29]

But let's leave operative magic to conclude with this summary on love and its natural incantations. We have seen that love can be envisioned or felt as a harmony, the "attunement" of the rhythms of two beings, then sometimes their "unison" with a vaster, more general rhythm as amplitude, which they will perceive as a great enveloping and penetrating harmony—the "music of the spheres," the voice of the planetary sirens, of the "love that guides the sun and other stars." It is in these moments of perfect resonance with the Great Rhythm that the three kinds of ecstasy or "natural" magical states that I listed at the beginning of this chapter—"cosmic" euphoria, earthly love, and divine love—can be combined and commingled.

(cont. from p. 208) was called the "Lady of the Green Stone," after the pale-green turquoise of the Sinai. Robert Eisler, *Die Kenitischen Weihinschriften der Hykososzeit* (Freiberg: Herder, 1919).

29. Éliphas Lévi, *Dogme et rituel de la haute magie.* This is not a love spell but a necromantic working, the raising up of a dead person.

The path to divine love chosen by Plato is not that of initial ascetic renunciation, but the one that travels through earthy love, and the love of beauty, including that of the physical. It is of course not the search for easy carnal pleasures, but the ardent pursuit of the buckskin-clad mystes through the dark kingdom of Persephone, among the sweet or terrible images that hang over his road, extending their arms with the smiles of sisters, bacchantes, or gorgons, the calls of brothers-in-arms, or the snickering of monsters—until the final revelation reserved for those who have believed in love, a revelation that Plato, in fact, compares to the illumination of the mystes who has finally become a "seer."[30]

---

30. As I have chosen Dante as the prototype of the mystic lover, it is a good idea to stress the fact that the fervent worship of the physical beauty of the "chosen being," in order to be "fixed" in some way on this individual's eyes or smile, played a no less dominant role in this type of love. In the *Divine Comedy,* which, as a metaphysics of love, is moreover a very strict transposition of the *Symposium,* itself a Platonic condensation of everything that in the Eleusinian Mysteries and their very gradation concerns love, Beatrice in the previously mentioned passage in which she commands Dante to avoid gazing at her, puts it this way:

> "If I were to smile,"
> She told me, "why then you would share the
> Fate of Semele when she was turned to ash,
> Because my beauty becomes more inflamed the higher we ascend
> Up the steps of the eternal palace
> And burns so hotly, if it were not tempered here,
> Your earthly forces would shatter at its flash,
> Just like a branch struck by a lightning bolt."

And in the thirtieth canto, upon arriving in Empyrea and already enlightened by the Divine Beauty, just before his companion leaves him, Dante turns toward her and lets loose a cry of admiration for her beauty:

> Love made me turn anew my gaze toward Beatrice. If all I have said of her until now were combined in a canticle of praise, it would still be too weak to do her justice. The beauty I saw exceeds all measure, I even believe its Creator alone can fully enjoy it.
>
> Here, I acknowledge defeat. . . . Like the sun erases the pale image, remembrance of her sweet smile has erased memory itself from my memory. If from the first day I saw her, in this life, until this vision, my poem has been able to follow here, I must now abandon all future attempts to describe her beauty as the artist must stop before the realization of his supreme goal.

I can think of no better way to bring this chapter to a close than by quoting the words attributed in the *Symposium* to Diotima of Mantinea:

> "These are the lesser mysteries of love, into which even you, Socrates, may enter; to the greater and more hidden ones which are the crown of these, and to which, if you pursue them in a right spirit, they will lead, I know not whether you will be able to attain. . . . He who has been instructed thus far in the things of love, and who has learned to see the beautiful in due order and succession, when he comes toward the end will suddenly perceive a nature of wondrous beauty—a nature which in the first place is everlasting, not growing and decaying, or waxing and waning; secondly, not fair in one point of view and foul in another, or at one time or in one relation or at one place fair, at another time or in another relation or at another place foul, as if fair to some and foul to others . . . [beauty which is not this word or that science, and which resides in no other being but itself . . . but which remains by itself eternally identical to itself. . . . So, when one has risen from these particular beauties to these perfect beauties, and begun to contemplate them, then one has almost attained the full vision of the mysteries of love.]
>
> "And the true order of going, or being led by another, to the things of love, is to begin from the beauties of earth and mount upwards for the sake of that other beauty, using these as steps only, and from one going on to two, and from two to all fair forms, and from fair forms to fair practices, and from fair practices to fair notions, until from fair notions he arrives at the notion of absolute beauty, and at last knows what the essence of beauty is. This, my dear Socrates," said the stranger of Mantinea, "is that life above all others which man should live, in the contemplation of beauty absolute."[31]

(cont. from p. 211) This already divine beauty that Dante contemplates here in Beatrice is obviously not earthly physical beauty, or at the least she is its archetype. The ratio of one to the other is, if you like, comparable to that of "pure numbers" to scientific numbers in Pythagorean mysticism (chapter 1).

31. [Translated by Benjamin Jowett, except for the sentences in brackets.—*Trans.*]

It is easy to see that the "frozen mysticism of pure number" hardly prevented its Mediterranean adepts from ascending to a more vibrant vision of love and one that was both more human and divine than the one glimpsed by all the others on less arid paths. The arithmological underpinnings of the *Divine Comedy* have just recently been studied by René Guénon,[32] and in the *Vita Nuova* itself, one is surprised by the veritably haunting presence of the number nine (the "ennead"), in particular. The great visionary who ended his life's Great Work with the invocation to "*L'amor che move il sole e l'altre stele*" ("the love that moves the sun and other stars") and the author of the *Timaeus* were both, like Jacob following his long struggle with the angel at Penuel, capable of contemplating the dazzling face without dying, and like Jacob could have proclaimed:

> I have seen God
> Face to face,
> And yet my life is spared!

For this face was the face of love.

32. *The Esoterism of Dante.*

PART 2

Rites

Introduction to Part 2

In the first part of this book ("Rhythms") we have tried to define and analyze the various categories of rhythm envisioned as aesthetic expression. For example, we examined the "irreversible" rhythms that develop in duration (music, poetry)—direct emanations of living experience, from the rhythm of the "pneuma"—and we studied them by means of suitably conceived notations, depicting in series of simple numbers those rhythms that reflected affective undulations parallel to the two physiological cadences that can be framed in numbers: the fundamental weave of the heartbeat and the respiratory rhythm, an even more direct reflection of the flow of emotions.

We similarly examined from the perspective of measure, and with even greater precision, the categories of the rhythms of space (architecture and plastic arts), domain of the reversible and continuous and of proportion as such. All this was viewed through the particular lens of the creative cycle I dubbed Mediterranean, in which the sense of proportion and a spatial rhythm based on the sequences and combinations of proportions, overseen by a mathematical desire, was exhibited in a specifically characteristic manner.

We saw that this conscious eurhythmic composition and this rigorous orchestration in space (what could be called the "Pythagorean" aesthetic) bestowed a distinctive harmonious quality on the monumental or plastic works produced during this Mediterranean cycle in the broad sense (Egypt, Greece, Byzantium, the Gothic age, Renaissance); one of their characteristics being the ability to reflect in a completely unique manner everything that in the morphology of life could obey geometric laws: proportions of the human body, harmonious growth, and so forth.

This led us to not merely contemplating and comparing aesthetic results or successes, but to look at converging trajectories, innate tendencies, and questions of origins and transmission. It is this historical aspect that we are going to examine more closely here; its interest and difficulty come from the pains taken to cover its trails and conceal its keys by the successive keepers of the tradition.

For example, having excessively analyzed the strict "regulating designs" of the plans of Egyptian and Greek temples and Gothic churches, I demonstrated that these various kinds of plans generally culminated in variants of the same "theme," that of the golden number, or golden section, theme of the living pulsation, which, moreover, the Renaissance rediscovered under the spiritual aegis of Plato, the preeminent Mediterranean "thinker." Now, the pentagram, the geometric sign of this golden number, was the emblem of harmony and health for the Pythagoreans, then their secret rallying symbol. By striving henceforth to prove the continued transmission of this symbol through the ages, as well as its variants and related geometric patterns, I hope to clear the ancient royal way that we have only marked out up to now.

The chains of transmission that we are going to bring to light are not, moreover, formed solely by the secret techniques of architects; we will find other currents—Kabbalah, magic, secret societies—but beneath all of them we can detect the same origin and the same inspiration: the Master of Samos, who formerly proclaimed this law of number of which the golden number is the most remarkable invariant.

Regarding this tradition, this fraternity devoted to proportion, harmony, and beauty that distinguishes our Occidental aesthetic, Plato was the herald whose voice still echoes down through the centuries. And, as we have said so much about Pythagoras and Plato in the first part, yet barely touched on their lives, it is now high time to approach them head on. The following part will start with Pythagoras.

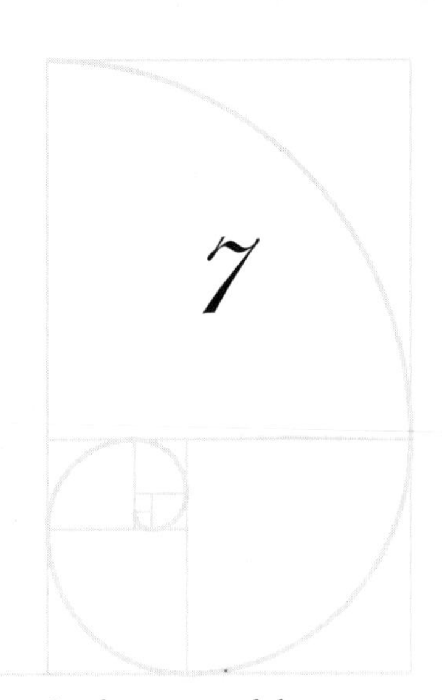

7 Pythagoras

I am a son of the earth and the starry sky, but I am of the race of heaven, know that you are too!

Inscription on an Orphic-Pythagorean funerary lamella found near Rome (first or second century CE)

On April 23, 1917, the weakening of a portion of the ballast under the rail line from Rome to Naples in close proximity to the Porta Maggiore and the eccentric "baker's mausoleum" revealed the existence of a crypt that, once systematically excavated, appeared at first glance to be an underground Christian chapel, to which the faithful gained entrance through a gently sloping narrow tunnel. All the specific features of a primitive basilica were found there: nave, side aisles determined by two rows of imposing quadrangular columns carved in the very tuff, a "cathedra" or bishop's throne against the apse wall, facing the congregation. However, once the walls and vault had been scrubbed, delicate stucço moldings appeared, depicting scenes that did not correspond with either Christian symbolism or the standard Greek or Roman mythological subjects.

The technique and the style of the moldings and the other details of the crypt, as well as the stratigraphic indications, made it possible to place the establishment and then intentional destruction (it was systematically filled in) of this temple or meeting place of a mysterious sect to the time between the advent and death of Emperor Claudius (41–54 CE).

In 1918, a passage from the *Annals* in which Tacitus tells of the disgrace of Statilius Taurus and his suicide (53 CE) following the senatus consultum that, at the request of the emperor, exiled from Italy all the *mathematici* (that is to say, the "mages" and neo-Pythagoreans who had been growing steadily in numbers in Rome for the previous one hundred years[1]) was connected by Formari with the destruction of the "basilica" of the Porta Maggiore.

From another angle, the eminent archaeologist Franz Cumont, by examining the moldings mentioned above, recognized that they depicted episodes of the soul's journey through its successive ordeals, a theme that, with many variants, was the basis of the teaching and the symbolism of all the initiatory ritual cults (Egyptian mysteries of Isis and Osiris, mysteries of Demeter and Dionysus in Eleusis, the Orphic mysteries, and the Pythagorean cult). However, the extremely pure and musical accent of these symbolic scenes, in which neither the masked sensuality customary to the allegories of the other mysteries nor any Mithraic or neo-Egyptian intrusions are displayed (this impression of a starkly abstract harmony is intensified by the admirable proportions of the basilica and its uniform and gentle lighting, falling from a "hole of light" at the top of the atrium) induced Cumont, from the beginning of his examination, to opt for a hypothesis of a "den" or "house of philosophy" for a Pythagorean association. Porphyry[2] and Iamblichus[3]

1. The first known neo-Pythagorean lodge in Rome was founded by a friend of Cicero, the senator P. Nigidius Figulus. He was exiled by Caesar as a supporter of Pompey and more specifically as grand master of one of those secret societies that first the Roman state then the Church always held in dread and that they would always suppress whenever they could. "*Nigidius Figulus Pythagoricus et Magus in exilio moritur,*" Saint Jerome tells us in his chronicles. What the state feared in these Pythagorean groups was not their religious and philosophical teaching—the members of the exotic cults of Isis and Mithras enjoyed greater tolerance—but their organization based on secrecy. It was for this same reason that Caesar tried to suppress the "colleges" of craftsmen.

2. *De antro nympharum.*

3. Porphyry and Iamblichus probably citing Aristoxenes of Tarentum.

say, in fact, that in Samos, in addition to the meeting room the Master owned in town, he had installed in the countryside an underground cavern, symbol of the world of appearances in which moved about the souls who had not yet been brought to the light by death or initiation, a symbol as well of the prison of the body, and memorial of the underground site in which, according to Clement of Alexandria, Pythagoras was initiated into the mysteries. It was this cave that was his true "house of philosophy."

This immediately brings to mind the splendid myth of Plato's cave (the *Republic*), in which only the shadow puppets of reality, projected by a still inaccessible sun, appear to us, and the association of these ideas is correct, for Porphyry explicitly tells us (*De antro nympharum*) that this comparison was of Pythagorean origin.

Cumont's hypothesis was provisionally adopted, then subsequently verified by the scrupulous checking of the moldings by authorities like Eugénie Strong and by the discovery of other, cumulative clues. For example, there is the fact that only the bones of suckling pigs were found in the pits reserved for the remains of animal sacrifices. As it happens, Pythagoras only accepted young pigs and goats as sacrificial animals. These animals, moreover, formed the main course of the "sacred banquet" or the meal eaten together after the morning's religious observances (cf. Porphyry, Diogenes Laertius, Aulus Gellius).

But it was Carcopino who provided a new bundle of evidence, presented in his very fine book titled *La Basilique pythagoricienne de la Porte Majeure*,[4] culminating in the interpretation of the most important molding, the large stucco one of the apse. This depicts a young woman holding a lyre, who under the eyes of Eros is diving into the waves from atop a stone cliff; a siren appears to be waiting to receive her and guide her to an island on which a solar god is enthroned.

Cumont sees this as an allegory of the human soul who, compelled by Love and holding the vibrating heptachord of the harmony

4. L'Artisan du Livre, 1927.

of the world, is not afraid to entrust herself to the transitory ordeal of death, in order to rediscover, beyond the swells of imperfect matter, the revelation: Apollo, god of Pythagoras.

Densmore-Curtis, for his part, believed this image depicted the leap of Sappho, lover of Phaon, from atop the Leucadian cliffs. According to legend (Ovid, *Heroïdes*), she was taken in by Apollo.

It so happens that Pliny's text introduced by Carcopino explicitly says that the Pythagoreans took possession of this legend[5] (Sappho's love for Phaon), probably in echo of Cumont's intuitive interpretation.[6] Carcopino thus seems right in viewing this explanation of the major stucco image as the "key" that confirms his theory that had already been almost irrefutably proven by a convergent set of conjectures,[7] and thus concludes:

> The proof is there, irrefutable and peremptory, that this religion of Pythagoras—which several texts that are, alas, too rare, elliptical, and disjointed, have informed us existed at the end of the first century BCE—possessed, in imperial Rome under the rule of Claudius, a church. . . . And from this fact, henceforth established on foundations that seem unshakeable, proceeds the greatest consequences.

5. Carcopino notes that the date of Pliny's text is in accord with that of the basilica's destruction.

6. "Pythagoreanism consciously, systematically identifies life (life in this world) with death . . . it makes a play on words between σῶμα and σῆμα (body and tomb)."

Plato presents this idea in *Cratylus,* then in *Gorgias,* in which he cites Socrates as asking: "Who knows if life is not death, and death life? . . . our body a tomb? (σῶμα ἐστιν ἡμῖν σῆμα)." Carcopino, *La Basilique pythagoricienne.*

7. Another example, the location of four marble tables for the sacred supper has been found in the basilica as well as 28 stucco constructions on plinths, each representing a funerary enclosure guarded by a god or goddess (the symbolic equation of σῶμα and σῆμα in the previous note). Strong and Carcopino deduced from this that the brotherhood likely had 28 = 4 × 7 members. In a first-century neo-Pythagorean dialogue (*Palatine Anthology*), Pythagoras, questioned by Polycrates of Samos about the number of his disciples, responded with an arithmetical riddle whose solution is 28. In the atrium, to the contrary, the same motif is repeated ten times, probably in honor of the decad, and so on.

The completely accidental discovery of the basilica of the Porta Maggiore has thus restored to topicality this little-known domain of the religious doctrine and ritual of the Pythagoreans. The studies made following this discovery and the books successively published by Delatte, Méautis, and Isidore Levy[8] now make it possible, after a cross-checking of the information that was as laborious as that required by a cryptographic decoding, to lift a corner of the veil and measure the prodigious influence of Pythagoreanism on the world of antiquity and the subsequent development of European thought.

The historical data we have on Pythagoras himself is fairly scant and wrapped in the golden fog of the legend that had already crystallized by the fourth century BCE.

Pythagoras was born on Samos between 592 and 572 BCE, in other words, during that same sixth century that saw the lives of Gautama Buddha, Zoroaster, Confucius, and Lao-Tzu. Together with the Master of Samos, they form a sparkling pentad of supermen, demigods, or "daimons," to use the term that become dear to the disciples of Pythagoras.[9]

The long-haired, purple-garbed adolescent Pythagoras appeared at the games of the forty-eighth Olympiad where, contending against "heavyweight" adults, he won the inestimable olive branch for pugilism.

---

8. Delatte, *Études sur le littérature pythagoriennes;* G. Méautis, *Recherches sur le Pythagorisme* (Neufchatel, 1922); Isidore Levy, *Recherches sur les sources de la légende de Pythagore* (Paris: Ernest Leroux, 1926), and *La Légende de Pythagore de Grèce en Palestine* (Paris: Champion, 1927).

9. Starting in the fifth century, the "arch-secret" formula of the Pythagoreans, transmitted by Iamblichus, inserts Pythagoras as a veritable harmonic "mediety" between the divine and the human. In the fourth century, he was even attributed with the status of divinity. Pythagoras is the Hyperborean Apollo, whose gold thigh (or gold-like, rather) dazzled the eyes of the spectators in the stadium of Olympus. Later in Sicily, the Nordic mage Abaris, who had been his priest in the inaccessible "temple of clouds," recognized in Pythagoras, encountered with his disciples who were oblivious of their master's true nature, the very god he had once served; and to the mage prostrating himself before him the god, smiling, showed this same sign (the golden thigh) as confirmation of their mutual secret. Fragment from the *Abaris* of Heraclides of Pontus, a disciple of Plato.

He then departed on long journeys. His sojourn in Egypt is confirmed by all sources, as well as his initiation into the Egyptian mysteries, on the one hand (the three sanctuaries named in this instance are Memphis, Diospolis, and Heliopolis), and, on the other hand, into geometry. His travels in Phoenicia (retreat on Mount Carmel) and Chaldea (Pythagoras was taken prisoner at the time of the conquest of Egypt by Cambyses and taken to Babylon) may have been invented for symmetrical ends at a time when, less than a century after his death, his legend was being woven together, but it is possible to maintain that here the legend, at least partially, does conform with the truth.

All the sources are in agreement on the fact that his studies and pilgrimages lasted quite a long time and that he was more than fifty years old (fifty-six, Iamblichus specifies) when he returned to Samos. The success of his lessons drew a growing crowd of adepts, and the enmity of the tyrant Polycrates forced him into exile. Hence his arrival in Italy (Croton), where he settled once and for all in Magna Graecia. Here also soon appeared the radiant force of his teaching that embraced religious doctrine, ethics, and a scientific "corpus," all connected by mathematical keys that condensed the unvarying relationships and principles common to these three domains.

From chaos, through creation, order is born: the word *cosmos,* which Pythagoras was the first to apply to the perceived universe, means "order."[10] Order can, and should, become harmony (being perceived as consonant harmony in ourselves).

Souls are subject to successive reincarnation until the liberation of those that have proven themselves worthy over the course of this palingenesis[11] (succession of life cycles). They then become "daimons," semi-divine "spirits" who no longer return to this plane of existence

10. And Plato: "The sages, o Callicles, say that friendship, order, reason, and justice hold heaven and earth together, and gods and men; this is why they call this grouping the Cosmos, in other words the good order." *Gorgias,* cited by Méautis.

11. Παλιγγενεσία (*paliggenesia,* palingenesis): technical term of Pythagorean initiates for designating metempsychosis; (γενέσθαι πάλιν: "they receive a new existence").

but dwell in the starry gardens of the blessed "beyond the Milky Way."

The Milky Way returns often in the Pythagorean myths concerning the afterlife. Carcopino notes on one of the funerary lamellae of Thurii the twin inscription "password" and response:

> Young goat, I have fallen into the milk!
> Young goat, you have fallen into the milk!

As a symbolic correspondence: on one of the stucco surfaces of the basilica of Porta Maggiore, a bacchante is holding a young goat toward another who is offering her breast to nurse it. Like the young goat reaching toward the breast of the bacchante, the spirit of the mystes, surviving the death of the body, soars toward the regenerating river of the symbolic Milky Way, toward the soothing bosom and felt presence of the Great Goddess.

All lives, all souls, including those of animals and plants, derive, in reality, from a great universal soul: a fraternity, a true kinship (συγγένεια), reunites animals with humans, and humans with gods (we find here that Panpsyche,[12] Plato's world soul).

The foundation and completion of the normal Pythagorean path, the consequence of real fraternity, of authentic community and the interpenetration of the living world, and of the law of harmony in which sister-souls seek to find harmony and join together, is love; love for animals (and plants as well) and humans culminating in the grand illumination of divine love as we glimpse it in the final flash of Plato's *Symposium.*

An important part of the apprenticeship and transmission of the

---

12. "*Audiebam Pythagoram Pythagoreosque . . . nunquam dubitasse qua ex universa mente divina delibatos animos haberemus.*" Cicero, *De senectute (On Old Age),* cited by Méautis.

This doctrine of palingenesis was not borrowed from the Egyptians. Herodotus (who lived during the fifth century BCE, thus a century after Pythagoras) even states that the opposite was true: the Egyptians borrowed this belief from the Pythagoreans.

Pythagoras could have had contact with Hindu metaphysics, from which this notion appears to originate, either in Egypt itself (where the occasional presence of Hindu gymnosophists is mentioned in the ancient texts), or during his hypothetical sojourns in Asia.

doctrine of the Master rests on the use of symbol.[13] Iamblichus mentions the Egyptian source of symbolic technique (Clement of Alexandria and Plutarch even say that Egyptian symbology also holds the key to the "acousmatic" questions of the Pythagorean catechism and to hieroglyphics). The symbol can be a phrase, a word (the "words of power" mentioned in part 1), a geometric "sign," or a number. As we have seen in chapter 1, the geometrical sign and number share in the nature of the paradigms or models that preceded creation, and form the specifically Pythagorean contribution to initiatory symbolism; they are eternal principles, symbols and agents of harmony, condensing agents acting through suggestion, liberation, and incantation, hence their essentially "magical" nature.[14]

The sign and its corresponding number are interchangeable: pentad and pentagram, "punctual" decad and tetractys (ten points arranged in a triangle representing the units of the first four triangular numbers), and so on.

We already know that the pentagram, the symbol of life, health, and love, was the rallying symbol of the Pythagoreans.[15]

---

13. Androcydes (fourth or first century BCE) wrote a treatise *On Pythagorean Symbols,* of which only scattered fragments have come down to us.

14. The importance of number in Pythagorean metaphysics and ritual appears in both the legend and history of Greek science. Thus in the first "novelized biography" of Pythagoras, that of the Platonic disciple Heraclides of Pontus (author of the *Abaris* cited above), portions of which have come down to us through Iamblichus and Diogenes Laertius, it is said that Pythagoras, to reward the Hyperborean traveler for having recognized his god, taught him divination ("prognosis") through number, the purest form of divination because of its connection with the "divine numbers" (the "pure" numbers that we find in the work of Nicomachus of Gerasa). Under the name of Kabbalah we find the Hebrew daughter of this prognosis founded on "divine numbers."

15. Lucian of Samosata, *Pro lapsu.* Lucian also tells us of the tradition concerning the dream in which Alexander appeared to Antiochus and showed him a vexillum [Roman military standard—*Trans.*] bearing a pentagram as its emblem. Alexander's association with the major Pythagorean symbol here is suggestive. We are very familiar with the role played by the eldest daughter of his philosophy and will—the city of Alexandria, Egypt—in the development of not only Greek mathematics and neo-Pythagoreanism, but of an extremely rich cluster of philosophical schools, of sciences and beliefs, which the Mediterranean culture could simply call Alexandrian culture.

We have also seen in the first chapter that this esoteric mathematics, this mysticism of number, is a pinnacle of crystalline abstraction where the metaphysics of the harmony of the Great Whole is joined with the theory of musical harmony and eurhythmy in general. In this synthesis the directing mathematical concept is the geometrical proportion (the *analogia* whose mathematical theory was explained by Archytas of Tarentum in terms that Plato reemployed in the *Timaeus,* and whose simplest and most fruitful type par excellence is precisely the "analogy" or continuous proportion, the golden section embodied geometrically in the "stellar" diagram of the pentagram). Its metaphysical parallel is the great principle of analogy (with its different logical levels: identity principles of the same and the other, similarity, unity in variety), which is itself reflected in harmonic correspondences,[16] among which that between the universal soul and the human soul stands out as a fundamental chord intended to resonate through the centuries, endlessly echoing the solemn promise of the Master to his disciples: "You shall know, to the extent it is permissible (for a mortal) to know, that nature from all points of view is similar to itself."[17]

This doctrine of harmony, the concordance between the great rhythm of universal life and that of the human soul (and secondarily

---

16. Another example of the role of harmonic correlations in Plato's work, which in this instance concerns daimons or spirits, as he says in the *Symposium:* "Filling the interval that separates man from God, the daimons unite the Great All with himself. It is from them that comes all divinatory science, the entire priestly art of sacrifice, initiations, incantations, all high magic, and all goety. God mingles not with humans."

17. Γνώσῃ δῆ θέμις ἐσεί, φύσιν περί παντός ὁμοίην, an extract from Pythagoras's *Ieros Logos* in Delatte, *Études sur la littérature pythagoricienne.* Delatte also cites another very important fragment of the *Sacred Discourse* in which the Master invites his disciples to be reassured: the mortals to whom sacred nature has revealed all things can rely on their kinship with the gods. He notes that this idea can also be found on Pythagorean-Orphic funerary tablets (lamellae, thin metal strips) of Petelia and Thurii. As the epigraph to this chapter, I cited a passage from the Petelia tablet (Petelia is near Rome). Here is a passage from its counterpart from Thurii: "I come purest of the pure, o Queen of the underworld. . . . For I, too, glorify myself as belonging to your blessed race." We should note that Leopold and Carcopino share the belief that the last Pythagoreans of Sybaris were buried in these tombs of Thurii in the fourth century.

with the harmony of the body itself, the soul's material projection) is, as we have seen (part 1) one of the major themes of the *Timaeus.* But the expressions *macrocosmos* (or *megacosmos*) and *microcosmos* are not found anywhere in Plato. Mario Meunier has pointed out that he found these expressions used for the first time in a fragment of Democritus of Abdera (among others: "man is a microcosm," Diels, *Fragmente der Versocratiker,* frag. 34).

This contemporary and rival of Plato (he lived between 460 and 360 BCE) was in contact with Philolaus and other Pythagoreans (cf. Diogenes Laertius and Apollodorus of Cyzicus) and adopted their mathematical notions, including the theory of harmony, while firmly rejecting their spiritualist monism (to the contrary, through his notion of the indivisible atom, he was the spiritual father of Lucretius and materialistic determinism). He had, moreover, during a five-year stay in Egypt, studied Egyptian geometry and natural sciences for his own benefit (the first treatise on chemistry has been attributed to him): hence the famous witticism that Clement of Alexandria put in his mouth (*Stromata*): "He said he never met anyone who surpassed him in the art of drawing lines in figures, not even among the *arpedonaptes* (geometer-surveyors) of Egypt."

But let's return to the life of Pythagoras at the time of his arrival in Croton (set at 529 BCE by Aristoxenes of Tarentum). The success of his teaching in Calabria then in Sicily was such that the number of disciples increased rapidly. Nicomachus and Iamblichus mention a great discourse[18] of the Master, which consequently led to the founding of a society whose members committed, among other things, to communal sharing of property, and, by meditating on the new revelation (or "philosophy," the word itself is attributed to Pythagoras, many of whose words definitely went on to enjoy success), to extend themselves toward realization through the knowledge of love, inner harmony, and accord with the

18. Discourse whose contents may well have been flowed into the Ionian verses of the *Ieros Logos* by the immediate disciples or even the son of the Master, as reported by Iamblichus.

great harmony. The knowledge or "gnosis" for which mathematics (recall the "all is arranged according to number") is the sole path of access is essential. This is what distinguishes Pythagoreanism from other ethical systems or dogmas founded on harmony.[19] Hence the meaning of a passage of Heraclides transmitted by Clement of Alexandria (*Stromata*): "Pythagoras places the supreme happiness (literally, 'the eudaimonia of the soul') in the contemplation of the harmony of the rhythms of the Universe (τῆς τελειότητος τῶν αριθμῶν, literally: 'on the perfection of numbers,' number being here rhythm and proportion)."

As the Pythagorean society or confraternity attracted many minds curious about the sciences as well as mystics with a hunger for the ideal, it quickly acquired, as I mentioned in part 1, a majority and then almost absolute political power over the largest part of Magna Graecia (the Crotoniate League). This was a sort of "esoteric fascism" made up of three categories of initiates: the contemplative philosophers (the mathematicians), the *nomothetes* (those philosophers who directed the social and political activity of the brotherhood by giving their instructions to the third category), and lastly the "politicians" (who have not yet attained perfect purity), who are executive and liaison officers. A novitiate of three years (exoteric stage) was required for admission to the first degree of initiation. This stage lasted five years, only at the end of which would the initiate move into the category of full initiate who could then see the Master (see Iamblichus, *Life of Pythagoras,* and so forth; the "politicians" mentioned earlier were novices[20]). After the death of Pythagoras, which

19. For example, the Zen Buddhist sect that in its Japanese form realized (fifteenth and sixteenth centuries) a communist feudal state based on the cult of honor (absolute devotion to clan and emperor), the scorning of wealth, and the cult of beauty in its most serious and subtle forms. The ideal follower of this noble branch of Buddhism could paradoxically combine the frugal stoicism of the ancient Cato, the Albanian concept of clan honor, that of the Prussian Junker for military romanticism, the serious love of a Leonardo da Vinci for the beauty of forms in art and nature, and the hypersensitive aestheticism of an Oscar Wilde.

20. We can also apply the term of first-degree initiate to novices, although this concerned a preparatory stage that assumed the nature of a test. We then have three stages of initiation.

took place between 510 and 489 BCE, the political hegemony of the society continued until about the middle of the fifth century, when, following a great popular revolution that captured the cities of the Crotoniate confederation one after the other, the leaders of the brotherhood perished in a final massacre (the conflagration of Metapontum) around 450. Only Lysis and Philolaus managed to escape, with a small number of novices. Among them may have been Hippocrates of Chios—who settled in Athens at around this time—Hipparchus, and Hippasus, all three immortalized by the fact that, for divulging certain mathematical secrets to the public, they were excommunicated by cells of the brotherhood that reformed some time after the catastrophe in Sicily and southern Italy. Because the sect reformed in the shadows and in small local brotherhoods, political activity was completely abandoned (with the brilliant exception of Archytas of Tarentum), but the bonds of mutual aid and secrecy sealed by oath remained.

It is precisely this rule of secrecy that made it so difficult to research the formation of the old Pythagorean society and its occult continuation until the time of its brilliant rebirth in the first century BCE as Alexandrian, Roman, and Syrian neo-Pythagoreanism.

For the first point, we had the good fortune that in antiquity itself (during the fourth century BCE) two diligent researchers were concerned with reconstructing through authentic sources these inward-looking "establishments." The first, Aristoxenes of Tarentum, was a friend of a group of Pythagoreans of the old tradition and thereby obtained the elements for a *Life of Pythagoras.* The second individual, Timaeus of Tauromenium, traveled specifically to Sicily and Calabria to search the archives and local traditions concerning the Crotoniate era. The book that he wrote, like that of Aristoxenes, has not come down to us, but, transmitted through the intermediary of also-vanished Alexandrian texts, we can find many fragments of them in the three lives of Pythagoras by Diogenes Laertius, Porphyry, and Iamblichus.

We know though Aristoxenes of Tarentum that Philolaus was considered by Pythagoreans as the earliest among the "traitors." He was accused of not only divulging philosophical and mathematical secrets in his writings but also—tempted by the enormous price offered—selling to Dionysius of Syracuse or his brother-in-law Dion three books containing the secret doctrine, books that Plato would have learned of during his first sojourn in the court of Syracuse. We actually know through a text by Plato himself (the *Seventh Letter*) of the enthusiastic friendship, illuminated among other things by the shared practice of a "philosophy" unknown to the common folk, that bound him beyond death to this same Dion, who was also the son-in-law of Dionysius the Elder.[21] We also see here the regenerative ideal founded on this philosophy was also that of the Pythagorean regent of Tarentum, and the same document confirms the influence Archytas had on the thoughts and even the actions of Plato.

Indeed, here is the way Plato describes in the aforementioned letter the insistence of Dionysius II (who shortly after his succession to power, and during the philosopher's second visit to Syracuse, exiled his uncle Dion, whose reformist tendencies he feared) when inviting him to return to his court.

As during his first stay at the court of Dionysius II, he (Plato) had formed bonds of friendship between the ruler and "Archytas and the Tarentine circle," and Archytas and other "philosophers" had, after forging this relationship (and following Plato's departure), personally gone to Syracuse. The young tyrant, who confessed his shame to them for not having gained a deeper knowledge of Plato's "philosophy," sent him an invitation by way of a warship carrying his best Sicilian friends, especially Archedemos, "from Archytas's intimate circle," with a pressing message as well as a letter from "Archytas and the Tarentine circle" certifying Dionysius's new zeal for seeking the truth, and underscoring their request by expressing their wish that this recent friend-

21. "I expressed my point of view on this subject to him . . . and he (Dion) then decided to devote himself forever to a kind of life completely different from the one that had become customary (in Syracuse), in other words to devote himself to a life not of pleasure but of actions based on our ideal." *Seventh Letter.*

ship between Syracuse and the Tarentine circle not be compromised by Plato's refusal. Plato then acceded "in the interest of Dion and of their friends and brothers in doctrine."

After describing his new failure to convert the young tyrant, he again repeated that it was "the friends of Archytas" (in other words the Pythagoreans) who had urged him to undertake a third voyage "in the desire to be useful to the 'philosophy' of his friends."

Seeing that Dionysius was still seeking to deceive him and, far from considering recalling Dion from exile, wished to keep him in "a cage," and feeling he was even in danger of being killed, Plato sent a message to "Archytas and his other friends of Tarentum."[22] They then sent a fifty-oar war galley carrying as ambassador extraordinary Lamiskos, who requested the philosopher be set free. The tyrant yielded to the command of the powerful regent of Tarentum, and Plato could embark, leaving Syracuse forever.

A short time later Plato met Dion at the Olympic games, where the exile asked him to take part in a military expedition against Dionysius. Plato refused to take part in a violent action against someone who, despite all his perfidious behavior, had been his host and with whom he had shared "house, table, and altars." But while advising Dion to seek a more amicable arrangement between uncle and nephew, he did not oppose Dion's request to appeal to "his friends."

This document of capital importance confirms, as if there was any need, that Plato's entire "harmonic" philosophy, in which "pure numbers," geometry, the theory of proportions, the correspondences between the human and the divine aspects in the world soul (Panpsyche) are inextricably entwined, is essentially Pythagorean, as is his apology for love and the very special note of his belief in immortality.[23]

---

22. This expression recurs like a leitmotif throughout the whole text.

23. "The body of man is certainly not immortal . . . good and evil can only touch the soul—incarnate or disincarnate. And I believe in the immortality of this soul that is judged in the beyond on its merits; so it is better to suffer injustice than inflict it on others." *Seventh Letter.*

Heinrich Gomperz, who has recently published[24] an appealing analysis of this letter by Plato, of which I will later cite an even longer passage, has also reached the conclusion that this document, written by the philosopher at the age of seventy-five, provides the key to his entire life and work. His interpretation is the same as mine.

Plato's central doctrine, the one that fed his meditations since his first journey to Sicily, and that inspired the books of his maturity—*Parmenides, Theaetetus, Timaeus, Philebus,* and the *Laws*—is the Pythagoreanism of the initiates, founded on the metaphysics of numbers.

I said at the beginning of part 1 that as Plato, outside of the compilers, forgers, copyists, and exploiters of the Pythagorean ranks, was the sole man of genius to have given us a direct and potent reflection of the doctrine, he could be considered as an unsworn initiate, an "honorary" Pythagorean, a sort of "corresponding member" of the school.

But by reflecting on the number of riddles scattered throughout his work (some of which have barely been interpreted) and especially by rereading a certain passage of that magnificent *Seventh Letter* that is—and allow me to emphasize this point—one of the most important documents in the history of human thought, an autobiographical confession, the testament of a "daimon" or demigod, if ever there was one, one could be tempted after all to believe that not only had Plato sworn the oath of secrecy, but had firmly kept it, only allowing a few sparks of the great light to escape in order to mark out, through the ages, the path for those who would be worthy of handing on the torch.

Here is that passage:

> If there should be someone to write a book in which he claims to explain the points of my doctrine that I hold most dear, whether he believes to have learned them from me or someone else, or had managed to discover them on his own, know that this man under-

24. Heinrich Gomperz, *Platons Selbstbiographie* (Berlin & Leipzig: M. de Gruyter and Co., 1928).

> stands nothing about it. There exists no text by me dealing with these points, and there will never be one. This knowledge cannot be passed on like a series of theorems; it is only after long meditations, and personal habituation with it that, as if by the kindling of a light, a flame springs forth . . . and its light sustains itself with no need of any exterior element.

Immediately following is this passage: "For he who at one time grasped this teaching, there is no danger that he would ever forget it; and, moreover, it consists of only several brief phrases. . . . A scant few of existing men have any knowledge of it."

This brings to mind several of the symbol-dense verses of the *Ieros Logos* that have come down to us, the "passwords" on the gold lamellae of Thurii, the mathematical condensations of tetractys, decads, pentads, and their geometric symbols that the flood of texts and traditions periodically offer us in the cresting of one wave to the next in the swell of the centuries, of the beautiful, evocative phrases (Pythagorean as well, as we have seen) gathered up into metaphors that Plato himself set in his works like gems or beacons intended to illuminate human thought eternally: the body-tomb, the cave, the song of the cosmic sirens . . .

I have spoken several times about the law of secrecy that with purity of heart was the essence of the Pythagorean rule.

Silence and secrecy, the secret sealed by sworn word. Absolute silence (*echemythia,* ἐχεμυθία) was required during the full period of the first degree of initiation, strictly speaking (the five years during which the initiate heard the words of the Master but did not see him), at the same time as the *catartysis* (not to be confused with "catharsis") or spirit of submission. ("Better to perish than speak" was an aphorism dear to the Master, which enjoyed great success among the Stoics and found one last echo in "La Mort du loup" ["The Death of the Wolf"] by Alfred de Vigny.)

The law of secrecy is mentioned by all sources, Pythagorean and outsider alike (Lysis, Dicaearchus, Aristoxenes, Timaeus, Plutarch). It forbade, on penalty of excommunication (and this excommunication was identified as spiritual death[25]), the divulging of either the philosophical doctrine or the rites (secret, as were those of all the ancient mysteries,[26] Egyptian or Greek), or the mathematical teachings because they formed part—as the reader has realized from reading part 1—of the central metaphysical nucleus of the doctrine. The secrecy naturally extended to the signs of recognition. We know from the passage by Lucian cited earlier that the fact the pentagram was the great geometrical Pythagorean "rallying sign" was already known.

But the most interesting ancient document concerning the law of secrecy that has, remarkably, survived into the present is the letter of Lysis to Hipparchus, dating from the fifth century BCE, which Delatte cites at length in his aforementioned book.[27] We have already seen that Lysis was one of the leaders to escape the massacre in Metapontum (around 450 BCE) and that he settled in Thebes with the family of Epaminondas; therefore, at the very latest, he would have died at the end of the fifth century. In this letter that Timaeus (born some thirty years after the death of Lysis) probably brought back from his journey as a student in Magna Graecia with those fragments of the *Ieros Logos* that have come down to the present,[28] Lysis communicated to his correspondent that despite the Master's instructions and the oath

25. Clement of Alexandria (*Stromata*) writing on Hipparchus even reports that funeral steles were erected to confirm this spiritual death.

26. As the great night of the Eleusinian Minor Mysteries fell, when the neophytes, introduced by the *hierokeryx* (great herald, disguised as Hermes) into the sacred grove bordering the temple of Agrae, could hear the choir of the white-clad *hierophantides* (priestesses) evoking "real" life, the life that preceded birth and followed death. At the end of the chorus the *prophantide* that directed the songs called down terrible curses on any initiate who dared divulge the mysteries. During the Great Mysteries in Eleusis itself, the mystai, after the arrival of the torch-bearing procession that brought the statue of Dionysus from Athens, and before their entry into the cave symbolizing the underground home of Persephone, would renew their "sacred oath" before the *hierokeryx*.

27. *Études sur la littérature pythagoricienne*. The text of the letter, transmitted by Timaeus and Apollonius of Tyana, can be found in Iamblichus.

28. See their reconstruction by Delatte, ibid.

of secrecy he had sworn he was teaching the philosophy publicly without any concern for choosing his students from those who had gone through the novitiate and other essential preparations. Lysis went on to say that he felt compelled to criticize him severely and threaten him with excommunication.

Encompassed in the ritual of secrecy and silence is the superstitious respect that forbids speaking the Master's name aloud. He is called: "That one, the immortal Genius, the Divine." The Αὐτὸς ἔφα ("He said it himself!") of the acousmatics has remained famous.[29] This can even be seen in the old gossip Herodotus, who merrily handed down all the fifth-century (BCE) scandals but lowered his voice and whispered, "He whom I shall not name . . ." when referring to Pythagoras.

I have already noted that the Pythagorean pact of secrecy was sealed by a solemn oath, the integral text of which has come down to us by virtue of Iamblichus, after Timaeus.[30] ("No, I swear by he who has transmitted to our soul the tetractys in which is found the source and root of eternal nature," see chapter 1.) The oath is mentioned, moreover, in the first two verses of the *Ieros Logos,* in which Pythagoras, addressing his disciples to decree their religious duties, stated as his first

29. Members of the Pythagorean popular cells in which were transmitted, especially in Greece, with the exclusion of the scientific philosophy and the metaphysics of the school, practices of communistic asceticism, vegetarianism, the respect for all life—both human and animal—and so forth, the whole accompanied by a meticulous liturgy in which, through an intangible "catechism," the wreckage of an ancient ritual was crystallized. The doctrine of Pythagoras the redeemer, a god or spirit who came down to earth to reveal the true doctrine, complete with miracles, a descent into hell, and so on, must also be traced back to this catechism that also contained a summary of the Pythagorean "gospels." The existence of this Pythagorean "Salvation Army" is categorically confirmed by Aristotle, who seems to have taken from it his ten categories of oppositions. The persistence of this democratic branch of the brotherhood into the Christian era and beyond is proven by the mockery of the poets of the Middle Comedy period (Cratinus the Younger, Alexis, Aristophon—the first two each wrote a play titled *The Pythagorean*—cf. G. Méautis) who were exasperated by their plebian puritanism and their affectations, and by the references of Lucian.

30. This is still Timaeus of Tauromenium, who is not the same as Timaeus of Locris whose name furnished the title of Plato's dialogue.

commandment to honor "the immortal gods, the oath, and the heroes" (Iamblichus).[31]

These oaths, in addition, were usual not only in the sects of initiates affiliated with the mysteries (Eleusinian, Orphic, and so on) but even in the guilds or professional brotherhoods (Hippocratic oath). We shall see that this tradition was not lost.

We have spoken of the law of mutual aid binding all the members of the brotherhood that derives from the general dogma of charity and universal love. The Pythagoreans attached an extreme importance to friendship that for them took the form of an intermediary sentiment between what we call friendship and love, a tender camaraderie in which even the sacrifice of life for a friend is an incident as normal as the absolute sacrifice of a perfect knight in the Middle Ages for his "lady."[32] Plato's love for Dion of Syracuse, which is a tragic reflection of this notion (tragic because the assassination of Dion occurred after the misunderstanding that Plato tries to explain in the aforementioned letter addressed to the family and friends of the murdered

31. The commandment related to the civic duties of the Pythagoreans is concise and complete: νόμῳ βυηθεῖν, ἄνομια πολεμεῖν—respect the law (in time as order) and combat illegality (because of anarchy, chaos).

I must once again here pay homage to the infinite patience of Delatte, who in the book I've been citing repeatedly, through sorting and cross-checking sources repeatedly, has pulled from their raw ore a respectable number of the verses of the *Ieros Logos.*

"We find nothing in the form of these fragments," he says, "that prevents us from attributing the *Ieros Logos* to Pythagoras himself. A Samos native could use an Ionian dialect, a sixth-century religious reformer could prefer poetic expression."

But all that one is permitted to assert is that the composition of the poem is earlier than the third part of the fifth century. There is nothing improbable about the tradition mentioned by Iamblichus according to which it was written by a son of Pythagoras (Telauges) from notes kept in the family.

32. The military "fraternities" were already explicitly mentioned by Homer (Nestor asks Agamemnon to parade his army in review by ranking them by clan and fraternity. Walt Whitman praises his barracks brothers, his "camaradoes" (we know he was a stretcher-bearer in the Civil War). The adoptive fraternity (with a blood oath on the cross) still exists in eastern Europe. It can be found, without any ritual confirmation but with extremely strict duties, in the Foreign Legion. Cf. Shakesepare's immortal "We few, we happy few, we band of brothers."

nobleman), is the culmination of a friendship of this nature; Plato's enthusiastic fervor for his young disciple and spiritual brother, and the filial tenderness he retained for his teacher, Socrates, are the inspiring and illuminating focal points of the *Symposium,* the book Plato devoted to love.

The two tragic poles of his emotional life were thus the cruel and sublime death of Socrates, which sealed an end to his adolescence (and whom he also mentions in still moving terms in his *Seventh Letter*), and the murder of Dion, which occurred, as noted earlier, because of a total misunderstanding.

But getting back to the cult of friendship among the Pythagoreans: Dionysius the Elder tells (the anecdote was collected by Aristoxenes) of the experiment he performed in vivo on the two Pythagoreans Damon and Phintias. He imprisoned Phintias on a capital offense, then allowed Damon to take the prisoner's place by having Phintias guarantee with his own life that he would return before sundown. Phintias returned, and the tyrant, moved by this, restored him to liberty.

An interesting aspect of this Pythagorean solidarity is its international character: the Carthaginian Pythagorean Miltades recognized the Pythagorean Argien Possides among the mercenaries who had been condemned to death, and saved him; the Etruscan Nausithous saved the Messenian Eubulos, and so on.

Here is the proper place to cite this anecdote recorded by Iamblichus (*Life of Pythagoras,* chap. 33):

After a long illness contracted during a journey, a Pythagorean died at an innkeeper's for whom this illness had caused great expense and much bother. Before dying, the Pythagorean inscribed a certain sign on a tablet and begged the innkeeper to hang it over his door following his death. This he did. A great time later, a Pythagorean passing by recognized the symbol and fully reimbursed all the innkeeper's expenses.

Méautis, who cites this passage of Iamblichus, concludes from it that "the Pythagoreans must have had some recognition signs similar to those of Freemasonry." And with respect to this same anecdote,

Perdrizet[33] rightly wonders if this recognition sign might not be the Pythagorean rallying symbol mentioned by Lucian: the omnipresent pentagram.

Another important source of information on the rites and spirit of the Pythagorean congregations that still existed in the first century CE are the works by Plutarch in which worship of the hero is in close conformity with the first commandment of the *Ieros Logos* mentioned above. It is superfluous to recall the influence that his *Lives of Illustrious Men* exercised—especially since the Renaissance—on the ideals of so many children and youths who became "heroes" in turn for having read the lives of Leonidas, Alexander, or Caesar. But Plutarch's Pythagorean sympathies are revealed in his book *Isis and Osiris,* and even more specifically in his *Quaestiones conviviales* (a collection of short "topical" dialogues exchanged at the table) and his *De genio Socratis* (which concerns the "daimon," the guardian spirit of Socrates).

The Pythagoreans who appear in Plutarch's dialogues are not fanatics and unsophisticated "acousmatics" but representatives of the sect's philosophical tendencies.[34]

They were young, benevolent, but reserved patricians who practiced the silence of the *echemythia.* They would appear unexpectedly surrounded by pomp and mystery, like the noble Theanor, who (*De genio Socratis*) arrived in Thebes following the death of Lysis (in 379 BCE) to supervise his funeral services in conformance with the "secret practices" and to reimburse the family of Epaminondas for all the expenses generated by the long stay of the venerable exile.

We can easily see the motif similar to the one in the anecdote from

---

33. *Negotium perambulans in tenebris.*

34. This was also continued in a distinct branch whose members called themselves the "mathematicians" and who passed down the metaphysics of the true initiates (of the second degree) of the ancient brotherhood, whereas the "acousmatics," despite their claim of representing the true tradition, only possessed knowledge and phrases belonging to the old reserves of the novices and first-degree initiates. The prototype of the initiated "mathematician" was Archytas of Tarentum, and the Pythagoreans frequented by Aristoxenes came from this same rootstock.

Iamblichus summarized above: the moderation, gravity, discreet abstinence, the reserved gentleness, and the love of harmony from which flow respect for family and established institutions;[35] charity toward all living things and the cult of friendship are thus the "outside" characteristics of the Pythagorean. His inner harmony, for which mathematical knowledge and love are the sources, is governed by a ritual discipline whose basis, at the time of the "Great Brotherhood," is a communal life (with shared ownership of property).

After a strict entrance examination, the candidates were trained in the observance of the rule by a physical and moral educational system with physical purifications, ablutions,[36] and lustrations that were both ritual and mental (we have already looked at how music, perfumes, and dance played a major "cathartic" role by the harmonization and "liberation" of the feelings).

Iamblichus has passed down, following Aristoxenus, a fairly detailed account of the "day of a Pythagorean" inside one of these communities.

"What shall I do today?" the faithful disciple asks himself upon arising, and at evening he does not fail to review the day's activities, transgressions, and omissions.[37]

We should not overlook the mnemotechnical exercises[38] (particularly the recitation of the beautiful "euphonies," as well as mental gymnastics, catharsis, and incantation). Finally the common sacred feast

35. When they are respectable and harmonious; I mentioned earlier the summary of the duties of the citizen according to Pythagoras (passed down by Aristoxenus): respect the law and fight disorder. This is a charming variant (before the fact) of "Render unto Caesar . . ."

36. The most scrupulous physical cleanliness was rule and symbol. When the Pythagoreans were in their "phalanstery" or when they met "officially" they were dressed all in white (toga included) like the priests of Isis (Isis λινόστολη, the queen "in the linen vestments"). They were obsessed with purity. Diogenes Laertius and Iamblichus.

37. Delatte identifies five verses from Porphyry (according to Timaeus) as belonging to *Ieros Logos* that prescribe this "examination of conscience" and that through a curious reincarnation can be found in the Boy Scout code ("one good deed a day").

38. Plato admits he tried mnemotechnical procedures on Dionysius II during his novitiate that turned out so badly.

with libations and as main dish only the flesh of animals eligible for ritual sacrifices: white roosters, suckling pigs, and kid goats, all topped by a brief sermon and a grace recited by the "presbyter," the eldest member of the community. Before sleep, certain chords on the lyre and a certain perfume (the χύφι for which Plutarch gives the recipe[39]) "soothe as if by incantation the sensitive and irrational part of the soul . . . soothing and untying like knots, and without intoxication, all cares . . . polishing and purifying like a mirror that which, in the soul, is imaginative and receives dreams" (*De Iside*).

It was only after his elevation to the first grade that the disciple was initiated into the mathematical metaphysics (the law of number) and the interpretation of symbols, through whose knowledge he would arrive, having become "sighted," to perception, through love, of the Great Harmony that he had approached, now prepared fearlessly to attempt the Great Adventure: deliverance from the tomb, from the long passage of ordeals, the abrupt exit from the cave of shadows into the full light . . . to join—beyond the Milky Way, beyond time and the cycles of palingenesis—the "daimons" and sister-souls finally found or found again.

Thanks to the patient and underappreciated work of researchers (Delatte, Carcopino, Levy, Méautis) whom I have named in this chapter, a labor demanding the examination and verification of an immense number of texts and allusions—I almost said "alluviums," as it sometimes brings to mind the grinding and washing of diamantiferous conglomerations, and sometimes the scientific decryption of messages sent from a master of cryptography—we have been able to lift the veil and glimpse how the "Great Brotherhood" in the white togas functioned, and, filling in an apparent four-century darkness, rediscovered the unbroken chain that connects it to Alexandrian and Roman neo-Pythagoreanism. To conclude this examination, I am going provide

39. *De Iside.* Plutarch says that it was composed of sixteen ingredients: honey, wine, raisins, galingale (a sedge), resin, myrrh, rosewood, seseli, mastic, bitumen, aromatic bulrush, dock, large and dwarf juniper, cardamom, reed.

some more details on these two parallel movements, starting with, since we began this chapter under the auspices of the Pythagorean basilica of the Porta Maggiore, a return to the Eternal City.

We should note right off that in Italy the Pythagorean tradition was never interrupted. Carcopino notes that not only was it never at risk of being extinguished, but that the Pythagorean "disapora" caused by the revolutions of Croton and Metapontum had diversified and extended rather than slowed its secret propagation. We have already mentioned many times the government at Tarentum of the Pythagorean Archytas. The lodges of the Rhegiums of Calabria and Philonte are explicitly mentioned, and at the end of the fourth century Aristoxenes of Tarentum (cited by Diogenes Laertius) placed the center of Pythagoreanism in Etruscan territory.

In Rome itself at this same time, we see the ancient patrician house of the gens Aemilia attach its origins to an alleged son of Pythagoras, Mamercos (hence the name of Mamercus held by many of this family's members between 376 and 270 BCE, according to Plutarch). According to Cicero (*Tusculan Disputations*), the famous censor of the year 312, Appius Claudius Caecus, was a Pythagorean.

Pliny the Elder (*Natural History*) tells us that between 298 and 290 BCE, the statue of Pythagoras was erected in the Forum following a command of the Pythia as "being the wisest of all the Greeks." Cato the censor, who had been a guest of the Pythagorean Nearchus in Tarentum in 293 (Plutarch), was classified by Cicero as a member of the sect (*De senectute*). The affinity between Pythagorean asceticism and the harsh discipline that was the pride of the republic's patrician class provided an encouraging environment for these influences. Cicero finished by convincing himself that the word of Pythagoras had never stopped echoing in Rome, and that a large number of Roman institutions were models of his. Like Timaeus of Tauromenium three hundred years earlier, Cicero even visited Metapontum in search of remembrances of the Crotoniate era. Waxing lyrical, he attributes Pythagoras with the discovery of the "sublime truth of the immortality of souls."[40] In his *De republica,* he quite rightly makes Plato the spiritual heir of Pythagoras, and repeats the tradition maintaining that Plato secured the secret books owned by the disciples of the master for their weight in gold.

It was during Cicero's own era that the Roman elite was seduced by a second wave of Pythagoreanism, a parallel wave to the simultaneous Greco-Alexandrian or neo-Pythagorean awakening (Posidonius of Apamea).

"In the century that frames the beginning of the Christian Era," notes Carcopino, "Pythagoreanism attracted from all points of the intellectual horizon those who, eager for certainty, felt as smothered in the emptiness of the state sanctuaries as by the swirling of Lucretius's atoms." This is an infatuation similar (and for the same psychological reasons) to what was known as Bergsonism at the beginning of the twentieth century. "From 60 BCE to 50 CE, all . . . academy members, Stoics, Peripatetics, Eclectics . . . all were more or less Pythagorizing."

Varro, the most famous scholar of Cicero's era, was a Pythagorean and requested that his body be laid to rest in accordance with Pythagorean rites on a bed of myrtle, olive, and black poplar leaves (Pliny, *Natural History*).

We already met at the beginning of this chapter an interesting figure who was another friend of Cicero's and embodies in history and legend Roman neo-Pythagoreanism: Publius Nigidius Figulus, senator, astronomer, mathematician, diviner,[41] mage, and grand master of the most important Pythagorean lodge in the city. We have seen that he was exiled by Caesar (in 45 BCE) and that "*Nigidius Figulus Pythagoricus et Magus in exilio moritur*" [that is, he died in exile] (Saint Jerome).

Despite the strict measures taken against secret societies, Pythagorean ways continued under imperial Rome. Seneca confessed that Sotion had

---

40. In his *Dream of Scipio,* it is to the Milky Way that the soul of Scipio the African, like a Pythagorean daimon, descends to converse with his grandson in dream.

41. Dion Cassius and Suetonius reported a famous prophecy by Nigidius: he announced to Octavius on his way to the Curia that the son just born to him (63 BCE) would become master of the earth. As Octavius's republican sentiments were affronted by this, he spoke of returning home to slay the newborn. Nigidius forestalled him by telling him that Fate was stronger than he.

inspired his love of Pythagoras.[42] Moderatus of Gades taught the doctrine during Nero's reign, and we have seen, in connection with the time the Porta Maggiore basilica was destroyed, the accusation of magic that included Statilius Taurus among "the mages and mathematicians" exiled by Claudius.

Pythagorean ways even won over exotic kings like Juba II, king of Numidia and husband of Cleopatra Selene, daughter of Antony and Cleopatra.

Then, concurrent with the success of the apostles Peter and Paul, Apollonius of Tyana and Simon Magus appeared in the capital of the empire. They were the first leading players of Gnosticism, the alarming bastard child of the young church of Christ and Alexandrian neo-Pythagoreanism, conceived in the hot twilight of Egypt, the mother of all magic.

This is an opportune moment to look toward the East.

One hundred years after the death of Plato, Alexandria had become the intellectual and scientific capital of the world. Euclid, Eratosthenes, and—much later—Diophantus completed, in the context of Archytas, Eudoxus, and Plato's theory of proportions, the crystal palace of Greek geometry with its overly neglected annex, the theory of figurate numbers.

We have specifically seen the theories of proportions and figurate numbers, with their most subtle refinements, occupy the place of honor in the treatise of mathematical popularization written by Nicomachus of Gerasa (see chapter 1), thereby proving that around the first century CE the Pythagorean notions of numbers and geometry formed part of the scientific legacy of the educated circles of the Greco-Roman world. We will find these same notions in the only treatise on building that has miraculously survived into the present: the book by Vitruvius in fact shows us that the theory of proportions and harmonic convergences, with the directing terms and ideas found in the *Timaeus,* had given to architects and sculptors a doctrine (linkage of proportions, "analogies," and a

42. *Mihi amorem Pythagorea iniecit Sotion.* Cited by Carcopino.

"symmetry" of concordances culminating in eurhythmy) as well as practical procedures of harmonic composition passed down most likely in the form of guild secrets in the families of architects and craftsmen colleges.

Neoplatonism and neo-Pythagoreanism harmoniously combined to form—on the old Syrian-Egyptian lands—a humus ready to allow strange metaphysical flowerings to germinate, some of which crystallized into religious sects. The most audacious among these, founded like early Pythagoreanism on knowledge and love, and believing in palingenesis, was Gnosticism.

In this milieu of intense mathematical culture, the personal contribution of Pythagoreanism, the mysticism of numbers, found, moreover, enthusiastic adepts in one of the most brilliant communities of the Alexandrian "intelligentsia," namely, the elite of the Jewish diaspora.[43]

The reciprocal visits of Greek and Egyptian gods noted by Herodotus continued. Hermes, creator of the word, number, and music, who, as we know from Plato, was none other than the ancient Thoth, had resettled once and for all in the Nile valley under the name of Hermes Trismegistus (ὁ μέγας, μέγας, μέγας καὶ μέγας, τρὶσμέγας τρὶσ μέγιστος was the inscription to Thoth on the temple of Dendera; Ερμῆς μέγας καὶ μέγας δεῷ μέγιστος Ερμῆ was the inscription on the Rosetta Stone—the adjective *τρὶσμέγιστος* as one word is first found in the work of Tertullian) and had brought back online the barely cooling laboratory where talismans and words of power were once crafted.

This is how both the Kabbalah and Hermeticism, inseparably connected to Gnosticism by a common ancestry, were born; they are, as in a *trimurti,* the three faces (Hebraic, Egyptian, Hellenistic) of one same deity.

Like an instrument for formulating metaphysical equations for all

43. "The conquest of Judaism by the Pythagorean doctrine began long before the Roman era. All the masterpieces of the Jewish literature of Roman Alexandria were connected to the Pythagorean tendency." And "Alexandrian Judaism, Phariseeism . . . and Essenism, offer, compared to biblical Mosaicism, new characters, signs of the conquest of the Jewish world by notions for which the legend of Pythagoras was the narrative expression and vehicle." Isidore Levy, *La Légende de Pythagore.*

their problems, the three disciplines employed by choice the theory of the analogy of the macrocosm and the microcosm, brilliantly reemployed by the choir of Neoplatonic commentators of the *Timaeus* (Posidonius of Apamea, Chalcidius, Theon of Smyrna, and so on—see chapter 1).

In addition to the metaphysical speculations—cycles of palingenesis, mysticism of numbers—we see the revival in Egypt, then in Syria, of contemplative communities of "brothers" of the ancient "society." The Jewish Pythagorean Philo of Alexandria (first century CE), in his *De vita contemplativa,* describes the societies of "Therapeutae" established in the isolated regions of Lake Mareotis. They would leave the *monasterion* where they led solitary lives to celebrate together, in the main phalanstery, on the seventh and the fiftieth day, a choice motivated by the nature of the ever-virgin number 7,[44] and the number 50, the holiest and most natural of numbers, because it was equal to the sum (9 + 16 + 25) of the squares constructed from the "sacred" (3-4-5) triangle of Pythagoras and also to the product 5 × 10 of the pentad and decad, numbers of generative life and the world harmony, of the microcosm and the macroscosm.

In Palestine, the Essenes or "Silent Ones" already existed before the preaching of Christ. Their meeting hall, a dependency of the temple of Jerusalem, was called *Hassa'im*, hall of the Silent Ones (hence Εσσαῖοι, Εσσηνοι, Esseni). Josephus (*The Jewish Wars*) says when speaking of these meetings: "No shout, no tumult pollutes the common house at any time; each is given the chance to speak in turn. . . . They swear to never reveal to any outsider of matters concerning the members of the sect."

Oath of initiation, precept of silence, communal sharing of meals and

44. The choice of the number 7 as symbol of "virginity" is even more appropriate as not only is it a prime (indivisible) number, but while it is easy to divide a circle into 3 or 5 equal parts (3 and 5 being the other prime numbers of the decad 10), it is impossible to divide it (by a strict euclidean construction) in seven. This was only demonstrated by Gauss at the beginning of the nineteenth century. The allusions to the virginity of the number 7 are frequent in arithmological texts, and in the works of the Church fathers. In a manuscript at the Bibliothèque Nationale, Delatte found: ὁ πέντε γάμος / ὁ ἑντά παρθένος.

Let's add that by a combination of analogous ideas, 7 was also "the number of the oath."

housing, successive stages (here one, then two years) before being admitted to the communal sacred feasts and at the status of full initiate—we would recognize the line of descent even if Josephus had not explicitly stated in another book (*Jewish Antiquities*): "Those we call Essenes practice a lifestyle that conforms with the principles of Pythagoras."

The Essenes displayed a stoic courage during the course of the final battle against the Romans. We have no certain proof of their survival after the destruction of the Temple. But on the recently unearthed Galilean synagogue of Capernaum (beginning of the third century BCE), the pentagram of the Pythagoreans appears instead of the ritual Hebrew hexagram (Solomon's seal).[45]

Thus we can again see—at the threshold of the Christian era, in this Garden of the Hesperides of metaphysics and religions that is the sparkling Alexandrian microcosm, dominating the flowering of systems and rites, in this center from which radiated broad avenues of thought and the obscure arbors of sects—standing tall, more tutelary than ever, the royal tree of the "philosophy" par excellence, that of the Master of Samos and Metapontum.

Of course, some strange vines are coiling around its main branches. The dark scarlet color of the fruits whose use the Master did not advise suggests, among the green foliage, their "secret architecture"; these are the pomegranates of Persephone.

For the perfume of the mysteries floats in this garden; like Thoth-Hermes, Demeter-Ceres has returned to the land of the black silt and become anew Isis, the "perfumed queen clad in linen." Like her companion, she has been oddly rejuvenated, and like him has recovered her dispensary of charms and "words of power."

"My veil, never has any mortal yet lifted it," said the Isis of antiquity, goddess of the mysteries and initiations. The Alexandrian Isis holding in her open hand the heartlike fruit of the persea, lets slip the black veil, and appears, amber-colored fruit in the tight envelope of immaculate linen,

45. Levy, *La Légende de Pythagore.*

goddess also of fecundating life, source and receptacle of all generation.

Overlying the supremely pure harmony of the spheres once heard by the initiates of Metapontum is a minor cantilena; the far-off voices of the planetary sirens are drawing near.

Didn't Plato cite Diotima as saying in the *Symposium:* "The object of love . . . is therefore not love of the beautiful . . . it is the love of generation and creation in the beautiful . . . and the procreation by the union of man and woman is also creation, and a divine work"

Why read any further?

Why should what is static in the ideal analogy of the "little world" and the "large world," not become dynamic? Why should the man-microcosm not become closer to the macrocosm, living in continuous creation (*natura naturanda*) of the Creator himself, by fervently repeating his fecundating activity?

Hermes Trismegistus appears to approve of this shortcut to heavenly love; his hand engraved this summary of it on the Emerald Tablet:

> *Id quod inferius*
> *Sicut quod superius!*
>
> (That which is below / Is like that which is above.)

And "Above the heavenly things, below the earthly things; through the male and female the work is achieved."

Heavenly love, cosmic love, earthly love, seeking to find harmony; the cycles of palingenesis undulate to the rhythm of pangeneration.

On a large alexandrite gem in the Vienna Museum, the Gnostic Hermes himself, a muscular beardless youth with no other costume or insignia save the large serpent comfortably coiled around his left arm, is contemplating, with a young goat by his side, the flame dancing above a vase, be it cauldron, crucible, or incense burner. Above his head is radiating the ritual pentagram, with wreathed sides, pentagram of harmony transformed for good to the Πέντε Γαμος of Hathor-Aphrodite, fecundating goddess of love.

8

The Lamp under the Bushel

Ah, the dazzle on my brows blindingly gilded,
Eyelids overborne by a night of riches,
Gropingly I was praying in your golden glooms

Paul Valéry, "The Young Fate"

In the presence of the unsettling avatar of the sign of harmony that we have seen radiating above the Gnostic Hermes, another star, or another reflection, not far from here, appeared in the firmament of symbols.

Magi are also guided by this star, leaning over, like official monitors responsible for an observation or a transmission of powers, over the cradle of a new god, who, too, will say he is love. He needed barely two centuries to be installed as master, at least as acknowledged suzerain of the Alexandrian mystical garden grown into a world. He is acclaimed as such by thaumaturges, mages, and Therapeutae—Gnosticism becomes Christian.[1]

But Christianity did not become Gnostic.

1. The greatest doctor and patriarch of Gnosticism was Valentinus (circa 100–170 CE), a disciple of Theodas, who was himself a disciple of Saint Paul. He continued the tradition of Simon Magus. Other Gnostic luminaries include Basilides, Menander of Samaria, Bardesane the Syrian, and Marcion.

The father of the Church who was most greatly attracted by Gnostic symbolism and Pythagoreanism was Clement of Alexandria. The Ophites (snake worshippers) were also a Gnostic sect, with a subdivision in the Luciferian Cainites.

In Isidore Levy's *La Légende de Pythagore de Grèce en Palestine,* we find the account of an Egyptian tale known as the story of Siosiri or the "Duel of the Magicians." This is a cyclical tale in which a black magician from the South reappears at the court of the reigning pharaoh in Memphis after long intervals between visits. This magician, who has challenged the official mages of the royal house to a contest of spells and divination, is on the verge of defeating them, when one of the royal magicians, sometimes even a stranger whose "magic" qualities are suddenly revealed, unmasks the identity and dangerous evil spells of the dark sorcerer. He routs him at his own game by using more effective spells, and delivering Egypt and the pharaoh from the perilous stakes of the occult duel, forces the demonic wizard to leave after promising never to return. As I said above, he always returns, under the same name, moreover: Hor, son of the Negress. The centuries (sometimes a thousand years) have erased all memory of his former avatar, but the providential figure who accepts his challenge also proves to be the reincarnation of the "white magician" of the previous encounters, which permits him to thwart the wily "son of the Negress" and exorcise him by the terms of the periodically renewed pact that is always forgotten and always broken. It is in the court of one of the final indigenous pharaohs, great initiates in the occult sciences, that the story places the last "accredited" encounter of the two magicians.

It was an enemy like this malefic riddler that the young Christian church immediately recognized in Alexandrian Gnosticism, with its seductive riddles and its charms bearing the mark of a good craftsman.

The Church refused to accept magic—or at least the "desire for magic." It did not accept the predominant role awarded to "knowledge," and the desire for knowledge, and even less the tendency to reserve this knowledge for a circle of the elite. Finally, while openly accepting love as light and supreme purpose ("Simon, son of Jonas, do you love me?" the man-god of Galilee asked three times the one who would be the cornerstone of his Church), it denied any blend of divine love with earthly love, or at least with the love that seeks in the creature, next to beauty of the soul, that of intelligence and form. It commanded its faithful to seal their ears, like the companions of Ulysses, to the song of the cosmic sirens, and to ratify the verdict of the mysterious voice that cried out at night over the Phoenician sea, during the reign of the same Tiberius who had witnessed the torment of Christ, the yell that astounded the pagan world: "Pan, the Great God Pan is dead!"

It was in vain that Marcion, Valentinus (first patriarch of Gnosticism), Basilides, and Bardesane deployed the blandishments of their Word to save the scintillating mixture of Neoplatonic metaphysics and Syrian tenderness—the words of power of Isis no longer worked, Isis who had become the feminine Holy Spirit, the Gnostic Sophia—in vain that Simon Magus[2] tried to prove with his miracles that he was the true keeper of the words of Christ. Patriarchs, bishops, and magicians of Gnosticism were excommunicated, like Arius, and the sect, with its strange branches that were equally Syrian, Coptic, Cainite, Ophite, and so forth, was vanquished and expelled from the Roman bosom, doomed, it seemed, to destruction. We shall see later that it led a hardy life, and with its sisters, the Kabbalah and Hermeticism, it was content to wend its way through the esoteric shadows, which was more in keeping, moreover, with its origins, passing down from century to century its ritual and ideological heritage, of which certain Pythagorean rites and symbols formed no small part.

In this chapter we are going to try to see exactly how among these symbols, principles, or even procedures, those that were of specifically Pythagorean essence were handed down: through geometrical symbols and designs.

2. Simon of Samaria, known as "Magus" or "the Magician," who preached in Rome during the reign of Claudius. He had met in an "infamous place" of Tyr the woman he made his life companion and introduced her as a reincarnation of Helen. His disciples maintained that she was the sorrowful symbol and living image for him of the fall of thought into matter, followed by the new ascent toward the light, love, and revelation.

* * *

We saw in chapter 3 how Gothic architecture was "expert" from the perspective of the strict geometry[3] of its plans. We saw how nothing was left to chance in the overall plans or in the design of the details, and that especially in the drawings of the roses, rose windows, and stained glass of the Gothic cathedrals, we found an entire graphic encyclopedia of regular polygons inscribed in the circle as well as the polar segmentation of this same circle (Mössel's *Kreisteilung*).

I already noted in my book *Esthétique des proportions* how often the pentagon and pentagram recur in the designs and the motifs of Gothic roses and rose windows. The inscribing of these figures in the circle (probably one of the geometric "secrets" of the school revealed to the profane by the Pythagorean Hippocrates of Chios[4]) was handed down from antiquity according to the method of Ptolemy in the *Almagest*,[5] based, as noted by Ptolemy himself, on the division of a straight line into mean and extreme ratio, which is to say, following the golden number (divine proportion or "golden section") that indeed governs the play of proportions in every regular figure with pentagonal or decagonal symmetry.

The old Pythagorean symbol for harmony radiates especially at Notre Dame, where we find it inscribed inside a pentagonal stained-glass rose, as well as at the heart of the north rose of Saint Ouen in Rouen,

---

3. And also from the perspective of the perfect dynamic balance of the forms culminating in the tips of the ogives, vault keystones, flying buttresses, and so on. In modern times only the builders of large steel bridges have rediscovered a dynamic architecture. In the living world, the armature-supports of plants sometimes resolve in similar fashion the problem of obtaining the maximum height and resistance with a minimal amount of substance.

4. An empirical variant of the construction that has been given his name, that of a pentagon with a given side, can be found in the first geometry manuals to be printed during the Middle Ages and in Dürer's treatise on proportions.

5. Claudius Ptolemy (90–168 CE). This refers to his manual of astronomy and mathematics called *Μεγάλη Σύνταξις*, then *Almagest,* according to the name given it by Arab scholars.

and in the magnificent north rose of Amiens cathedral (plate 49: the rose window itself is pentedecagonal, with fifteen points).

Notre Dame also possesses two beautiful pentagonal plain roses (no stained glass), whose flexible and perfectly proportioned designs evoke a floral calyx or some sea creature.

Pentagonal rose windows can also be found in the Saint-Chapelle, in Strasbourg (at the choir and at the tip of each point of the large rose window of the cathedral), and on the entire perimeter of Westminster Abbey, and so on.

In the architect Villard de Honnecourt's sketchbook, preserved at the Bibliothèque Nationale, we often find the use of the pentagram as the regulating design of human figures, animals, and plants (as we do later in the botanical sketches of Leonardo da Vinci). It was this same time (thirteenth century) that Campanus of Novara made his observation on the role of the golden section in the organization of transcendent geometrical symphonies (cf. chapter 3).

Before leaving the rose windows of the cathedrals, I would like to recall the analogy sketched out in part 1 concerning the Celtic-Nordic contribution to the European concept of love and relations between the sexes, between what I have called "Gothic" love and Gothic architecture, and I will note another correspondence here, a purely verbal one, if you like. At the same time the rose of stone and glass appeared in decorative symbolism, the rose as the flower of love replaced the Egyptian lotus and Greek narcissus. These are not the frivolous roses of Catullus, those whose petals were plucked over the beds of the poets of the Latin decadence, but the vital, imperious Celtic roses, which were not lacking in thorns and were heavy with a sweet symbolism: this is the rose of the *Roman de la Rose* from which Guillaume de Loris and Jean de Meung made the mysterious tabernacle of the Garden of Love of chivalry, the *rosa mystica* of the litanies of the Virgin, the gold roses that the popes gave to deserving princesses, and lastly the immense symbolic flower that Beatrice showed to her faithful lover who had reached the final circle of Paradise, both rose and rose window, and in which René Guénon[6] sees an initial apparition (and a connection to the white militia of the Templars) of what would become the Fraternity of the Rose-Cross:

Nel giallo de la rosa sempiterna,
che si dilata ed ingrada e redole
odor di lode al sol che sempre verna,
qual è colui che tace e dicer vole,
mi trasse Bëatrice, e disse: "Mira
quanto è 'l convento de le bianche stole!"
....
In forma dunque di candida rosa
mi si mostrava la milizia santa
che nel suo sangue Cristo fece sposa . . .[7]

Into the yellow of the Eternal Rose,
which outward spreads in tiers, whose fragrance praises
the Sun which makes an everlasting spring,
was I, like one who, fain to speak, keeps silent,
led on by Beatrice, who said to me:
"Behold how vast the white robed Convent is!"
....
In semblance, therefore, of a pure white Rose
the sacred soldiery which with His blood
Christ made His Bride, revealed itself to me . . .

6. *L'Ésoterisme du Dante.*
7. *Paradiso,* cantos 30 and 31. The possibility of an allusion to the Templars (a "secondary" allusion since the white-clad phalanxes who form the petals of the "eternal rose" constitute the heavenly army of all the elect) is curiously underscored by Beatrice's prophecy in the final verses of canto 30, concerning the damnation of Clement V. This was the pope who supported Philip the Fair in his battle to wipe out the Templars and who was cited by Jacques de Molay, when about to burn at the stake, to appear before the divine tribunal. Pope Clement died that very year (1314) as did the king (see chapter 9).

A flower with five petals like its wild sister the sweetbriar, whether heraldic like the red rose of York and the white rose of Lancaster, or mystical like the gold rose and the Rose-Cross, the medieval flower of love is graphically a floral image of the "cinquefoil," the five-part rose window, a gentler version of our old pentagram of harmony.

The geometrical esotericism of the Pythagoreans, with the ideological form given it by Plato in the *Timaeus* and which we see again in Nicomachus of Gerasa and Vitruvius, was in fact passed down after the fall of the Roman Empire in the West and the conquest of Egypt by the Arabs through two underground tracks: the designs of the architects and the "pentacles" of magic, which we shall now analyze.

TRANSMISSION OF THE PYTHAGOREAN DIAGRAMS THROUGH ARCHITECTURE

The exclusively Pythagorean-Platonic aspect of Vitruvian mathematics (were it not for the fact that Vitruvius wrote his book at least a half century before Nicomachus,[8] we could easily believe he copied it word for word from the latter's book) is visible in the fact that Vitruvius, the son of an architect, spoke in the name of a solidly established professional tradition and that in the art of building, as in the majority of others,[9] the tradition was handed down as family or guild secrets (though the simple fact of the absence of printed books would have imposed this

---

8. Cf. chapter 1, the extracts from the treatise on numbers by the neo-Pythagorean Nicomachus of Gerasa.

9. The wording of the "Hippocratic oath" has preserved proof of this kind of transmission of professional secrets, which only came to an end with the general implementation of "university" teaching, and especially the invention of printing. Here is this Hippocratic oath:

"To give a share of precepts and oral instruction and all the other learning to my sons and to the sons of him who has instructed me and to pupils who have signed the covenant and have taken an oath according to the medical law, but no one else."

This oath came down to us through the writings of Scribonius Largus, doctor of the emperor Claudius and of Messalina.

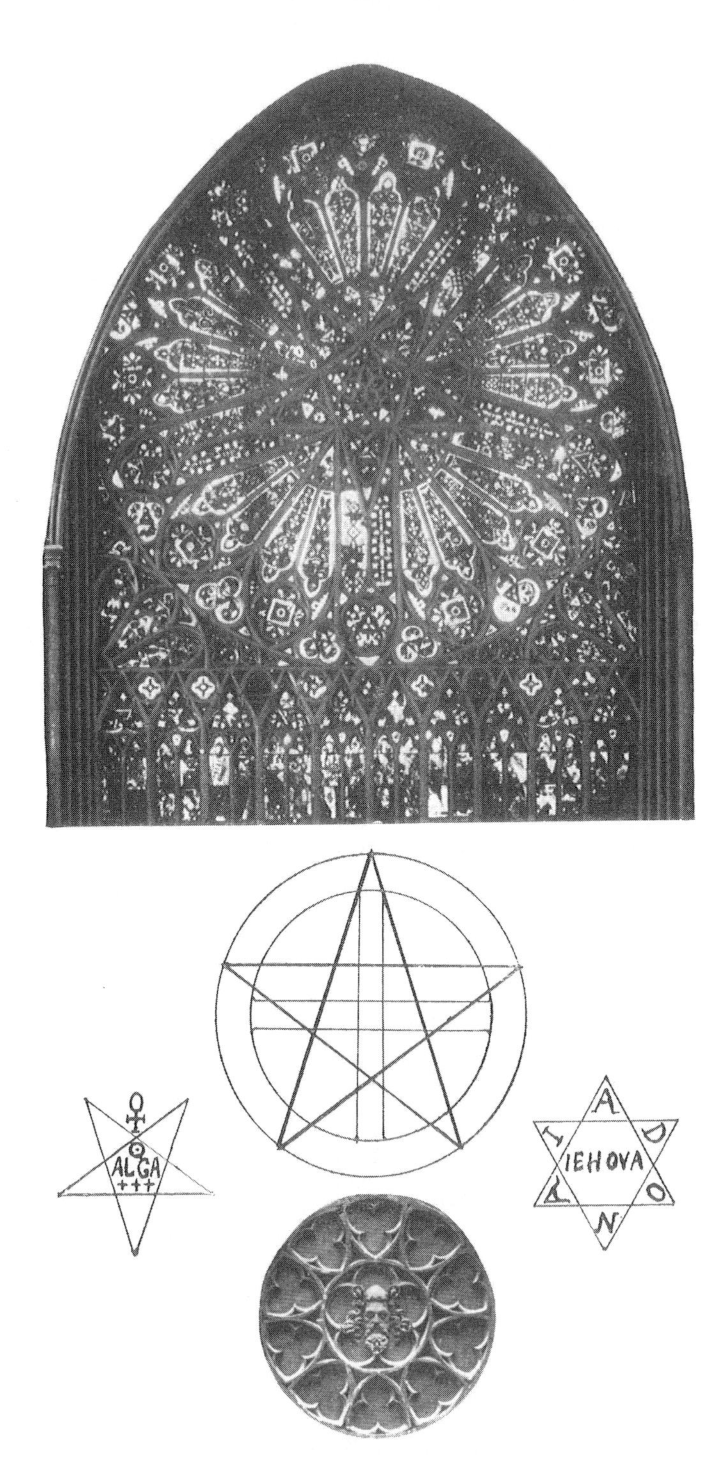

Plate 49. *Top,* pentagram in the great rose window of Amiens (photo Régnaut); *middle,* magical pentacles from a German manuscript of the *Faust* cycle; *bottom,* cinquefoil from the Saint-Chapelle, south portal (photo L.P.)

"esoteric" tradition to a certain extent). Let us already assume that this tradition would include the graphic procedures for establishing proportion, the geometrical directing diagrams[10] that consisted of drawings whose lines, inside a directing circle, represented varieties of individual or combined projections of the regular polyhedrons inscribed within the sphere; as these same drawings would have automatically provided series of proportions, reflections of "analogies" conforming to the desiderata of the "symmetry" (commensurability, *commodulatio,* whether as lines or areas, which in this last case Plato had already called "dynamic symmetry"), it clearly seems that they should resemble the logical drawings later identified in the plans of Egyptian, Greek, and Gothic buildings by Lund and Mössel. Basically, they all come down to variations on several very simple themes of which the most fruitful in eurhythmic combinations were based on the pentagon and decagon being inscribed inside the circle.

As confirmation of the transmission of the principles of composition in the families of architects we have, among others, a text from the Theodosian Code saying that architects are exempt from all obligations in order that they may teach their children the practice of their art more easily.

A parallel path of transmission of professional techniques, also in the form of hereditary secrets, was formed by the guilds or "colleges" of masons and stone carvers, whose existence and operative methods in the Greco-Roman world are proven not only by mentions in the chronicles or annals, but in legislative texts.

Tradition credits Numa with the founding of eight craftsmen guilds. What is certain is not only that these guilds (*collegia opificum*) were in existence in the third century BCE, but that they multiplied and degenerated by adding to their purely corporate character electoral

---

10. The lost plates from Vitruvius's work would probably have confirmed this kind of regulating diagram, which we can imagine (and even draw exactly in the case of Greek and Roman theaters) based on the text. Perhaps professional secrecy had a role here, too; in this case we would have found specific examples rather than true keys.

agencies and secret political clubs. We know from Suetonius that Caesar dissolved all the guilds "except those of ancient foundation."

Under the empire, in any case, hereditary enrollment in the *collegia* of craftsmen had an obligatory character. Adrian (117–138) registered the artists. We found numerous laws concerning guilds in the fourth century. The Theodosian Code even provides for the forcible return of *collegiati* who illegally left Rome back to their guilds, and the same arrangement was promulgated anew by Honorius in Milan. The last ancient regulation concerning the "colleges" was an edict signed by Emperor Majorian (died 461 CE) in Ravenna.

We have the membership list for the boatmen guild of Ostia in 152 CE, which was preserved in an inscription found during excavations in Rome's ancient port.

In Gaul, these guilds were established with all the administrative and social mechanisms of Roman municipal life of which they were the primary actors. The "water merchants" of Paris appear to have unbroken ties to the old college of the "Nautes Parisienses" that was established in the city during the reign of Tiberius, but it was especially in southern France that the Roman *collegia* resisted the "barbarian" invasions and opposed with their communal liberalism the feudal spirit brought down from the north by the clans of Frankish warriors, Visigoths, and so forth. It so happens that these corporate guilds of the Roman Empire displayed, in addition to their technical side (organization and execution of work, enterprises, reciprocities) a religious and social aspect. They sometimes had as appendages, and sometimes as nuclei, religious fraternities with annual festivals, feasts, and rites, and these religious brotherhoods in turn were affiliated with funerary associations with a cooperative and ritual basis that played a major role in the lives of the artisans.

We know by the few inscriptions found on this subject that there existed among crafts guilds, funeral guilds, and religious fraternities (*sodalitates*) not only these bonds of affiliation, but also a close similarity from the point of view of their initiatory ritual character: the kinds of ceremonies (feasts, sacrifices), names of officials, elections (and probably admissions) of honorary "patrons." We also know through the fairly plentiful inscriptions going back to the college of the "Arvalian Brothers" found in their small temple[11] that the ritual was excessively complicated and was able to be passed down for six centuries without any modification.

Similar guilds existed in Greece and in the Hellenic East. The mason guilds that interest us specifically spread throughout the Eastern Empire during the entire lifespan of Byzantine civilization, and we have numerous legal texts referring to them. The sole modification after Constantine was that the Christian worship replaced that of the fickle tutelary deities from the beginning of the empire, but the technical traditions and the ritual itself, including the principle of professional secrecy and initiation, were passed down with no changes.

What is even more interesting is that the fall of the Byzantine Empire after the Turks took Constantinople, instead of destroying these guilds, saw them—like other local institutions that the conquerors respected almost to the present day (I imagine that the new nationalist regime in Turkey* and population exchanges have brought an end to these institutions that were more than two thousand years old)—strengthen their autonomy and keep both traditions and organizations fully intact. This is how in the second half of the nineteenth century, Auguste Choisy[12] was able to find in the mason guilds of Constantinople and the other large cities of the Ottoman Empire with Christian craftsmen (such as Salonica) the structure, ritual, and even the names of the guilds of Justinian's era, themselves a continuation of the guilds of pagan antiquity.

Choisy was thus able to find the "protomaster" or project manager

---

11. In the sacred grove of the "Dea Dio," local goddess of the nurturing earth who was especially honored by their brotherhood. It was on the walls of the temple that this carved statement was found: *Acta Fratrum Arvalium.*

[*This book was written shortly after Kemal Ataturk took control of Turkey and radically altered its society—*Trans.*]

12. *L'Art de bâtir chez les Byzantins* (Paris: Libraire de Société Anonyme de Publications Periodiques, 1883).

(a sixth-century fragment speaks of the "πρωτομαίστωρ"of the Hagia Sophia worksite), the "μαίστορες" or masters, the religious or funerary brotherhoods, with annual feasts, and the name of "χῆρυξ" (for the protomaster's assistant bailiff), which is reminiscent of the title of the heralds assisting the mystai during the high holy days of Eleusis.

The initiatory symbolism extended to the tools of the trade, and for architects and masons this technical symbolism acquired a very special importance due to the geometrical "secrets" handed down by the masters, which conferred a doubly initiatory nature to their use. These instruments—compass, square, plumb line—are sometimes depicted on the sarcophagi of Roman-era architects,[13] as well as those of the Gallo-Roman era (one can be seen at the Alyscamps outside of Arles) arranged exactly as they would be later on the tombstones of the fifteenth-, sixteenth-, and seventeenth-century master builders (for example, that of Master Franciscus Hietz, dated 1675, on the façade of Saint Stephen's in Vienna—the "mason's mark" of the master is carved in the escutcheon above the inscription). We find it again holding a place of honor in the symbolism of the "speculative" Freemasons.

The "masons' marks" I just mentioned, which could take up an entire chapter by themselves (a chapter that could easily extend into volumes, a labor taking years) form a several-thousand-year-old chain of the corporative procedures and rites of masons and stone carvers. These signs are sometimes obvious monograms, and sometimes more or less complicated geometric symbols that can be found on the stones of some ancient monuments and on the majority of Roman and Gothic buildings.

For ancient monuments, Didron, who fancies himself the first to have discovered, recorded, and assessed this kind of hieroglyph, in a report made to Guizot after a trip to central and southern France,

---

13. Lund mentions one (in *Ad Quadratum*) on which is depicted a ruler divided into four segments forming a geometric series with the "golden section" as ratio between consecutive terms. I am reproducing in this book (plate 57) a funerary mosaic from Pompeii on which the square and plumb line appear.

mentions that they can be found on the Egyptian pyramids, and in the arenas and amphitheater of Nimes. Choisy and Rziha mention their existence in Pompeii and on the Palatine Hill, and a large number at the palace of Diocletian in Spalato [now the city of Split in Croatia].

I found no mention of them in Greece for the monuments of the classical era,[14] but there were many, to the contrary, from the Hellenistic and Byzantine eras: the theater of Salonica [the modern Thessaloniki], the theater in Nicea, the aqueduct of Ephesus, the Hagia Sophia (all the stones of the decorative exterior surface are "signed"), and so forth.[15]

For medieval buildings, to the contrary, where these masons' marks or "sigils" are much more numerous and also crafted with such a precise and mysterious geometry (whose mathematical and logical key the Austrian architect Franz Rziha appears to have discovered at the cost of an immense labor), we know from the ancient documents of the stone-carver lodges that each journeyman at his reception into this second grade of the hierarchy received a "mark" that belonged to him for life (unless he proved undeserving of it). This was his signature on the important pieces (such as keystones) for which he was responsible, his sign of recognition, his password in his travels and contacts with members of his lodge or affiliated lodges (in these instances he would have "to place" and "read" his mark, which means provide the geometrical construction and symbolic meaning).

Studies of the *Bauhütte,* a federation with secret rites formed from an autonomous association of all the stone carvers of the Germanic parts of the Holy Roman Empire (including the lodges of Switzerland and other neighboring countries with a shared language or Germanic tradition) that lasted until the end of the seventeenth century, have allowed us to reconstruct the history of these lodges. In the Holy Roman Empire, as in France and England, they were continuations of

---

14. Since I wrote this, Paul Le Cour, editor of the magazine *Atlantis,* has shown me photographs of signs that he took of the stones of Erechteion (the double Cretan axe) and a temple of Eleusis (circle with eight radii).

15. Choisy, *L'Art de bâtir.*

the "colleges" of builders that after the fall of the Roman Empire continued in tandem with Roman municipal institutions, with southern France and the Rhineland as conservation then propagation centers, when the era of constructing great religious buildings began during the Carolingian period (we can find confirmation of this in medieval documents dealing with English "masonic" guilds).

During the Carolingian and the Romanesque era (that is, between the eighth and ninth centuries), the blossoming of religious architecture, in which the construction of the magnificent Benedictine abbeys played a major role, first brought together around these abbeys specifically the workshops or lodges of masons and stone carvers that went on to become veritable schools of architecture directed by Benedictine monks. It was the disciples of Saint Benedict who, in fact, at Saint Gall, Monte Cassino, and so forth, not only conserved or rediscovered the mathematical texts of Greek and Alexandrian antiquity that have come down to us, as well as Vitruvius's architectural treatise (at Monte Cassino Abbey), but most specifically transmitted the Pythagorean mysticism of numbers—through the lineage of Nicomachus of Gerasa, Martianus Capella (fifth century), Boethius and his friend Cassiodorus (sixth century), Isidore of Seville (beginning of the seventh century), the pope Sylvester II (tenth century)—and the geometry of Platonic solids and their harmonic correlations (Walter of Speyer, Campanus of Novara).

These monk-architects, their master masons and journeymen stone carvers, also revived the ancient tradition; in addition to long apprenticeship journeys and individual pilgrimages (replacing journeys to Eleusis, Delphi, and other initiatory centers of antiquity), there was the movement of work crews and whole workshops of builders. The slow reconquest of Spain from the Arabs taking place during this same period (Toledo was won back in 1088) led to new contacts, technical this time, with the traditions and procedures of Hellenistic and Byzantine architecture (especially that of the Egyptian and Syrian regions), contacts that were formed by the constant exchanges in Spain with Arabic architects and foremen who brought architectural solutions and forms (among others, the Gothic arch) that had evolved in the eastern half of the Mediterranean basin under the threefold influence of Greece, Iran, and Egypt.

During this same period, the Crusades were creating another zone of direct contact between Western builders—both ecclesiastics and laymen—and the architects, procedures, and corporative centers of the eastern Hellenistic tradition, with its Iranian-Arabic and Coptic branches, on the very soil of Syria, on the Phoenician coast from where the king of Tyr once sent to Solomon the master Hiram with his crew of masons and carpenters, and on the very parvis of what was the temple of Jerusalem (1099).

This was when Western masons and architects, while retaining close ties with the Christian church and holding on to their devout attachment to its God and saints who during this quasi-monastic period had definitively replaced the gods or tutelary spirits of the *collegia* of antiquity, regrouped into semisecret societies that were purely secular[16] and formed in the Holy Roman Empire the powerful *Bauhütte,* a federation of stone-carver lodges, attached to the four great lodges of Strasbourg, Cologne, Vienna, and Bern,[17] with that of Strasbourg acknowledged as the Supreme Grand Lodge, where the master builder (or *Meister vom Stuhle*) of the Strasbourg cathedral exercised authority over all the local lodges that were dependencies of the aforementioned four great lodges (*Haupthütten*).

---

16. Certain documents of the *Deutsche Bauhütte* specifically mention the "Brotherhood of Saint Guy (Veit)" existing in 1088 in Corvey as the secular transformation of the monastic architecture school that had been a dependency of the old Benedictine abbey of this name. All the great Benedictine abbeys (Cluny, Saint Gall, and so on) founded similar schools, but we do not know if during this Carolingian monastic era any guilds of lay builders had also survived in the West. It should be noted in this regard that in the questionnaire for the apprentice candidates to the *Bauhütte*'s second grade, this question appears: "What was the first 'honorable' building" (which means was built in conformance with the rites of the *Bauhütte*), and the correct answer was the cathedral of Magdeburg, built in 876 during the reign of Charles II. [Construction on the existing cathedral of Magdeburg was started in 1209—*Trans.*]

17. Replaced in the sixteenth century by that of Zurich.

The first supreme grand master (*Obermeister*) of all the lodges of the Holy Roman Empire was, according to tradition, Master Ervin von Steinbach,[18] architect of the cathedral of Strasbourg. The supremacy of the lodge of Strasbourg had been recognized earlier in 1275 by Rudolph of Hapsburg, then in 1278 by Pope Nicholas II.

It was the privileges acknowledged or granted to these lay guilds of builders by emperors and popes that earned their members the name of *Freie Maurer,* or "free masons."

The first modern work on the history and traditions of the German *Bauhütte,* especially with regard to the Grand Lodge of Strasbourg, is that by Abbé Grandidier, canon of the cathedral, who published his *Essai historique et topographique sur la cathédrale de Strasbourg* (Historical and Topographical Essay on Strasbourg Cathedral) in 1782. His book is valuable because it was written before the interruption of corporative traditions and the dispersal of numerous documents by the French Revolution.

But the fundamental work on this subject is the one by Rziha (*Studien über Steinmetz-Zeichen* [Studies on the Stone Carvers' Symbols], Vienna, 1893). Rziha began to take an interest in the *Bauhütte* by way of the masons' marks that appeared in such infinite variety on the important buildings of the Romanesque and Gothic eras. I noted earlier that these mysterious symbols, which are obviously the continuation of the similar marks carved on the stones of antiquity and Byzantium, are much more complicated than these latter, generally (but not always[19]) simple initials or monograms. In the Byzantine lapidary seals (in the Hagia Sophia, for example), the monogram, in accordance with Byzantine style that gathers together all the letters of the name, already displays a fairly complicated geometrical design. In the masons' marks, seals, or sigils of the Romanesque and Gothic eras, on the other hand, the monogram has vanished entirely and been replaced by a purely

---

18. Tradition attributes to his daughter Sabina the magnificent sculptures *Church* and *Synagogue* in the cathedral's south portal.

19. Purely geometrical designs are also found on the stones of the Palatine Hill.

geometrical design, one that is sometimes quite simple and sometimes quite complicated.

If they are a delight to behold, it is because beneath their apparent whimsy there is always a very skillfully concealed geometric composition. Rziha has recorded some nine thousand of these masons' marks throughout Europe (one thousand of which are reproduced in his book), and has found, after years of patient research, the key to their geometry. All these signs in fact exhibit the impression of an "organic" geometrical kinship; in particular, each of them has a center of symmetry. They are all, without exception, composed of fragments (generally joined together) from one of four "matrix" types. Each matrix is a geometric diagram corresponding (for the masons' marks conferred by the lodges affiliated with the *Bauhütte*) to one of the four grand lodges. Here are the specifications of these grids, each of which is inscribed inside a directing circle:

1. *Quadrature.* Obtained by the orthogonal and oblique divisions (by means of diagonals) of two squares superimposed at 45° (forming a pseudo-octagonal star) inscribed inside the directing circle. Each square is divided into sixty-four little squares whose diagonals also form part of the complete matrix.[20] It is the center of the directing circle that is naturally the center of symmetry

20. The partition of each square is made by means of the following independent operations:

a) Division of the grand square into 4, 16, or 64 small even squares (by the division of each side of the principal square into 2, 4, or 8 parts);

b) Drawing of a square inscribed inside the principal square by connecting the middle of its sides, then the inscription of a third square inside the second using the same procedure;

c) One can direct all the diagonals of the squares obtained by *a* and *b* (many of these diagonals already are the result of the drawings *a* and *b,* especially from the following operation *d;*

d) The grid obtained by *a, b, c,* is moved by a 45° rotation around the center of the circle, and it is the superimposition of the two drawings that produces the final grid or matrix type.

(which we discussed earlier) in this matrix, as it is in the three others.

The masons' marks of the "quadrature" type were given exclusively to the stone carvers who had been initiated as masters in the territories affiliated with the Supreme Grand Lodge of Strasbourg and the secondary lodges directly connected to it.

2. *Triangulation.* Obtained by divisions in a triangular grid of two equilateral triangles superimposed head to tail (in the form of a hexagram), inscribed inside the directing circle.[21]

   A very rare variant of this triangular grid, and which encircles it, is obtained by rotating a second grid 20°, twice in succession, around the center of the directing circle. Rziha only found one example of the application of this second grid. It was in the mason's mark of the master of the pulpit of Saint Stephen's in Vienna. G. Bals has found another example in a Moldavian church (Ruseni).

   The masons' marks of the "triangulation" type were exclusively conferred by the lodges that were under the authority of the Grand Lodge of Cologne.

3. *Quatrefoil* (or four-lobed rose, "*Vierpass*"). All the circumscribed circles and inner squares resulting from the various division operations are dawn over matrix 1 (quadrature). Circles with the diameter R, $\frac{R}{2}$ and $\frac{R}{4}$ (R being the radius of the directing circle circumscribing the principal square) are sufficient to obtain this matrix.

   The masons' marks derived from the quatrefoil type were

21. The division of the equilateral triangle is done by the following independent operations:

a) Division of the principal triangle into nine inner triangles (by dividing each side into three);

b) Drawing of a triangle inscribed inside the principal triangle by connecting the middle of its sides, then the inscription of a third triangle inside the second using the same procedure.

exclusively conferred by the lodges affiliated with the Grand Lodge of Vienna.

4. *Three-lobed rose* (or trefoil, "*Dreipass*"). Circles circumscribing all the inner triangles are drawn on matrix 2 (triangulation).

   The masons' marks derived from this matrix were conferred by the Grand Lodge of Bern (but the rule was not exclusive as in the three previous cases) and by the Lodge of Bohemia (Prague), which had acquired in the fifteenth century autonomy comparable to that of the four grand lodges.

The examples in plates 50 and 51, on which the matrices have been drawn in fine lines, illustrate the above explanations.

It will be noted that it is not always necessary to draw the complete grid; for example, for the "quadrature" (signs derived from matrix 1), as for the "quatrefoil," the 45° rotation (with superimposition of the two diagrams) sometimes does not come into play. It is also fairly common for the division into 4 or 16 small squares (instead of 64) to be sufficient.

An extremely interesting observation made by Rziha is that not only the Romanesque and Gothic masons' marks, which are so numerous[22] (finding them and placing them in a grid is a very interesting sport), but also the ancient marks (except for those that are simple nongeometrical

---

22. Rziha notes that on the older Romanesque constructions built by monastic workers no marks are found. They only reappear around the beginning of the eleventh century, the date when the lodges became secular again. The great Gothic cathedrals (Regensburg, Vienna, Prague, Strasbourg, for example) contain many masons' marks. The keystones and capitals are almost all signed in this way. The *Wandergesellen* or journeymen making the "three journeys" necessary for initiation to the grade of master thereby left evidence of their passage. We sometimes find (in Regensburg, for example) group stones, *Sammelsteine,* on which are carved a large number of journeymen and masters' names.

Among the *Bauhütte* blueprints for Saint Stephen's Cathedral that can be found in the archives of the Viennese Academy of Fine Arts library, there are two plans of vaults (the ribs are projected horizontally) on which, next to each keystone is marked in red pencil a different "mark." They are the marks of the masters or journeymen responsible for carving the corresponding stones.

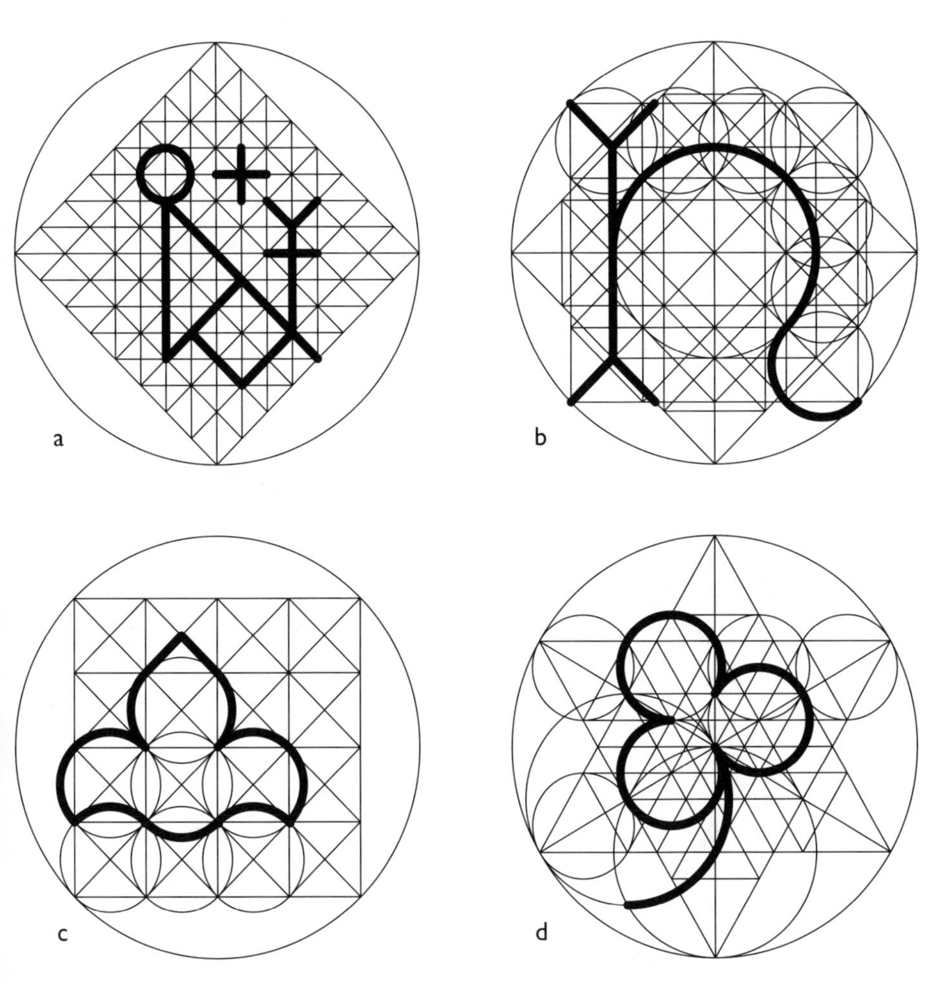

Plate 50. Masons' marks: (*a*) Hagia Sophia (Byzantine); (*b*) Saint Stephen's Cathedral of Vienna (Gothic); (*c*) Barbarossa Tower at Gelnhausen (Romanesque); (*d*) Hradschin cathedral in Prague (Gothic) (from Rziha)

Plate 51. Gothic masons' marks (from Rziha)

initials or monograms) and the Byzantine marks (even the complicated monograms of Hagia Sophia) can all be placed in one of the four matrix types. Rziha also notes that the masons' marks from the Romanesque era, independent of their provenance, most often are derivatives of grids 3 and 4—those that contain arcs of a circle—and sees in this distinctive feature a reflection of the Roman arch. The Byzantine marks are also in the key of the "quatrefoils" (group 3) and in this way exhibit an affinity with Romanesque marks.

Among the Romanesque masons' marks, it is necessary to highlight, first, a carved stone from the Naples Museum that has greatly excited the curiosity of archaeologists; it contains four geometric signs, each placed in one quarter of a swastika. One of them appears to be the isosceles triangle of the pentalpha (the top corner is 36°, the long side and base are in the ratio of the golden section) decomposed into a similar small triangle and its gnomon.

And second, the mason's mark on a flat marble surface from ancient Rome, the Capitoline Hill, which is a classic seal of the triangular type: a hexagram with all the radii leading from the center to the six vertices of the small hexagonal nucleus.

The diagrams represented by these grids offer this interesting feature in that they also graphically contain what the stone carvers called the "fundamental grid" or the base (*Steinmetzgrund*), which is to say, the majority of geometrical designs used for, among other things, establishing the horizontal outlines of the capitals, steeples, bell towers, for the drawing of Gothic rose windows, and so forth. If we add a fifth diagram to these four "lapidary" grids, one in which the directing polygons are two pentagrams (the vertices determining those of the regular decagon inscribed inside the directing circle), we obtain all the designs used by Gothic architects not only for the details and the rose windows but for the overall compositions, the synthesizing establishment of proportions of the buildings in such a way that they seem to emerge from the analyses and "canons" presented in part 1, especially that of Mössel (*Kreisteilung,* the polar segmentation of the circle). It is the fifth pentagrammatic drawing-grid that, to the best of my knowledge, is never revealed in any mason's mark, strictly speaking (since we have seen they are all of square or triangular symmetry[23]), that perhaps constitutes the geometrical "major arcana" as such of the Gothic builders, accessible only by the "masters" (second degree of initiation or third if we count the novices). This would explain the lack of any pentagonal symmetry in the lapidary signs that the journeymen (first degree of initiation) had to know how to "prove," that is, to explain geometrically.

It is this geometry of the circle and the inscribed regular polygons that *Bauhütte* documents call "the very noble and very straight fundamental grid of the stone carvers."[24] To "prove" or "place" (*stellen*) his mark was to place that sign precisely in the directing circle, in conformity with an old *Bauhütte* saying, whose meaning is thereby made quite clear:

> A point that goes in the circle,
> And that sits in the square and the triangle:
> If you find the point, you are saved,
> Pulled from torment, anguish, danger.[25]

This point, this pole whose marking is so important, is, in the reading of the "marks" as in that of the "plans," the hidden center of symmetry, that of the directing circle. It is still the idea already embryonically present in the passage of the *Philebus* that struck Mössel so forcefully (see part 1) and that is found in Vitruvius, when, in comparing the sym-

23. The three pentagrammatic marks noted by Rziha in Diocletian's palace, in Vienna's Saint Stephen's, and in Regenburg cathedrals, share this oddity in that they are pseudo-pentagrams obtained by removing a vertex from a hexagram and linking the other five together.

24. "*Den fürnembsten und gerechten Steinmetzgrund.*"

25. *Ein punkt der in dem Zirkel geht,*
Der im Quadrat und Dreyangel steht,
Trefft ihr den Punkt, so habt ihr gar
Und kommt aus Noth, Angst und Gefahr.

(For a variation of this saying, see chapter 3.)

metry of a perfectly proportioned temple to that of the human body, he notes that the navel is the human's center of symmetry,[26] and when he provides the design types for theaters in general (four equilateral triangles inscribed inside a circle) and Greek theaters in particular (here the twelve points of the circle are provided by three inscribed squares), which we lastly find in a phrase in the Como edition of Vitruvius (printed by Gotardo da Ponte) analyzing the façade-elevation of the plan for Milan cathedral[27] (see plate 38). It is important to note that the construction of this cathedral took from the thirteenth to the nineteenth century, and that the commentator Caesar Caeseriano, himself an architect, who brought out this magnificent edition of Vitruvius (dated 1521) had at his disposal the cathedral archives of the builders who, at the height of the Renaissance, were working on the Gothic plan in a Gothic spirit. Moreover, this plan proved to be—thanks to the uninterrupted transmission by the builders' guilds and the lay and religious architects of the procedures for establishing proportion that were honored in antiquity along with Platonic harmonic ideology—in complete conformity with the precepts and terminology of Vitruvius, whose work had just recently been rediscovered.[28]

---

26. "*Similiter vero sacrarum aedium membra ad universam totius magnitudnis summam ex partibus singulis convenientissimum debent habere commensuum responsum. Item corporis centrum medium naturaliter est umbilicus.*"

27. I will quote the entire passage despite its length: "*Idea geometricae architectonicae ab iconographia sumpta ut peramussineas possint per ortographiam ac scoenographiam perdurcere omnes quascumquae lineas non solum ad circini centrum sed quae a trigono et quadrato aut alio quonismodo perveniunt possint suum habere respunsum tum per eurythmiam proportionatam quantum etiam per symmetriae quantitatem ordinariam ac per operis decorationem ostendere uti etiam haec quae a germanico more perveniunt distribuentur pene quedmadmodum sacra cathedralis aedes Mediolani patet.*"

28. The same idea appears in the very final verses of the *Divine Comedy,* when Dante, contemplating the center of the "eternal rose," strives, "like the geometer," to understand "how the image unites with the circle and finds its place there."

> *Veder voleva si convenne*
> *L'imago al cerchio vi s'indova.*
> (*Paradiso,* canto 33)

Rziha, moreover, wrote in 1883—thus well before Lund and Mössel established their syntheses—a perfect summary of the idea that he drew from his analysis of thousands of these marks: "All understanding of masons' marks rests on this knowledge of the 'fundamental grid' (*Steinmetzgrund*), or in short: on the realization of the utility of geometric designs for construction."

Once more, the mason's mark that had been conferred on the journeyman—or rather his "proof"—helped, along with other ritual signs of recognition,[29] to validate his legitimacy during his years of wandering, especially during the "three journeys" that, as a *Wandergeselle,* were required by guild rule.

The oldest official mention of the *Bauhütte*'s masons' marks can be found in the *Hüttenordnung* (lodge ordinance) of Röchlitz of 1462 (the original of which still exists), which refers to the congress (*Hohe Morgensprache*) of Torgau. It speaks of the masons' marks (*Zeichen*) conferred by the master to the journeymen after their internship and required during a solemn ceremony. This document also shows, as does the ordinance of Basel of 1563 (another official document that mentions them), that these signs are strictly personal and that the journeyman does not have the right to change them or transfer them to another person.[30]

We will now leave these masons' marks to revisit the overall history

29. There were ritual forms of address, of dress (the last three buttons buttoned, and so on), of looking, standing (feet at a right angle), walking (by three-step measures), knocking, greeting, picking up a cup, drinking, carrying the cane (*Wanderstock*), holding one's arms, taking the hand of a brother (*Griff*). All of these ceremonial attributes appear in French *compagnonnage* about which we will speak later.

30. In the dispute between the lodges of Magdeburg and Annaberg, the latter lodge had rebelled and reduced the required years of apprenticeship from five to four. The master of Annaberg, Jacob von Schweinfurt, was threatened, following a congress of masters that met in Halle and was chaired by the supreme grand master (Hans Hammer) of Strasbourg, to have his mark carved on the "table of felony" (*Schelmentafel*). This punishment was actually inflicted in 1718 by the Grand Lodge of Strasbourg (when the supreme master was Michael Erlacher) on two master stone carvers and a journeyman of Regensburg and Kelheim. We should note that at this time, Strasbourg had been a French city for more than twenty years.

of the *Bauhütte* and the authentic documents that mention it. The oldest, found in Trier by a Dr. Reichensperger, is the order (*Hüttenordnung*) of October 22, 1397. It begins as follows:

"*Hic incipiunt constitutiones artis Geometriae . . .*" We can see that geometry for the master stone carvers and master masons of the Middle Ages was clearly the fundamental science, and that the "*Ars sine scientia nihil,*" spoken in Milan by the master architect Jean Mignot of Paris was not an isolated quip. This document already mentions the "Quatuor Coronati"[31] as patron saints of the fraternities of stone carvers and masons. The other legal records that mention the *Bauhütte* found within the borders of the Germanic parts of the Holy Roman Empire are the three Viennese documents of 1412, 1430, and 1435, then the major order of Strasbourg of 1459, in which mention is found of the division of all the lodges into four zones (*Hüttengaue*) under the suzerainty of the Supreme Grand Lodge of Strasbourg (with Josse Dotzinger of Worms as master builder), and a certain number of orders and congress minutes dated from Regensburg (1514, 1555, 1559, 1616, and so on). The *Bauhütte*'s privileges were confirmed by the majority of the emperors, from Maximilian[32] to Charles VI (1713).

31. The "Quatour Coronati" are the saints Castorius, Symphorian, Claudius, and Nicostratus, architects or master masons who were martyred during the rule of Diocletian. The medieval builders' guilds claimed them as honorary brothers. They are honored in Rome itself (by construction workers) in the Santi Quatro Coronati church. They are depicted in Orsanmichele church of Florence, and on the tomb of the master stone carver Tenk (died 1513) in Steyer, and so on. We should note that a large number of masons' marks have been found on Diocletian's palace in Split.

32. Like many other guild traditions, the one that records the affiliation of this emperor to the *Bauhütte* can be verified. The curious *Theuerdank* (printed in Nuremberg in 1517), which traces the life of Maximilian under the allegorical name of *Weiss König* (White King) shows us "how the young king learned the art of building with stone." He had to legitimize his initiation like any other journeyman by answering the master's questions, for example, "What are the three fundamental principles of the work?" The young king answered: "Pleasure, Necessity, and Force," from which the master builder understood that the candidate had learned and clearly grasped the bases (*Grundt*) of the art (the symbology of the aforementioned *Steinmetzgrund*). In the ritual of the *Bauhütte* we find that on November 8 (feast day of the "Four Crowned Martyrs")

The decline of the *Bauhütte* in Germany was gradually brought about by the disappearance of the Gothic style, the influence of the Italian Renaissance (which marked not the end of Platonic influence, but the end of "Hermeticism" in architecture), the Reformation (the separation of the lodges into Catholic and Protestant), and the Thirty-Years War. The Reichstag of August 12, 1671, took away the suzerainty of the Strasbourg Supreme Grand Lodge, but restored it in 1707 and confirmed it again in 1727 and 1731 (we can see that the incorporation of this city into France played no role from this perspective). In 1771 (July 15) the Reichstag eliminated all the actual privileges of the *Bauhütte,* but it continued to exist formally. In 1883 it was still possible to count around one hundred *Hüttenbrüder* (lodge brothers) striving to maintain the traditions and rituals of the once powerful *Bauhütte.*[33]

Several of the original documents mentioned above confirm the traditional division of the stone-carver "brothers" into apprentices, journeymen, and masters. Journeymen themselves were divided into journeymen in the strict sense (*Gesellen*) and *Parlierer* ("talkers"). These documents also sometimes mention, without providing any details, not only masons' marks but the other means of recognition—passwords, gestures, and so forth, listed above—as well as other secrets or *Heimlichkeiten* (the most important of which was the *Grundt* or the art of the drawings) that were handed down through the generations, in parallel with those of their brothers, the masters of the stone-carver lodges and freemasons in France, England, and so forth, (the term *Freie*

(cont. from p. 273) the brothers honored "the three pillars of the lodge": Beauty, Science, and Strength. A variation discovered by Rziha in an ancient legitimization formula calls them "Verity, Science, Force."

33. Masons' marks did not die with Gothic architecture. They can be found on monuments of the German Renaissance and baroque era. The forms they take are equally "baroque" but still obey the control of the "placement in a grid." The marks found in the stucco and mortar constructions of this era prove the affiliation of stone-carver lodges with those of masons in the strict sense. Even in the contemporary era, the use of "masons' marks," although rare, has not completely ceased.

Maurer had been bestowed on the members of these privileged guilds as early as the fifteenth century).

I have gone on at length regarding German corporative masonry because of the fairly complete documentation we possess concerning it. But the parallel to this perpetuation of the Roman-era builders' colleges, their collaboration or temporary merger with monastic worksites and workshops (of the eighth and ninth centuries), then the resumption of their autonomy and secular rituals, can be found in France and in England as well. Despite the ruination and material despoliation of Italy and Gaul and the Roman Western world in general as a result of the invasions and pillaging of the fifth and sixth centuries, documented proof of the persistence of craft organizations survived. In seventh-century Naples, in Comacchio, in eighth-century Ravenna, then in Saint-Omer, Corbie, Soissons, and so forth, we can find trade groups, including masons. They sometimes formed, like the ancient *collegia* or *corpora opificum* from which they descended, "societies of mutual aid or fraternities of a religious nature (*confratriae, gildes, geldoniae*), which the authorities banned when they took the form of unions organized on the faith of the sworn oath."[34] We have proofs of the initial hostility of the Church against these first *jurandes;* for example, the Synod of Rouen declared in 1189:

> There are clerics and lay persons who form associations in order to assist each other mutually in all kinds of affairs, especially in negotiations and in filing suits against those who oppose their statutes. The Holy Church has a horror of such associations of brotherhoods, because it exposes their members to perjury. Consequently, we forbid, on penalty of excommunication, that anyone form similar associations, or maintain those that have already been formed.

34. P. Boissonade, *Le Travail dans l'Europe chrétienne au Moyen Age* (Paris: Félix Alcan, 1921).

Plate 52. Gothic symphony (Burgos cathedral)
(photo L. L.)

Following the coordination by royal authority and the provost of the guilds of Paris to enact the statutes of the guilds recognized as official trade bodies (Étienne Boileau's *Livre des métiers,* 1268), also later called sworn trades, then *maîtrises, jurandes* [organizations with sworn members and the grade of master], and so on, the Church's hostility ceased, and the trade guilds and fraternities placed themselves under the protection of saints. But the "lay" rites were not abandoned. The trade hierarchy included three grades: apprentices, journeymen, and masters. The initiations of journeymen into the status of master were solemnized by a banquet, and the masons and stone carvers paid homage to an "initiatory" symbolism similar to the one we saw in the *Bauhütte* of the Holy Roman Empire. However, the fact that the master mason of the king, through his position, was the master of all the masons, mortar layers, and plasterers of Paris prevented the building guilds in France from acquiring an autonomy and power comparable to that of the *Bauhütte.* But from the perspective of what could be called technical esotericism (transmission of the "fundamental" geometry and other trade secrets), there is not only a community with shared roots and trajectory, but, thanks to the traditional long journeys especially in favor among the craftsmen of the building trade, exchange was possible between architects and masters in all the construction sites of Gothic Europe, and continuous and personal contact existed between all the lodges and workshops of Europe. Masons' marks can be found on the French cathedrals as well as on many castles and other civil buildings.[35]

35. Didron (*Annales Archéologiques* III, 1845, and II, 1858) reports the existence of masons' marks in the chateau of Coucy, in the palace of the popes in Avignon, the chateau of Vincennes, and in the Palais de Justice [law court] of Paris. He found 242 varieties in Strasbourg and 237 on the ramparts of Aigues-Mortes (in an hour and a half). Klotz, grandson of the twenty-third and last grand master of the Strasbourg cathedral workshop, reproduced (*Annales Archéologiques* V, 1846) the 28 masons' marks of the architects who met in counsel in Strasbourg on December 26, 1658, under the leadership of Andreas Schmidt, master builder in Colmar. Abbé Laran recorded masons' marks in England, Belgium, Switzerland, and Spain (in the Burgos cathedral and in the Monastery de las Huelgas, in particular).

They are like their Holy Roman Empire contemporaries and obviously derive from the same grid types. If, thanks to its centralized supervision by the crown, the admission rites and internal functioning the official French guild did not weave around its ceremonies the veil of mystery that was dear to the *Bauhütte,* France did see the formation, precisely in reaction to the meddling of the central power and the constant support it lent its managerial element (that is, to the "masters" and responsible officials), of an institution with a clearly "occult" nature that existed during the entire ancien régime outside the guild strictly speaking: the *compagnonnage.* This tradition, which even survived the abolition of the guilds and professions by the French Revolution, was given center stage by the attempt in 1848 to form a federation based on its principles by Agricol Perdriguier (journeyman carpenter "of the Devoir de Liberté," known as "Avignonnais-la-Vertu," which gathered together 10,000 journeymen on the Place des Vosges on April 10, 1848), and by the *Compagnon du tour de France* by George Sand. It has persisted into the present, outside of workers unions in the strict sense and the C.G.T. [*Conféderation générale du travail,* General Confederation of Labor] and has experienced a kind of renewal since the war [WW I].[36]

(cont. from p. 277) Didron and Klotz make the distinction between masons' marks as such, which are individual signatures, and stone-setting marks, which make it possible to place the stones at the desired story and location. These have been found in Reims and Cologne (in the form of soleplates, keys to a number of notches that varied depending on the height of the foundation).

36. We will revisit these precursors to the workers' movement in the next chapter, in which I will discuss the political activity of secret societies at greater length. Let me just say for now that what essentially distinguishes the "trade guild" spirit, despite its clearly pro-worker, democratic, and rebellious nature, from the "trade union" spirit as such, is, in addition to faithful observance of the rites and symbols whose origin is lost in the depths of time (the cliché is a reality in this case), pride not only in work and in the quality of the work achieved, but in the perfection of the work, of the "masterpiece" executed by the journeyman or the *cayenne* (a local group, which is a *compagnonnage*-inspired equivalent of the lodge).

We can cite as an example of a masterpiece the extraordinary lock (in the shape of the Legion of Honor medal) that can be seen at the Borély Museum in Marseilles, carved from the massive iron of an anchor by the master locksmith Ange Bonin (aka "Ange

This persistence and the renewed curiosity that had been triggered earlier in the Romantic era for this "chivalry of the people" (the expression is George Sand's) allows us to find in what the profane may know of the rites of the *compagnonnage* (still kept as secret as those that were once held by the Pythagorean brotherhood, and those still held by Freemasonry, and this double rapprochement is no accident), all the initiatory ceremonial attributes—passwords, feet at a 90° angle to each other, a defined number of steps in this or that direction, gestures, greetings (*topage*), all the way up to the major importance of the cane (adorned with ribbons), the insignia of the journeyman that I earlier mentioned was an attribute of the *Wandergeselle* in the great Germanic *Bauhütte.*

It is a curious thing that, although the *compagnonnage* does not only embrace masons or stone carvers, but the bulk of the trades represented in the regular guilds, it is the rites, symbols, and legends of the stone carvers, masons, and carpenters that were impressed on all categories of journeymen. This is similar to how, in the Middle Ages, the lapidary signs of the building artisans were often adopted as "marks," especially in central Europe, by other guilds, goldsmiths, illustrators (sculptors), metal casters, and so forth.[37]

There is another country in which guild life developed, as in the

---

de le Dauphiné) in response to a challenge. This was a professional duel intended to determine which of the two fraternities of rival rites, Compagnons serruriers du Devoir de Liberté, Children of Solomon, or "Gavots," and Compagnons serruriers du Devoir, Children of Master Jacques, or "Devoirants" [*gavot* can be taken to mean peasant; literally it means an inhabitant of the rural commune of Gap, and *devoirant* in this context would mean dutiful, as in obedience to the organization commonly known as a *devoir,* or "duty"—*Trans.*], would hold all rights to the city of Marseilles. What this means is that the members of the defeated rite would have "to travel the Tour de France without stop and not establish a headquarters in the city until a hundred years had passed" (cf. the organization's journal, *Les Muses du Tour de France*, published by the Fédération Compagnonnique de la Seine).

37. Many sixteenth-century German printers and engravers had symbolic marks of the same type as the stone carvers' masons' marks. These include several painters; I have found them on the paintings of Joos van Cleve, Frans Pourbus, and Bartholomäus Bruyn. A sixteenth-century engraving depicts a compass with the letter *G* between its points; the same symbol can be found on a seventeenth-century Italian engraving.

Germanic lands, in an initiatory fashion and where, thanks to an innate traditionalism paradoxically combined with an extremely progressive individualism, we have had the good fortune to find a series of recently discovered original documents that nicely top off what France and the Holy Roman Empire have given us on the life and origins of builders' guilds. It is England.

The oldest traditions connected to English masonic guilds (and in this chapter the word *masonic* applies to documents, rites, and so on, concerning actual masons or stone carvers, which is to say it concerns "operative" masonry or freemasonry, like that of the *Bauhütte,* and not at all the "speculative Freemasonry" we will look at in the next chapter) mention, in a document written in 1475 during the reign of Edward IV, the importation by the Anglo-Saxon king Athelstan (925–940) of a certain number of French master masons to whom he entrusted supervision of the work sites of the numerous important buildings constructed during his reign. As evidence of the affection that the makeup of this artisan guild and the honorable principles on which it was founded inspired in him, the king is said to have granted them a free charter.

This tradition probably contains a basis of truth; French influence is evident if not under the Saxon kings, at least during the beginnings of the Norman Dynasty, as shown by the very word *masons* that has been applied ever since in England to those who work on buildings.

But let's take a look, as we did for the *Bauhütte,* at the authentic documents that still exist.

1. The Ordinance of York (in Latin) of 1352 was issued by the chapter of the cathedral of York and was addressed to the workers employed in the workshops of the church. It commanded that "the ancient customs in use among the craftsmen of the building must continue to be respected." It also stipulated that the masters were expected to "swear before the chapter that they would ensure their strict application."

   A second ordinance in 1409 (also in Latin) confirmed the

earlier regulations and provides more explicit instructions concerning the discipline; no one is allowed to enter the "lodge" of the works without permission of the canons and the master; the master mason, the supervisors, and the former masons shall swear an oath of loyalty and diligence, and so forth.[38]

2. The Regius Manuscript or "Masonic Poem" from the British Museum, which dates from the end of the fourteenth century and consists of 794 verses. It opens with the epigraph: "*Hic incipiunt constituciones artis gemetriae secundum Euclyde,*"[39] and comprises nine parts. I will cite or summarize those passages that are relevant to the subject at hand.

   Masonry is the art derived from geometry, and is the noblest of the arts. Learned masters taught it to the sons of distinguished families. The apprenticeship of this art, among folk of this stamp, is done together, and those who study it called each other by the name of "companion" or even "dear brother," while they reserved the name of "master" for their teacher.

   There is mention of the statutes in 15 articles and 15 points given to the masons by King Athelstan, and a list of the 14 articles (no apprentice is accepted unless he commits to at least seven years of apprenticeship; no serf can be accepted as an apprentice, but only young men of higher position; the masters will maintain fraternal relations with each other). In addition, there is the list of 15 points (the apprentice must *faithfully observe the secrecy*[40] of

38. Here we can cite, in addition to the mention of the guides for London masons in the "Articles of London," crafted in 1356 by the lord mayor (for work regulations; no apprentice can be accepted for a period of less than seven years, and so on), the ordinance of the Norwich carpenters' guild, the members and wives of the affiliated families would be known as "brothers" and "sisters."

39. We find the exact beginning of the aforementioned Trier document from 1397.

40. The prevetyse of the chamber telle he no man,
Ny yn the logge whatsever they done;
Whatsever thou heryst or syste hem do,
Telle hyt no mon, whersever thou go!

his master and his journeymen; he must never talk of what takes place at the lodge or in the private home; a skilled mason, who sees one of his journeymen about to commit a professional mistake, should give him the necessary instructions for the work to be performed safely; and the composition of the general assemblies that judge the professional and social mistakes of the journeymen is detailed).

The fourth part, under the title *Alia ordinacio artis gemetriae,* deals specifically with the general assembly (we can see that the term *geometry* symbolically applied to the functioning of the entire masonic organization).

Next comes the legend of the "Quatour Coronati" and discussion of the honor of the guild.

3. The Cooke Manuscript is also housed at the British Museum. It is a copy, dating from 1430–40, of a late thirteenth-century text that is divided into two parts.

   In the first, the author, clearly a member of the mason guild, presents the history of geometry and masonry. Geometry is the principal of all the other sciences, and no work can be achieved that does not find its reason and cause in geometry. Similarly, masonry is the most important of all the arts, because it is nothing other than the application of geometry.

   Before the Flood, the three brothers Jabal, Jubal, and Tubalcain (the eldest is the inventor of geometry and masonry, the second, music, and the third, the art of the blacksmiths[41])

41. Robert Eisler's research on the smiths and miners of the Sinai (*Die Kenitschen Weihinschriften*) offers a curious reason for the biblical tradition that connects the smiths to Tubalcain. Since the time of the Third Dynasty, these smiths and miners had been working in the copper and turquoise mines of the Sinai on behalf of the pharaohs, under the direction of Egyptian officials. Thanks to the decryption of half-Semitic–half-hieroglyphic inscriptions found on the votive offerings of the temple of Hathor/Astarte (the "Goddess of the Green Stone"), located in Serabit el-Khadim in the heart of this mining district, Eisler was able to demonstrate that in the Semitic lands during this time the name of Kaïnites (Kajn, Kainum, Beni-Kenim) or Kenites was

carved the principles of the arts they had invented on two pillars, one marble and the other brick.

After the Flood, the two pillars, which had resisted the waves, were discovered, one by Pythagoras, the other by Hermes, who communicated their contents to humanity.

One of the most glorious kings of the world, Nimrod, demonstrated the power of the masonic art by building the Tower of Babel, for which he had the services of more than 40,000 masons. Some of these masons next went to the land of Assyria to build Ninevah, and it is from this era that the veritable art of masonry traces its constitution date.

Euclid taught the sons of Egyptian nobility geometry and the art of masonry. The sons of Israel were initiated into the art of masonry during their sojourn in Egypt. During the construction of the temple of Jerusalem, King David granted the masons new statutes, which Solomon confirmed. They have remained almost entirely unchanged through the centuries.

Once the temple of Solomon was finally built, the 80,000 masons that had collaborated on this work spread across the world. This was how masonry was brought to France and many other countries.

In France, Charles II (Charles the Bald, grandson of

---

applied to the semi-nomadic caste of smiths (*kajn* means "smith," or literally, "he who blows"). The Tubal or Tubala (from *tubalu,* "shavings," "filings of copper or metal"), or Tubal-Kainites, whose name even appears in the heart of Arabia, were the members of the smith caste that specialized in the smelting of copper or bronze. The two roots, *tubal* and *kaïn,* have an interesting posterity (*topor,* "axe" in Russian, Hungarian, Romanian, then "cannon," "cane," perhaps even the *cayenne,* the name for the lodge in the French *compagnonnage* system, and so on), and the proven two-thousand-year existence of this caste—both tribe and guild—of smiths and miners at the feet of this "international" sanctuary between Phoenicia, Judea, and Egypt, makes the tenacious yet converging guild and masonic traditions less absurd and improbable. Eisler was the first to highlight the importance of this Sinai sanctuary for blacksmiths, temple both to Hathor and the thunderous Yahweh of the desert tribes. He tried to elucidate its role in the life of Moses (rebellion, exodus, revelation), son-in-law of the high priest of the neighboring sanctuary of Midian.

Charlemagne[42]) was an honorary mason. He gave the masons his full protection and granted them a charter that remains in effect in this country.

A charter was granted to the English masons by the Anglo-Saxon king Athelstan following the intervention of his youngest son, who had taken an interest in geometry; these statutes are still in use (we can see that this tradition concerning Athelstan is well-established in England).

The second part of the manuscript is a book of the duties of masonry (this part is a copy of documents older than the historical summary of the first part). It is a variation of the "Masonic Poem" we recently looked at. There is a division into masons, journeymen, and apprentices (seven years), and articles on which masons were questioned.

It also mentions nine points; among others are these: the mason *will preserve the secrecy*[43] of his fellow masons, both with regard to the lodge and the private home, and with regard to any place where masons could be found. And *he will never betray the masonic art.*

Then there are four rules; one of them states that a rebellious mason or journeyman who refuses to appear before the general assembly to receive judgment will be obliged "to renounce his masonry."

4. The William Watson Manuscript, which was discovered in 1890 in Newcastle upon Tyne. This is a parchment scroll that is twelve feet long and seven and one-half inches wide. The document probably goes back to the second third of the fifteenth century. It opens with the arms of the guild of masons, and like the others is divided into two parts.

   The first part is a historical summary. It is almost identical to

42. Charlemagne invited Byzantine and Italian masons to come work in France.

43. Here is another citation concerning secrecy, found in a somewhat later manuscript: "You shall keep secret ye obscure and intricate pts. of ye science, not disclosing them to any but such as study and use the same." Ms. no. 2, Grand Lodge of London.

that of the Cooke Manuscript, except that the duties of the masons would have been brought already to France by Amphibal, who converted Saint Alban (fourth century) to Christianity. Athelstan's charter was only putting these ancient duties back into effect, as they had fallen out of use; this charter was later revised by Henry VI.

The second part describes general obligations. Again, each mason is expected to faithfully guard *the secrecy of the lodge* and *all the other secrets of masonry.* In addition, (1) masons should call their fellow masons their brothers; (2) no master should take it upon himself to initiate anyone into the guild unless it is with the agreement of six or at least five of his fellow masons; (3) no journeyman should roam through a town at night in which a lodge is located unless he has a fellow journeyman with him who can swear that he was in honorable company; and (4) every master will give foreign masons a warm welcome; if he has no work to give them, he will give them financial assistance to allow them to make their way to the closest lodge.

5. The Tew Manuscript marks the transition between the ancient texts of the "old charges" and the modern forms, and dates from around 1660. Its title is *The Book of Masons* and like the previous ones is divided into two parts. It also mentions the two pillars found again by Pythagoras and Hermes. With respect to the temple of Solomon, it mentions the 80,000 workers, which includes 300 masters directed by the geometer Hiram, son of the Tyrian king of the same name.

   In France, it was during the rule of Charles Martel (died 741) that this manuscript places the initiation of masons by a master from the East named Mamon Grecus.

*
* *

These distinct groups of official documents on the guilds of builders in the Germanic parts of the Holy Roman Empire, France, and England mutually complement one another by both their mentions of actual lodge organization and those of traditions connecting them to the lodges of antiquity.

We shall revisit those aspects of their rites that are obviously reminiscent of similar aspects of the ancient "mysteries"; these influences are completely natural precisely because of the link of the *collegia*. With respect to the symbolic insignia[44] of the masons' guilds (apart from the lapidary signs whose antiquity we have already seen), we have proof that they have been rigorously passed down from the Roman era through the centuries (compass, square, plumb line, arranged in the same way on the sarcophagi of the architects of imperial Rome and on the gravestones of the master stone carvers until the seventeenth century, and so on).

In addition to the quasi-divine roles attributed to geometry in these documents (in those of both English and German provenance), the mention of Pythagoras as having found one of the two "masonic" pillars on which the first geometers had carved their principles before the Flood is highly intriguing.

Two bronze columns are mentioned in the Bible as having been cast by Master Hiram for the temple of Solomon.[45] The thirteenth-century builders of the Würzburg cathedral also placed two pillars inside it on which were carved the same names as those of the Bible: "Jachin" and "Boaz."

TRANSMISSION OF PYTHAGOREAN DIAGRAMS THROUGH MAGIC

In part 1, I spoke of natural magic with respect to incantation in general, and technical or operative magic with respect to love spells or

---

44. "There is sevrall words and signes of a free Mason to be revailed to yu wch, as you will answer before God at the great and terrible day of judgmnt, yu keep secret and not to revaile the same to any in the heares of any p'son . . . so help me God." Appendix to the Harleian Ms. no. 2054, circa 1650.

45. Pacioli's *De Divina Proportione* already contained the assertion that the two twisted bronze columns of Saint Peter's Basilica (they were located before the high altar of the ancient basilica in his day) are precisely the two columns mentioned in the Bible, which were carried off by the Romans after the destruction of the Temple.

incantations and the origins of Gnosticism and the Kabbalah. I recalled there the influence of Pythagorean mysticism on the operative Kabbalah that specifically formed part of the technique of European magic as it evolved into a special discipline grafted onto Gnosticism, Hermeticism, and the Kabbalah, in parallel to its sister, alchemy.

We are once more going to dive into this murky domain on the borders of science and religion, without lingering there any longer than necessary, as it cannot be overlooked in an objective examination of the history of thought, or even history in general. What is involved here is what we could call "Gothic magic," which adopted in its techniques and superstitions many of the Pythagorean catchphrases and methods that were generally received secondhand from Hermeticism and the Kabbalah.

The effect, moreover, of many of the passages by ancient authors informed about Pythagorean rites (Plutarch in particular), and the label of "magus" that was unfailingly appended to that of "Pythagorean" in Roman-era texts, was to convey the impression that magical practices in the strict sense formed part of the ritual and even the daily life of the Pythagorean brotherhood. Our information on this subject is distressingly meager; all we can assume is that the initiates possessed a specialized technique in which musical incantation, the use of various herbs and incenses, and beverages held an important place and sought to produce the same kind of phenomena that are very much on the agenda of today's circles of metaphysical students: clairvoyance, divination, and the summoning of the dead. The regular use of children as mediums emerges from several texts.

The Gothic magic we are now going to examine is a blend of Alexandrian magic, ancient sorcery, the Kabbalah, and residues of this Pythagorean magic. It presents the same relation of feverish caricature and distorted projection as the Kabbalah with regard to Pythagoras and Plato's number mysticism.

It should come as no surprise that the geometrical aspect of the metaphysics of numbers, already introduced by Pythagoreanism into the philosophy of beauty, or aesthetics, was also incorporated into Mediterranean magic. We could even see that this magic distinguishes itself from its sisters of Asia and Africa by its essentially geometrical nature. Here, too, it is the star polygons inscribed within the directing circle—which has become the magic circle—that under the name of "pentacles" appear as the essential symbols and instruments of conjuration. The very word *pentacles* brings to mind the pentagram of love and harmony, the essential "sign" of the Pythagorean brotherhood. It is, in fact, the pentagram or star pentagon that stands out from the beginning as the preeminent pentacle, a diagram that conjures both good and evil spirits and compels their obedience, that confers to the individual who knows how to use it power over the world of elementals, truly even over the higher spirits. This figure, when used for evil purposes (especially as the black or malefic pentagram, reversed with its two points on top like the head of a goat), can unleash the demons of the astral plane.

Paul Perdrizet (in his *Negotium perambulans in tenebris*),[46] mentions several instances of pentagrams in the magic manuscripts of the late Middle Ages. I have shown that their true ancestor is the pentagram carved on the Hermes with the goat in the Vienna Museum (a Gnostic gem).

There is a fourteenth-century grimoire at Oxford (Ashmolean Library), one of whose geometrical figures is a symbol that makes it possible to acquire "supreme knowledge": it is a pentagram with the letter *G* in its center. It is (to the best of my knowledge) the oldest document bearing this particularity, which reappears in the pentagrams of the seventeenth-century English mason guilds, and lastly within the center of the "blazing star" (a pentagram, too) of the "speculative" Freemasons after their appearance (first in England) at the beginning of the eighteenth century.

In the British Museum, among the objects once belonging to the

46. Published by the Faculté des Lettres de Strasbourg, 1922.

famous magician John Dee,[47] who only escaped death at the stake during the reign of Queen Elizabeth by finding refuge in Prague at the court of Rudolph II (the emperor of the alchemists), there are two lead talismans for conjuring, each bearing a pentagram inside a circle on one of their faces.

Cornelius Agrippa von Nettesheim (1486–1535), the famous magician whose *De Occulta Philosophia* was studied with great interest by Descartes,[48] presented the pentagram as the symbol of the microcosm, and an engraving reproduced countless times (see plate 17) depicts a naked

---

47. The hero of the strange occult novel by Gustav Meyrink (author of *The Golem*) titled *The Angel of the West Window,* which prominently features a crystal talisman and a gleaming black dodecahedron. The talisman exists in London's Museum of Medicine and Magic, with a seventeenth-century manuscript detailing its disturbing properties. Next to them is a rock-crystal dodecahedron that comes from the same coterie of "scientific" alchemists and magicians of the time of Queen Elizabeth.

48. Agrippa von Nettesheim, himself a student of a famous scholar in the occult sciences, the Abbot Johannes Trithemeus, published his *De Occulta Philosophia* in Anvers in 1530. A substantial portion of this treatise on magic is a book on the Kabbalah, in which the influence of Pythagorean ideas is obvious. In it we find, brilliantly attuned to Plato's harmonic correspondences, Nicomachus of Gerasa's numerical metaphysics. Here are some passages:

"The Doctrines of Mathematicks are so necessary to, and have such an affinity with Magick, that they that do profess it without them, are quite out of the way, and labour in vain, and shall in no wise obtain their desired effect. For whatsoever things are, and are done in these inferior naturall vertues, are all done, and governed by number, weight, measure, harmony, motion, and light." Chapter 1 of Agrippa's *Cabala* (London: Gregory Moule, 1651).

"*Severinus Boethius* saith, that all things which were first made by the nature of things in its first Age, seem to be formed by the proportion of numbers, for this was the principall pattern in the mind of the Creator . . . there should be in numbers much greater, and more occult, and also more wonderfull, and efficacious, for as much as they are more formall, more perfect, and naturally in the celestialls, not mixt with separated substances; and lastly, having the greatest, and most simple commixtion with the Idea's in the mind of God. . . . And these are proper to Magicall operations, the middle which is betwixt both being appropriated by declining to the extreams, as in the use of letters. And lastly, all species of naturall things, and of those things which are above nature, are joyned together by certain numbers: which *Pythagoras* seeing, saith, that number is that by which all things consist, and distributes each vertue to each number. And *Proclus* saith, Number hath alwaies a being: Yet there is one in voyce, another in the proportion of them, another in the soul, and reason, and another in divine things. But *Themistius,* and *Boethius,* and *Averrois* the *Babilonian* [Babylonian], together with *Plato,* do so extoll

man with arms and legs spread wide, with the ends of the hands and feet corresponding to the ends of a pentagram inscribed within a circle.

The pentad, the number of the pentagram, also receives its due:

> The number five is of no small force, for it consists of the first even, and the first odd, as of a Female, and Male; for an odd number is the Male, and the even the Female. Whence *Arithmeticians* call that the Father, and this the Mother. Therefore the number five is of no small perfection, or vertue, which proceeds from the mixtion of these numbers: It is also the just midle [middle] of the universal number, *viz.* ten . . . it is called by the *Pythagoreans* the number of Wedlock. . . . It is also called the number of fortunateness, and favour, and it is the Seale of the Holy Ghost, and a bond that binds all things. . . . The heathen *Philosophers* did dedicate it as sacred to *Mercury,* esteeming the vertue of it to be so much more excellent than the number four, by how much a living thing is more excellent then a thing without life.[49] But now how great vertues numbers have in nature, is manifest in the herb which is called Cinquefoil, *i.e.* five-leaved Grass; for this resists poysons by vertue of the number

(cont. from p. 289) numbers, that they think no man can be a true Philosopher without them. Now they speak of a rationall, and formall number, not of a materiall, sensible, or vocall, the number of Merchants buying, and selling, of which the *Pythagoreans,* and *Platonists,* and our *Austin* [Augustine] make no reckoning, but apply it to the proportion resulting from it, which number they call naturall, rationall, and formall, from which great mysteries flow, as well in naturall, as divine, and heavenly things."

The reader may recognize the phrases of Nicomachus (quoted in part 1). Agrippa was a magician who was as erudite as he was a master of reason, and it should come as no surprise that Descartes and his friend Beckmann took him quite seriously.

49. The *Zohar* (*Book of Splendor*) that I cited in chapter 6 as one of the two major texts of the Hebrew Kabbalah, also mentions the microcosm in "Platonizing" terms:

"When the (divine) power made a center, it created a new universe, microcosm, and all the others moved to gravitate around it."

Further on we find an echo of the *Timaeus,* the Emerald Tablet, and Vitruvius:

"The form of man sums up all forms, both those above and those below. Because this form summarizes all that is, we use it to depict God in the form of the supreme old man. . . . The higher world fecundates the lower world, when man, mediator between the thought and the form, finally finds harmony. . . . All that exists is a body animated by a single soul."

of five; also drives away divells, conduceth to expiation. (Agrippa, *Of Occult Philosophy,* trans. "J.F.," 1651)

Paracelsus (Theophrastus Bombastus[50] von Hohenheim 1493–1541), who devoted much effort to talismans and magic symbols, stated that all the signs that compelled obedience from the spirits could be boiled down to these two: macroscosm of matter (*natura naturata*) or seal of Solomon (the hexagram), and the "most powerful sign of all," that of the microcosm or pentagram.

In plate 3 of the *Amphitheater of Eternal Wisdom,* a blend of theology, Kabbalah, and Hermeticism,[51] published in 1609 in Hanau with

---

50. The adjective *bombastic* is derived from his name. His magic speculations did not prevent Paracelsus from being a great physician. Just as the ideas on the "music of the spheres" drawn from the *Timaeus* a priori helped Kepler discover the astronomical laws that paved the way for those of Newton, Paracelsus worked using a "harmonic" theory of physiology. In his opinion, any pathological state was a rupture of harmony, a "dissonance." Another researcher that falls into this intermediate category between the man of science and the magician was the great mathematician Cardan (1501–76).

This is white magic; we also see in our own day great scientists (Sir Oliver Lodge and Sir William Crookes, for example) exploring a domain that in the sixteenth century formed part of magic, and even black magic (necromancy, the summoning of the dead). All magic is relative; the magic of one age can be the science of another. In any case, the word is convenient and replaces many paraphrases.

51. At the beginning of the book is a poem dedicated to the author by his friend Jenn Senssius, who thanks him for showing through the Divine Emerald Mirror of Wisdom "the mysteries of the Macroscosm and Microcosm." These two words recur repeatedly throughout this intentionally obscure book:

"Because of the world's ingratitude, I am regretfully compelled (as God is my witness) to seal my lips with my finger. However, I have been faithfully useful to the sons of the Doctrine by the second and third figures of this amphitheater."

The long tunnel through which the initiate advances toward the door of the "Eternal Wisdom" is depicted in plate 7 of Khunrath's book. In plate 9, the Great Two-Headed Androgyne is enthroned; the female half (left) is white, with *Laus Virgineum* around the head topped by a crescent. The male half on the right is golden and its head is crowned by a solar diadem. A peacock crown with outstretched tail is atop the two heads, with at the very top the punctual tetractys in the form of the "figurate" decad. The *solve* and *coagula* of the alchemists is also written on the arms of the androgynous individual (*solve* appears on the male arm and *coagula* on the female arm) beneath which the sphere of chaos is revolving.

the favor of Rudolph II (Prague 1598), Heinrich Khunrath lets us see the mage kneeling with arms outspread in his luxurious workshop before a half-opened tent inside of which a book is sitting on a table that the mage is contemplating. On the right page it shows a triangle, square, and a circle inscribed within each other, on the left page a pentagram.[52]

When it comes to pentacles, we cannot pass in silence the *Arithmology* by Father Athanasius Kircher. The most frequent magic signs, he said (fifth book, *On Magic Amulets*) are the pentalpha and the hexalpha (pentagram and hexagram regarded as formed by 5 or 6 interlaced *A*s), and he noted that the sign for the pentalpha was the Greeks' (Pythagoreans') υγιηα of health and harmony that Antiochus saw on Alexander's vexillum in a dream.[53]

Lastly, as we are on the subject of magic, it is appropriate to cite the treatises of the "black science" attributed to the famous Doctor Faust. The success enjoyed by Goethe's play ignited a search throughout Germany at the start of the nineteenth century for all the still-unpublished grimoires

52. The text related to this figure says: "Here is the SIGNACULAM that defeats and sends fleeing the adversaries. The marvelous PENTACLE of the five hieroglyphs." Also appearing in plate 11 with its highly complex alchemical symbolism is a pentagram on top in the form of a flaming star with Hebrew letters in each of its five points.

53. *Athanasii Kircheri Arithmologia* (Rome, 1655). In another chapter (*De Cabala Pythagorica*) the Jesuit scholar speaks of the "mysterious tetractys of Pythagoras," the decad (the number of perfect harmony over all others), and the ten *sefirot,* and returns to the pentad:

"*Mirium igitur non est, Pythagoraeos tantum virtuti tribuisse Quinario, ut euis ope compositonem animae deprehenderint.*" The attribution, in contrast to the senary (number six) of the inorganic world, is also spelled out: "*Sicuti Quinarius extimam sensuum cogntionem discernait atque numeral, ita et senarius materialem cogntionem dispertiendam numerandamque assumit.*"

Lastly Father Kircher was fully aware of the malefic power of the pentadic square

S A T O R
A R E P O
T E N E T
O P E R A
R O T A S

especially when it occupied the center of the hexagram.

dealing with occult matters. The most important, including an entire treatise on magical conjurations—the *Höllenzwang* and *Geisterzwang,* and the *Kabbalah Nigra* and *Alba* (an edition of which had already appeared in Passau in 1604)—claimed to be written in fact by the famous necromancer. It provided a wealth of details on his relationship with one of the seven infernal princes, Mephistopheles or Mephistophiel. The text is most likely apocryphal, written in the sixteenth century at the time when the Faust legend came into being, but with the help of "technical" documents that were truly representative of the white and black magic of the time. In any case, the additional diagrams (they have been reproduced several times in facsimile) form a complete repertory of pentacles—and a geometry that is as learned as that exhibited in the lapidary signs and Gothic rose windows. In both, the pentagram plays a leading role. What is most striking when closely examining the details of these figures and the inscriptions, monograms, and signatures that form part of them, is how much all caprice and innovation are absent. There is not a sign, not a spirit's name, not an invocation whose ancestral form cannot be found, often identical, in the repertory of Hermeticism and Gnosticism. Evolution and deformation are present but almost never any additions to the symbolism and terminology of antiquity. Mephistopheles himself, despite his red doublet, his samurai sword of Hades, his humanist gift of gab, is none other than the latest avatar of Hermes Trismegistus (who, as we have seen, is himself identical to the great Thoth, the god of ancient Egypt who condensed the creative breath into words and numbers).[54]

54. Mephistopheles = Megist-Opheles = Megist-Ophiel

Mephistopheles is specifically named Mephistophiel in the treatises of the *Faust* cycle. He is one of the seven great princes of hell as well as a planetary spirit. He is the spirit of the planet Mercury.

In Agrippa, the spirit of Mercury (Latin name for Hermes) is simply called Ophiel. He already bore this name as god of this same planet in the Hellenic texts, and carries the caduceus with the serpents (ὄφις, "snake"; the Gnostic branch of the Ophites were serpent worshippers). This filiation between the "Tris Megistos," Megist'Ophiel, and our Gothic and Romantic Mephistopheles was discovered by Julius Goebel (supplement to the *Allgemeine Zeitung,* Munich, no. 195); cf. also Franz Strunz, *Astrologie, Alchemie, Mystik* (Munich-Planegg, O. W. Barth Verlag, 1928).

The pentagram of Agrippa, Paracelsus, and Mephistophiel would earn its poetic pedigree in Goethe's *Faust.*

* * *

So there is legitimate reason to claim that the esoteric Pythagorean geometry was passed down from antiquity to the eighteenth century by the fraternities of builders on one side (who concurrently also transmitted from one generation to the next an initiatory ritual in which geometry played a preponderant role) and magic on the other, by means of the rose windows of the cathedrals and the pentacles of the magicians.

A third chain of the transmission of Pythagorean-Platonic ideas, especially the principle of the correspondence between the "great world" and the "little world," is subsidiary to what could be called Platonic monasticism, or even "Benedictine Platonism." As in architecture, it was the Arab world that restored to the Christian West (that after the restoration of security by the Carolingian Dynasty was again turning people toward study and metaphysical speculations) contact with Plato and the pattern of correspondences of the *Timaeus.* This doctrine was newly revealed by the works of the Jewish physician Sabbataï of Otranto (913–982), also known by the name of Donnolo, who was long held prisoner by the Saracens.

The monk Bernard Silvestris, professor of the school of Tours cathedral, wrote a *De mundi universitate sive Megacosmus et Microcosmus* entirely based on the metaphysics of the *Timaeus* and the Neoplatonic thinkers.

The Benedictine nun, Hildegard von Bingen (1098–1179), abbess of Rupertsberg (Saint Hildegard), two of whose illuminated manuscripts have come down to us (*Scivias* and *Liber divinorum operum simplicis hominis*), describes her cosmogonic visions in which the nous (world soul), which interpenetrates the entire macrocosm, dominates and brings harmony to the chaotic monster (hyle, primordial matter).

(cont. from p. 293) The planetary sign of Venus, a cross topped by a circle, which is also the alchemical symbol of copper sacred to the Cypriot Aphrodite, is the second "evolved" version of the ancient Egyptian ankh, the first being the "crux ansata" or Chi-Rho, which was destined to remain a Christian symbol, and which the effigies of the dead in Antinopolis held in much the same way as the Egyptian goddesses once held the "ankh" of life.

In one of her extraordinary illustrations, man is depicted naked at the center of the planetary universe, to which he is connected by rays that cross one another to form a star polygon.

In a miniature from her Hortus deliciarum, the Alsatian abbess Herrad of Landsberg, a contemporary of Hildegard, also shows us a naked man called "microcosmus" connected by rays to a circular macrocosm.[55] These are the precursors to the pentagrammatic microcosms of Agrippa and Paracelsus. We know, moreover, that the teacher of Paracelsus and Agrippa, Abbot Trithemius of Sponheim (Sponheim can almost be seen from Rupertsberg), attentively studied the writings of Hildegard, many of whose expressions were incorporated verbatim into the writings of his two students.

Magic, the art of building, and the science and philosophy of the Middle Ages are in fact connected by this common core of the theory of correspondences. The same holds true for another half-natural, half-Hermetic science, alchemy, whose medieval adepts, refusing to restrict themselves to the study of the alloys and transmutation of matter, sought to establish the basics for an overall technique of transmutations, united, by the intermediary of a suitable metaphysics, to the general science of forces, or magic.[56]

---

55. Charles Singer, in *From Magic to Science* (London: Benn, 1928), in which we find not only extracts from the works of Hildegard von Bingen but also beautiful reproductions of the miniatures that illustrated them, even cites a drawing that dates from around 1000, by Brythferd of Ramsay (for which a copy from 1100 exists at Saint John's College, Oxford) that depicts a zodiacal universe at the center of which the word *microcosm* is written.

We saw in part 1 that the first extant instance of the term *microcosm* is in a surviving passage from the work of Democritus of Abdera; however, an anonymous Greek biographer of Pythagoras attributes him with the inventions of the terms *macrocosm* and *microcosm* (citing in particular the phrase "ὁ ἄνθρωπος μικρὸς κόσμος λέγεται").

56. We know the importance of alchemy as a practical precursor for modern chemistry. Plutarch provided the etymology of the word *chemia* (χημέια) derived from the old name *Chemi* ("the black one," "the black land") as a designation for Egypt. Zosimos of Panopolis (circa 300 CE) was the first "modern" alchemist whose name has come down to us. In his *Ἱερά Τέχνη* (Sacred Art) he uses the word *chemeia* for the chemical art. The oldest alchemical documents we have are the Leyden Papyrus and that of Stockholm (found in Thebes in 1828), which date from the third century CE.

European magic has therefore retained the Pythagorean pentagram, which became that of the Gnostic Hermes, to make it not only the symbol of knowledge but also a tool for conjuring and for power—both good and ill. Brought back to the heart of magic and transformed into an independent discipline by the underground waterways of alchemy and the Kabbalah, the seductive metaphysical games of Gnosticism reappeared: Manicheans, Albigensians, Cathars, Waldensians spread its theology along with their claims to being the true light of Christ, the true doctrine of love. Now magicians, alchemists, and Rosicrucians took up anew the twin question of knowledge and love at the point where we had left it when under the anathema of Saint Peter's successors, this world of Alexandrian thought, which I have called the "Garden of the Hesperides" of the ancient philosophy, had appeared to sink beneath the waves of oblivion, like a shimmering, impenitent Atlantis to the song of its sirens. Here again echoes the Gnostic paraphrase of the tomb-body of the Pythagoreans, the deliverance from appearances by a "Leucadian leap": "I was not living, no matter how much I appeared alive; it is only now that I live . . . because I am dying." But also reechoing in a whisper are the words carved on the Emerald Tablet: "Through the union of man and woman . . . the work is achieved."

The candid love with which Saint Francis of Assisi embraced the whole of creation, that fraternal tenderness for humans, animals, and flowers, which is so oddly in agreement with an entire chapter of the legend of Pythagoras,[57] appears far too vague to the initiates of the new gnosis. He who wishes to realize the Great Work, whose material transmutation is only a first effort, an accompaniment on the physical plane, should also know the second stage, the work of the transmutation of life, the arcanum of generation through which man, by temporarily reconstituting the androgyne, truly participates in the creative spark of the demiurge. For the feminine principle is not only the "materia prima" and crucible of this cyclical operation; it forms one of the keys

57. Cf. Méautis, *Recherches sur le Pythagorisme.*

to the great mystery whose revelation permits the accomplishment and crowning achievement of the final Great Work: the re-creation of the soul in full light through intelligence, vision, strength, and love. This is why, beginning with the respect of the mystes of Eleusis for the Great Goddess of the black robe and that of the Gnostic in the early centuries for "Notre-Dame Pneuma-Agion" (his feminine Holy Ghost), that magicians, alchemists, and Rosicrucians—these latter emerging mysteriously at the beginning of the seventeenth century and claiming to be the spokesmen for a "college of philosophers," an "immemorial fraternity"[58] whose location and formation they are not able to reveal—will all touch on the mystery of the eternal feminine and gradually blend together the symbology and terminology of generative love with those of mathematics and the laboratory.

It was at the beginning of this same sixteenth century that saw the flowering of geometrical magic, which succeeded the long esoteric phase of architecture, the explicit "official" Pythagorean-Platonic period, and of the mathematical aesthetic that shaped the doctrine of the early Renaissance.

Studying Plato and Vitruvius by the light of the thoughts of Campanus of Novara[59] on the "transcendent" role of the golden section as governing the proportions of the five Platonic solids, the monk Luca Pacioli di Borgo (born around the middle of the fifteenth century in Borgo San Sepulcro in Tuscany, also the birthplace of his friend Piero della Francesca) wrote at the court of Ludovico Sforza his treatise on the divine proportion[60] (the name "golden section" was given to it by Leonardo da Vinci, who drew the magnificent plates illustrating this

58. Johannes Valentinus Andreae (1586–1654), who claimed he met members of the Rose-Cross brotherhood in the course of his travels, published the first "official" manifestos of the fraternity, *Fama Fraternitatis* and *Confessio Fratrum Rosae-Crucis*, in 1614 in Regensburg, then in 1616, *The Chemical Wedding of Christian Rosenkreutz.*

59. Cf. chapter 3. Campanus of Novara also studied star polygons and was the first to use the term *radix* to designate the unknown quantity in an equation of the first degree.

60. *De Divina Proportione* (Venice, 1509).

work[61]) that had such an immense influence on the scholars, painters, and architects of the early Renaissance.[62] I have recalled elsewhere that Alberti was the friend of Fra Luca. German art critics have discovered a passage in a letter of Albrecht Dürer that tells how he went especially from Bologna to Venice (where this monk "drunk with beauty" was then living) to be initiated "into the arcana of a secret perspective."

I found in the Bologna Vitruvius of 1532 a passage of commentary written by Gian Bastista Caporali of Perugia that also appears to mention Pacioli's direct influence on Dürer (also the author, we should not

---

61. "As you can see perfectly here based on the representations of all the regular solids, and their derivatives, which have been drawn by the most worthy painter, architect, musician, and man endowed with all the faculties, Leonardo da Vinci of Florence, in the city of Milan, where we then found ourselves in the service of the most illustrious Duke Ludovico Maria Sforza, from 1496 to 1499, after which we went with shared purpose to Florence where we lived together." *De Divina Proportione,* chap. 6. And further on:

"And . . . the pyramids in this book, like all the other figures . . . are also by the hand of my aforementioned named compatriot Leonardo da Vinci who in the science of drawing has never been approached by any other."

62. "It was his books (those of Pacioli) that served as the foundation for the work of all the sixteenth-century mathematicians." G. Libri, *Histoire des sciences mathématiques en Italie.*

Luca Pacioli is also known for his *Summa de arithmetica geometria*, published in Venice in 1494. In it we find the first attempts to calculate probability for games of chance, algebraic resolutions of geometry problems (embryonic form of analytical geometry), and even in an appendix, devoted to commercial accounting, the first mention of double-entry bookkeeping. We also find in this *Summa* all that remains of the treatise on square numbers by Leonardo di Pisa (Fibonacci), author of the first algebra treatise written by a Christian (*Abacus,* 1228, dedicated to Frederick II's astrologer, Michael Scott), who is of particular interest to us as having discovered the "Fibonacci series," the twofold additive series 1, 1, 2, 3, 5, 8, 13, 21, 34, 55, 89, 144, ... Among other properties this F series possesses is the fact that the ratio between two consecutive terms quickly approaches the ratio of the golden section, $\Phi = \frac{\sqrt{5}+1}{2} = 1.618...$ ($\frac{8}{5} = 1.6$, $\frac{13}{8} = 1.625$, $\frac{21}{13} = 1.615$, $\frac{34}{21} = 1.619...$). It is, in fact, the asymptotic model, in whole terms, of the ideal Φseries. I noted earlier that this F series, which nature tends to imitate, approaching the continuous by a discontinuous series, is none other than the tenth type of proportion of the neo-Pythagoreans (cf. chapter 1) and Nicomachus. The Egyptians appear to have known it, as the multiples of 55, 89, and 144 as units of length can be found in the dimensions of the Great Pyramid. Leonardo di Pisa studied Arabic mathematics and stayed in both Egypt and Syria.

forget, of the *Treatise on Proportions,* in which the study of the five Platonic solids and their derivative semi-regular polyhedrons plays the principal role). What it says in fact is that the *analogia* of Vitruvius, on which the sequence of commensurabilities is based (the *symmetria*) is not continuous geometrical proportion in general, but a definite proportion, to wit, "the divine proportion of Fra Luca and Albert of Saxony," which is to say, the golden section.[63]

But these secrets of Pythagorean-Platonic harmonic geometry, for which the golden section is, if not the keystone at least the symbolic instrument, after being loudly acclaimed publicly for half a century, were once again obscured. Palladio and Michelangelo (and perhaps Gabriel) were probably the last architects to still deliberately apply the proportions produced by the golden section and the Vitruvian concepts of symmetry and eurhythmy in their works. At the end of the seventeenth century, the very meaning of the word *symmetry* was forgotten and replaced by the definition it still holds today: division of identical corresponding elements on either side of an axis or plane "of symmetry." These elements are often equal among each other, which produces a static arithmetical balance that no longer has any relationship with the "dynamic symmetry" of the ancients. Architecture has been mechanized. Only the baroque, which is generally scorned and poorly understood, continues to "sing the cathartic geometry" in its waves of stone and stucco.

---

63. I am supposing that "Alberto de Sassonia" is Dürer. We should note the very significant role played in the crafting of the humanist aesthetic by the associations of artists, writers, and philosophers that formed during the second half of the fifteenth century in some Italian cities under the ancient names of academies and sodalities. The most famous was the Platonic Academy founded in Florence by Cosimo de Medici in 1442 that included among its members Marsilio Ficino, Pico dela Mirandolla, Machiavelli, Alberti, and the Greek Plethon. This latter claimed to be bringing to Italy the tradition of the Pythagorean-Platonic schools that had survived in the Byzantine Empire after the official closing of the pagan schools by Justinian. As these quattrocento academies, with their appearance of secret societies with initiatory rituals, claimed to be connected not only to the Academy of Plato but also with the ancient mysteries, they drew down on themselves papal thunderbolts from 1468 (under Paul II) onward.

Plate 53. Baroque catharsis (Melk Abbey)
(photo B. Reiffenstein)

Plate 54. Baroque catharsis (Melk Abbey)
(photo B. Reiffenstein)

Mathematicians also forgot the pentagram of harmony. Kepler was the last scientist to mention the "golden section," then night fell once more over the "torchbearers." However, in the stone-carver lodges of the Holy Roman Empire and England, and in the *cayennes* [lodges] of the French *compagnons,* the "geometrical" ritual and the name of Pythagoras are still passed on, distorted and blurred; and if the Gothic rose windows are dead, magic's pentacles continue tirelessly to blaze in the shadows, signs of recognition from one initiate to another, calls to good and evil spirits. They are calls for love as well; the erotic wave has been refined. Magic is less and less interested in heavenly love; it is generative love, the principle of universal fecundation, whose dynamism it is striving to harness. It is no longer even the tender search for the "wild strawberry" but an almost scientific eroticism toward which magicians and alchemists strive in common, and to which also succumbed, in thought at least, the servants of the Rose-Cross. Just as once in Thrace the chaste Artemis was transformed into Hecate or Bendis, so the crazed virgins of Hieronymus Bosch become the young witches of Baldung Grien, the central *G* of the pentagram seeks to assert itself as that of the pan-generating sensuality, of the "Gamos," until such time when, ever more feverishly, the *rotas,* the cycles of fecundation, turn to the sterile rhythm of satanic debauchery. The young goat of the Gnostic Hermes has become the priapic goat of Mendes to whom women once prostituted themselves in Egypt. Here it is the large black goat, in whose form Satan chooses to visit the Sabbat, and whose symbol is the upside-down pentagram.

Simultaneously, in this fluid margin of magic, the arithmological esotericism of the Kabbalah, the symbolism of alchemy, follow the rhythm of the chain of generations, "of man recreating himself ceaselessly in accordance with the flesh to recreate the material world in his image."

It is the "male seed" ("Red Lion") and the "female seed" ("Angel") that are the fire and primal matter of the alchemists, the recurring elements of the *solve et coagula.* Their "Alkahest" or universal solvent is "the formal and active principle of the seed in the state of effervescence

Plate 55. Baroque catharsis, monstrance
(photo Bayern Landesamt für Denkmalpflege)

to which it has been brought by the power of the magistery," in the functional union of the generative organs sometimes symbolized by the androgynous pentagram interwoven with itself and sometimes by the interpenetration of the white triangle and the upside-down black triangle in Solomon's seal.

In this symbolism haunted by the rhythms of generation,[64] the "inexpressible name" or "Shemhamphorasch" condensed into YHVH or Yod-Heh-Vav-Heh also becomes the union of the male Yod and the female Heh, of Aaron's rod and the cup or flower of Isis, and this idea can be seen again in all the imagery of the alchemists and Rosicrucians: dragons and serpents, rose bush in the hollow oak (Nicolas Flamel), mystic cave (Khunrath), and so forth.

In the town of magicians huddled around the Hradschin, the successors of Rabbi Loew and John Dee tirelessly juggled with sparkling, crystalline "words," the shimmering gems of the *sefirotic* wheel with the small spermatic flames of the yods and alephs, long before the century of the Encyclopediasts and the new rationalism; in the Bohemian "tarot" the marionettes of the great symbols of the Kabbalah—the crown, the androgyne, the virgin, the cup—were dancing their farandole of archetypes in miniature, vanishing like the cards in the hands of a stage magician, and the supreme old man, transformed into the number-letter Aleph-One, points again, a grimacing ideogram with its combined downstrokes, toward the heights as toward below: "As above so as below, the souls and things correspond":

Id quod inferius
Sicut quod superius!

And already in a time much closer to our own, in Goethe's *Faust,* we find the romantic caricature of the high sign of Pythagoras that the old Thoth-Hermes, become the picaresque Mephistopheles, is unable to pass beyond:

---

64. Cf. the highly documented articles by Henri de Guillebert in the *Revue Internationale des Sociétés secrètes* on "pentacles."

Mephisto:	*Gesteh ich's nur! dass ich hinausspaziere,*
	Verbietet mir ein kleines Hindernis,
	Der Drudenfuss auf Eurer Schwelle . . .
Faust:	*Das Pentagramma macht dir Pein?*

Mephistopheles:	Let me own up! I cannot go away;
	A little hindrance bids me stay.
	The witch's foot upon your sill I see.
Faust:	The pentagram? That's in your way?

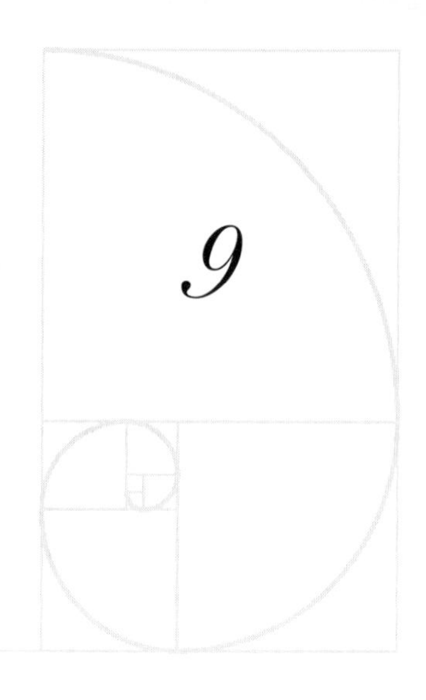

Esotericism and Politics: *From Plato's Cave to the Masonic Lodges*

Geb Er mir das Zeichen,
Gruss, Griff, Passwort!

He'll give me the sign,
Greeting, grip, password!

THE MASTER TO THE JOURNEYMAN,
ANCIENT STONE-CARVER RITUAL

We have seen that the republic of the philosophers, or the Crotoniate League, headed by the leaders of the Pythagororean brotherhood, after having controlled the political destiny of a large portion of Sicily and southern Italy for fifty years, saw its power collapse following a popular revolt whose final act—the conflagration and massacre of the leaders in an uprising of the lower class—tradition places in Croton and Metapontum. Among the few leaders to escape the catastrophe, some, like Philolaus, devoted themselves to the teaching of philosophy and mathematics. This was how, despite the protests and excommunications by the purists who remained faithful to their vow of sworn secrecy (see the letter of Lysis to Hipparchus, in chapter 7), the metaphysico-mathematical doctrine of harmony entered the intellectual public domain, the doctrine from which Plato borrowed the charm of his style and his own harmony, and several "geometrical" secrets of the school (construction of the dodecahedron, proportions between incommensurables, and so forth).

Other "brothers" who escaped the massacres regrouped in Magna Graecia itself (Locri, Rhegium, Philonte, Sybaris, and so on), in small initiatory communities in which the master's doctrine and the ritual of the "society" were passed down, but in which all actual political activity was excluded.

One sole but brilliant exception to this cautious attitude was the Pythagorean republic founded by the great mathematician Archytas of Tarentum in the city of the same name. The state governed by Archytas (he was regent and elected general seven years in a row) was prosperous and, as we have seen, respected even by the Sicilian tyrants. I have also shown how, in addition to the influence the regent of Tarentum's scientific teaching had on Plato at the time of his first journey in Magna Graecia (the theory of analogy and that of proportions in general were established by Archytas), there was the even more significant influence on his "philosophy" and his political ideas. Plato's political ideal is precisely the republic under the leadership of philosophers and geometers.[1] A major passage from the *Seventh Letter* (numerous extracts of which I already cited in chapter 7) shows precisely from the political perspective that Plato's ambition was not only to leave behind him a reputation as a pure theoretician: "And I was forced to say . . . there will be no cessation of evils for the sons of men, till either those who are pursuing a right

---

1. In the *Republic* (speaking as Socrates), Plato states that those who wish to attain high positions in the state should be mandated to meditate on the theory of numbers, not in the way suitable for a practical citizen, or great merchant or grocer, but in the way that leads to the perception of number in and of itself. The same is true for geometry, which is to say, the theory of Forms, not the surveyor's geometry. Plato lastly stresses the as yet unrecognized importance of three-dimensional geometry (the true "science of space"), and its "heightened charm" and the degree of power and civilization the state that appreciates its true value will achieve. Plato's prophecy has come true; it is geometers to whom Western civilization owes its temporal hegemony.

As a "college" for educating the elite based on the "geometrical spirit," we could say that the École Polytechnique of Paris (instituted by the Convention) corresponds fairly closely to Plato's ideal. It is the triumph of the Mediterranean synthesis. It is this spirit of synthesis that, with Joffre and Foch, former students of the Polytechnique, triumphed in 1918.

and true philosophy receive sovereign power in the state, or those in power in the state by some dispensation of providence become true philosophers." This was what Plato tried—unsuccessfully—on Dionysius the Younger:

> All our hope [his and Dion's] rested then on this attempt to win the ruler of a powerful city over to philosophy . . . it would be enough for me to win but one young man over to my ideas to always lead them to victory. . . . And what I feared most, if I stayed home [instead of returning to Sicily to attempt to convert the young Dionysius], was that I would have to deem myself as a man only good for thinking, but who would have retreated from the effort of converting thought into action. (trans. J. Harwood)

We have seen that this ambition of Plato was not realized, and that the murder of Dion at the moment he had just taken power over Syracuse not only put an end to a friendship that, like his earlier one with Socrates, represented in the affection of one human for another a supreme degree of fervor, but also, by cutting his ties to action, made him return henceforth to the domain of pure thought, far from this Sicily where he had hoped to find the promised land of his political dreams.

All that is left to us are the echoes of this dream in this emotional letter to Dion's supporters and in the *Republic.*

But getting back to the "Italian" period or the underground, second phase of "social" Pythagoreanism, we have already seen that the small Pythagorean communities or *hetairias* of southern Italy, after having continued their initiatory life for two centuries (often by merging with local Orphic sects), reemerged publicly and gained a foothold in Rome among the republican aristocracy (Cicero even attributes early Pythagoreanism with having a prominent influence on the development of Roman institutions). Toward the middle of the first century BCE, this infatuation assumed forms that the central authority could not abide. Senator Nigidius Figulus (as we saw earlier) had not only formed an academic circle or coterie, but a veritable Pythagorean "lodge" whose mysterious nature was mentioned by Aulus Gellius.[2] He was exiled by Caesar at the same time the majority of craft guilds were banned because of their tendency to meddle in political issues. In 26 BCE, another distinguished Pythagorean, Anaxilaos of Larissa (called "Pythagorean and magus" by Saint Jerome in his *Chronicles*) was expelled by Augustus.

This pronounced distrust of the Pythagoreans by the public authorities was not aroused by their doctrines but by the secrecy that ruled over their meetings; it continued under the first emperors (for instance, the expulsion of "mages and mathematicians" by the senatus consultum of the year 52 CE, and the disgrace and suicide of Statilius Taurus in the year 53). At the same time, and for the same reasons, the harsh measures wielded by the imperial authorities against the craft guilds continued (because they were hosts to rites based on secrecy).

One characteristic example is Trajan's response to the request by Pliny the Younger, proconsul of Bithynia, to authorize the founding of a "college" of firefighters in Nicomedia. The emperor denied this request stating that all colleges became secret societies too easily.

We also know, from a passage in Suetonius[3] and the famous letter

2. "Nigidianae commentationes non proinde in vulgus exeunt." Carcopino in his *Virgile et le mystère de la IVe églogue* (Paris: L'Artisan du Livre) reminds us that Nigidius Figulus wrote a commentary on the Latin translation of the *Phainomena* by the Pythagorean Aratus (Aratos of Soli, circa 50 BCE). This astronomical poem, the prevailing element of which is the "Great Year" and the cyclical recurrence of the Golden Age, had a great influence on Virgil. In Regensburg there is a mysterious bas-relief from the Benedictine abbey of Saint Emmeran, which dates at least to the eleventh century, that depicts a man kneeling on a small pillar whose head occupies the center of a large graduated disk divided into 360 degrees on which a diameter and several secants had been drawn. One of them supports the angle at the center of the pentagon. Around this stone diagram, which has greatly excited the curiosity of German mathematicians to no avail, there is carved a half-worn-away Latin phrase in which I was astounded to decipher the name Aratus.

3. The Roman chronicler mentions the expulsion of the Jews of Rome during the reign of Claudius (in the year 50) "impulsore Chresto assidue tumultuantes."

The emperor's letter dates from the year 41, and H. J. Bell published its complete translation from the original papyrus, held by the British Museum, in 1924. It confirms the cultural and other privileges of the Jews of Alexandria, but commands to neither

from Emperor Claudius to the inhabitants of Alexandria that the imperial police regarded the beginnings of Christianity as a subversive political activity fomented by Jewish secret societies.

We have no precise details on the existence of political Pythagorean lodges in Rome after the first century, because they were not persecuted. The secret-society phobia had vanished, religious ventures and "mysteries" were more in style than ever, and the guilds could, in both the West and Byzantium, pass on their rites in complete tranquility, rites that were heavily influenced by Orphic, Eleusinian, and Pythagorean initiatory rites. The adoption or overlay of new patrons (like the "Four Crowned Saints" for masons and stone carvers) soon gave them the external Christian cachet that would henceforth be essential.

In chapter 8 we looked at the persistence of these guilds through the Dark Ages, and discovered them again in the twelfth century forming an integral part of reconstructed municipal life.

Sometimes, despite the perfect adaptation of these guilds and brotherhoods to the Catholic rhythm of the city, the Church grew alarmed, and like the Caesars before it, thundered against those who "formed associations to offer mutual aid in all kinds of matters . . . carrying a penalty against those who opposed their statutes . . . exposing their members to perjury" (Synod of Rouen, 1189; see chapter 8).

After the official regulation, overseen in France by the central authority, of masteries and sworn trades, guilds could devote themselves entirely to their professional interests, and even the formidable *Bauhütte* (discussed in chapter 8) only concerned themselves with professional training, the technical supervision of the architects, masters, and journeymen who composed and erected their symphonies of stone to the glory of the sublime architect, the "Great Ordering One" of the worlds. And their secrets and rites, their "fundamental geometry," were piously passed down into

(cont. from p. 309) provoke nor even tolerate the influx of other Jews "as this would lend itself to serious suspicions of danger to the public. If any of them refuse to conform to my orders, I will be obliged to take the harshest measures in their regard, dealing with them as the carriers of a deadly germ that would contaminate the entire world."

the Renaissance and beyond until that time when, through their unanticipated confluence with the eddies of Rosicrucian ideology, they passed again from the hands of the builders of temples to the builders of systems.

However, before going there, we should examine two attempts—one abortive, the other successful—to place an organization founded on absolute secrecy and ritual discipline into the service of an idea, system, or a political or religious ambition: the Order of the Templars and the Society of Jesus.

The Templar Order, founded in 1118 in Jerusalem by Hugues de Payens and eight other knights, owes its official title (Pauperes Commilitones Christi Templique Solomonici) to the fact that its first headquarters was (as was the meeting room of the "silent ones" or Essenes) a building contiguous to the site of the former temple of Solomon (Baudoin II, the king of Jerusalem, in fact ceded to them the part of his palace built near the Holy Sepulcher, above this site).

We are all familiar with the rapid growth of the military, financial, and political power of this "militia of Christ" that at the beginning of the fourteenth century owned nine thousand fortified castles or manors, extraterritorial enclaves ("temples") in all the capitals of Christendom, and that, thanks to its material wealth, had also become a kind of "international bank of payments" that took deposits, gave credit, and granted loans to the majority of sovereigns. This went on until a king and a pope joined forces to strike down the proud "beauséant" (the order's standard on which figured the odd war cry "Long Live God Holy Love!").

On Friday, October 13, 1307, all the Templars of France were arrested on the orders of Philip the Fair. With the approval of Pope Clement V, the trials began that culminated on March 18, 1314, with the execution of the grand master of the Temple, Jacques de Molay, who was burned alive at the Place Dauphiné with Geoffroi de Charnai, preceptor of the province of Normandy (fifty-four other knights had already been burned at the stake in 1310). We know the main charges laid against the order (the official report of the accusations contained eighty-seven articles): clandestine admission ceremonies held at night with a ritual that included obscene kisses; renunciation of Christ with blasphemous insults to the Crucifix; and a solemn oath by the Holy Sacrament to never, under pain of death, reveal the secrets[4] of the order (the knights dared not even speak among themselves about the admission ceremonies); in addition, the members of the order were not permitted to make confession to any priests but only to those who were chaplains of the order (of which they were members), or even, in case of need, to its lay officials; in certain ceremonies they worshipped the head of an old bearded man carved in wood coated with silver called the "Baphomet"; and sodomy was encouraged; and finally, the supreme rule: the sole purpose of all endeavors without exception was to increase the power of the order.

Many of the knights confessed under torture, in all or in part, to the practices that were sacrilegious or simply "dangerous to the public order" of which they stood accused.

We also know that the grand master denied the charges lodged against him and cited, while at the stake, the king and pope to be summoned within the year to appear before the tribunal of heaven. Clement V died on April 20 of that year, and Philip the Fair followed him on November 29.[5]

The Templars were always suspected of having borrowed some of their secret rituals and vices from the religious and political sects of Syria and Egypt with whom they were in constant contact.[6] We know

---

4. The order comprised three classes: valets, squires, lay knights (*clientes, argeri, milites*). Chaplains held the same rank as the lay knights.

5. Dion Cassius reports a similar anecdote; when the magicians were expelled from Rome, one Vitellius responded by invoking the emperor's death within a certain period of time, which happened.

6. These include the Yezidis or devil worshippers, the descendants of the Gnostic Nasseenes or Ophites, and lastly the dreadful brotherhood of Assassins, headed by "the Old Man of the Mountain." As direct contact in the form of either implacable duels or suspect courtesy between the Templars and Assassins is vouched for by several documents (especially during the era of Saladin and Richard the Lion-Hearted), the following details have their value:

The Assassins were Ishmaelite Shiites, who revered Ismael, the seventh descendent

that until Jerusalem was retaken by Saladin, the residence of the grand master of the order was actually in that city, from where it next moved to Cyprus, then Paris.

What is certain is that despite their bad reputation from a moral standpoint (probably justified), a trend favorable to their memory was established in certain orders of knighthood (in Scotland, for example), in certain secret trade societies, especially those of the builders, in whose traditions the temple of Solomon has always held—for as far back as the documents take us—a major place. The motif of the iniquitous murder of the master by two jealous rivals is also ever present. Originally this was Adon Hiram (Master Hiram), Solomon's Tyrian architect who was betrayed and killed by three "journeymen"[7] (in the Bible, Hiram, "son of a widow," is the Tyrian master smelter who cast the two pillars of the temple and the famous "bronze sea").[8]

---

of Fatima, the Prophet's daughter. During the reign of Abdullah, a descendent of Ismael and first Egyptian caliph of the Fatimid Dynasty, the adherents of the Ishmaelite doctrines who, in the form of secret societies, were already quite numerous in Persia and Syria, founded their first Grand Lodge of an initiatory nature in Cairo. The envoy of the Persian Ishmaelites, Hasan-bin-Sabbah (who studied in Nishapur with the future high vizier Nizamul Mulk and the poet-mathematician Omar Khayyam), after initiation into the Cairo lodge, returned to Persia, where he founded a secret society in 1090 at the castle of Alamut. This society, which by and large used the great initiatory practices of Cairo as its model, added the regular use of assassination to eliminate its enemies as an original feature. It was headed by the Sheik-el-Jebal (Old Man of the Mountain), then, the three *dai-al-kirbal* (grand priors, one for each "province"), the *dais* (full initiates), the *refiks* (associates or companions), the *fedavis* (the "devoted," the guards or "assassins" strictly speaking), and the *lasiks* (novices). Koranic Islam was replaced for the higher initiates by a blissful pantheism combined with a political "will to power" that used murder as a standard tool. At the end of the twelfth century, the Syrian branch declared its independence under the *dai* Sinan.

The castle of Alamut was destroyed by the Tartars of Möngke Khan in 1280, but it appears that the Syrian branch still may have adherents in Jabal al-Druse.

7. A version of this legend accuses Solomon of initiating this murder because of the ardent feelings the architect aroused in Balkis, the queen of Sheba.

8. We saw that the bronze smelters and smiths of the eastern Mediterranean generally belonged to the tribe of the Kainites, Kenites, or Cainites of the Sinai. The name *Cainite* was also carried by the extremist "Luciferian" section of the Gnostic Ophites.

A funny thing, a tradition that appears historical,[9] combines a murder of this kind with the ceremony of laying the first stone of Strasbourg cathedral by the archbishop Conrad of Lichtenburg in 1277. Two master masons were disputing the honor of digging the symbolic foundations beneath the eyes of the bishop and, in the course of the ensuing quarrel, one was killed; work was interrupted for nine days and the foundations reconsecrated.

It can be seen in any study concerning the origins of the builders' guilds that legend, history, and symbol interweave in disconcerting analogous recurrence, and that historical fact sometimes follows the legend or symbol.

With respect to the grand master of the Templars, his memory is especially attached to that dissidence of the aforementioned French guild movement of *compagnonnage,* and, on the basis of a secret initiatic ceremony, signs, words of identification, and so forth, like the lodges of the Germanic *Bauhütte,* the journeymen were united in defense of their interests against the privileges and growing tyranny of the masters supported by royal and ecclesiastical authority.

It was precisely during the rule of Philip the Fair that the democratic movement in the guilds took shape. The king moved to suppress the "brotherhoods," but the urban guilds inflicted a bloody defeat on the royal army commanded by Robert d'Artois in the battle of the Golden Spurs (Courtrai 1302), and the 1305 treaty enshrined the temporary victory of the unions over the king and the bourgeois guilds (French feudalism would have its revenge in Cassel in 1328). This "dictatorship of the proletariat" almost succeeded in Florence, where the workers (*ciompi*) seized power for several months (1378). Again in Flanders at almost this same time, the weavers of Ghent tried again to shake off the tutelage of the masters, but were finally defeated, according to the extraordinary epic of Artevelde, on the fields of Rozebeke (1382).

9. *Essais historiques et topographiques sur l'Église Cathédrale de Strasbourg,* by the abbot Grandidier (1782).

In both cases, it was patricians or members of the upper bourgeoisie (Silvester de Medici and Philip van Artevelde) who organized the working proletariat. It was the bloody excesses and the divisions within the popular factions that led to the final triumph of the reactionary forces and the bourgeois masters.

Even in France, the *compagnonnique* brotherhoods of masons and carpenters were suspended by a decree of Parliament in 1501, and a decree of March 10, 1506, forbade journeymen from assembling, dining, or banqueting together on pain of imprisonment. But the *compagnonnage* held firm, and in 1665 Louis XIV's brother, the Duke of Orleans, was initiated as a journeyman hatter. Under the name of "Intrepide le Guépin" he had to answer the ritual questions, receive the watchword of the guild, and conform in every respect to the "duty" of the association.

> The companion sought in the rival organizations of the guild safeguards and means of defense. It was during the last century of the Middle Ages that associations of "brothers" or "companions" (*Brüderschaften* in Germany) began multiplying. Protected by secrecy or through the observance of mysterious rites, these worker unions broke through the narrow context of the city and spread into entire regions and countries . . . concluding treaties of alliance and reciprocity between each other . . . and organizing trips to various cities and countries, tours of France, tours of Germany (these would last five years). . . . They had their celebrations and banquets . . . their secret meetings, similar to those of the building workers, the "free masons," with initiations, oaths, and means of correspondence. . . . The first "workers' internationale," which coexisted with other groups . . . the brotherhoods, whose focus was mainly religious but whose framework also served the companions to organize agreements of understanding and defense despite the distrust and interdictions generously handed out by the Church and the public authorities. (P. Boissonade, *Le Travail dans l'Europe*)

The National Constituent Assembly vainly sought to eliminate the brotherhood-like associations that had resisted the revocation of guilds and trades; the *compagnonnage,* with the "worker romanticism" of the nineteenth century, took on a prestige following George Sand's *Compagnon du tour de France,* which ensured its entry into literature.

The secret of the rites of the *compagnonnage,* by very reason of the persecution and distrust of the public authorities, was much more absolute than that of official guilds or corporations.[10] A common rule of all the *devoirs* ["obligations," "duties"] (the traditional name of the *compagnonnage* association in a given trade that followed a given ritual) was: "All papers will be burned on this day of the year and their ashes will be mixed in wine that will be passed around in a circle and drunk." This topped off the other rule: "The secrets of the *compagnonnage* cannot be set down in writing." Some archives, but containing very few documents, would only appear after the French Revolution.

But to return to the memory of the Templars, we should note that an entire group of Compagnons du Devoir, or "Companions of Duty," also called itself "Companions (or Children) of Master Jacques," the patron thus designated being the executed grand master of the Order of Knights Templar. The Templars, according to a tradition held in honor by these groups, had, as the builders of "temples" and fortified castles in both the East and the West, maintained cordial relationships ("as one initiate to another") with the traveling builder associations. On the other hand, another association of the "duty-bound" called itself the "Children of Soubise," on the strength of its very warm relationship with the Church.[11]

---

10. Here is one example:

In 1811, a journeyman tanner of the *devoir,* Bavarois Beau Désir, who shared the guild secret of the tanners with two bakers who wished to start a *compagnonnage* for their trade (this was, moreover, the origin of the bakers' guild), barely escaped lynching and was only able to escape death by sailing for New York on an American ship in 1812.

11. According to another tradition, the "Jacques" and the "Father Soubise" of these *devoirs* would have been two masters (Jacques Molène and Soubise de Nogent) who supervised the building of the towers of Orleans cathedral (1401).

Finally, the members of a third group of *devoirs,* claiming to maintain the true

Later we shall revisit the "dynamic" motif or complex of the "vengeance of the Templars."

The other example of a "society" also founded on a principle of a community whose members are bound by obedience and secrecy, and have been recruited and selected much in the same way the members of the Pythagorean brotherhood were after long intellectual and spiritual preparations, is the Society of Jesus, or Order of the Jesuits, founded by Ignatius of Loyola in Montmartre on August 15, 1534, and confirmed by Pope Paul III in Rome in 1540.

The kinds of vicissitudes this religious association experienced are common knowledge, an order that from the outset was established on not a monastic but a political, scientific, and military model (its superior or "black pope" is still called the "general" of the Jesuits). Expelled successively from those countries where their influence was strongest

---

tradition of the ancient guild mysteries, called themselves "Compagnons étrangers du Devoir de Liberté, du rite de Solomon."

Not only were these federations (which in France appear to have most authentically retained the guild rites of antiquity and even what remains of the ancient mysteries, strictly speaking, and the ships and oath-sworn trades) regarded with distrust by the public powers and constantly suspected of subversive activities, but, especially in southern France, their bellicose spirit sometimes provoked, between *devoirs* of different rites, long-lasting conflicts, if not actual pitched battles, such as the bloody battle of La Crau in 1730 between *devoirants* (from the rite of the Devoir de Maître Jacques, who had as allies the *bons drilles* carpenters) and *gavots* (and their affiliate stone-carver allies), or companions of the Devoir de Liberté (rite of Solomon). The stone-carver companions added to their *compagnonnage* cognomen the name of their first *cayenne* or native town. This custom has been passed down from the Middle Ages to our own time. This is how Boucher Franc-Coeur of Avallon set the stone for the Grand Palais, Cornette la Franchise of Pont-à-Mousson laid that of the Gare Saint-Lazare, and so forth. The companions have three degrees of initiation: young man (aspirant, apprentice), journeyman, and first journeyman (or master for the stone carvers).

A Fédération Compagnonnique has existed since 1874; in 1899 it adopted the name of Union Compagnonnique.

The "master builder" of the stone carvers of France is currently A. Bernet la Liberté of Seméac, C.·. étranger D.·. D.·. D.·. L.·.

There is an extremely moving mention in a *compagnonnage* journal of the meeting in a restaurant, during the occupation of the Ruhr, between French and German companions, whom the first recognized from their style of greeting, and so on.

(from Portugal in 1759, from France in 1762 and again in 1880), and completely suppressed by Pope Clement XIV in 1773 then restored by Pope Pius VII in 1814, they have always emerged unbeaten from storm and persecution, and still form today a formidable political and intellectual presence, one that is occult in method if not in purpose.

The reading of the *Constitutions* of Saint Ignatius and of the internal regulations governing the order is most instructive. In it we find the most profound documentation of a politico-mystical organization since the rules, of which we have only fragments, of the Crotoniate "society." It is possible, moreover, that their memory inspired Saint Ignatius, just as it formerly influenced the ascetic associations of the Therapeutae of Egypt, the Essenes and their immediate successors, and the first monks of the Thebaid.

A parallel with the Templars[12] has often been made, as it has with another secret society, one whose foundation is initiatory and not religious, and which we shall now examine: Freemasonry.

This would be the "speculative" or political Freemasonry founded in London at the beginning of the eighteenth century by a group of "honorary masons" (accepted masons) affiliated with the "operative" masons' lodges we looked at in the preceding chapter. These speculative masons were scholars interested in the complex of ideas and tendencies stirred up for about a century on the continent and even in England by the adepts of the mysterious fraternity of the Rosicrucians.

The custom of electing honorary members already existed in antiquity for the secret societies formed by the affiliates of the "mysteries" as such. We have proof that, primarily during the imperial era, it was the fashion for the emperors and other prominent individuals to be initiated into the higher grades of the rites of Eleusis[13] on the occasion of their travels

12. Like the Pythagoreans and Templars, the Jesuits have three grades that in their order are: novice, approved coadjutor, and the professed.

13. Initiation into the mysteries of Eleusis appears to have consisted of the following grades:

1. Initiation into the Lesser Mysteries or purification (this took place near Athens at the temple of Demeter and Kore on the Ilissos).

in Greece (cf. in Victor Magnien, *Les Mystères d'Eleusis,* the inscriptions found on the statues of the Eleusinian *hierophantides* who had initiated as *mystipoli* the emperors Hadrian, Antony, and Commodus). It is likely that this custom existed as well if not for the original Pythagorean society at least for the neo-Pythagorean brotherhoods whose success we have seen during the early years of the Christian era. We find this custom in any case for the "colleges" of craftsmen (for example, those of the masters—

---

2. Initiation into the Greater Mysteries. Conferred on the neophytes on the *telete* or the *myesis* (they became mystai or "silent ones"); this initiation, like those that follow, took place at Eleusis.
3. *Epopteia* or Mystery of the Seal. The mystes becomes an epopt or "seer" (he "saw" the synthesis directly without any further need to travel the path of reason). In theory this grade was conferred five years after initiation into the Greater Mysteries, but in practice one year was enough.
4. *Holoclere* initiation, Mystery of the Circle, the *anadesis,* or crowning. The initiate leaves the circle of carnal desire (the cycle of generations) and wins the crown; he has gained his unity and sees the Unity. He has become a guide for others (poet, military or political leader, physician).
5. Priest initiation. Confers the duties of the *dadouches* (torchbearer), *keryx* (herald), and so forth; the adept has become a philosopher-poet. Insignia: headband (*strophion*).
6. Hierophantic or royal initiation. The initiate has become king through his illuminated thought, which is illuminating to that of others. He has quit humanity. He now only loves Beauty in itself (cf. the end of Diotima's speech to Socrates in the *Symposium*), for he has made contact with divinity itself; he has become a "daimon."

   The hierophants and *hierophantides* adopted a new secret name; only their grade (which the king magistrate always possessed) conferred the right to enter the sanctuary. Their insignias were royal ones (diadems and so forth).

   Victor Magnien, *Les Mystères d'Éleusis,* infers from some texts (the "Sixth Ennead" by Plotinus, among others) the existence of an even higher degree: Supreme or Divine Initiation, through which the initiate becomes one with God through celestial love.

   Initiation into the degrees above *epopteia* was reserved for the descendants of certain noble families and the initiates adopted by these families. Women could not only be initiated (*thesmophoria*) but attain the highest degree of the priesthood (*hierophantides*).

   Some children—both girls and boys—of the powerful families of Athens were initiated at a very tender age so they could figure in the ceremony of the Mysteries; they were called sacred children or "initiates since Hestia."

architects and masons) who, as the reader has already seen, borrowed all or part of their rites from the initiatory religious brotherhoods, and were especially influenced by the ideology and geometrical and technical symbolism of the neo-Pythagorean lodges, whose god was the "τεχνιτης θεος," the preeminent artist and artisan of Nicomachus of Gerasa.

The honorary patrons (*patroni*) are mentioned on the inscription (*ordo*) concerning the constitution of the college of the Ostia boatmen in 152 CE. We even have the same proof for the funerary colleges (the patrons were called *pater* or *mater* here) and for the religious fraternities in the strict sense, or Roman sodalities organized for a special cult, like that of the Arvalian Brothers, whose minutes were deciphered on the walls of a small secret temple in the sacred grove of Dea Dio.

Feasts and ritual played a great role in the internal life of the technical *collegia* (and we know from Plutarch's text on guilds that each of them had the "assemblies, feasts, and ceremonies that suited them"), and the similarity of the rites for the purely religious or initiatory fraternities and the technical "colleges" is confirmed by the name "brothers" that members of each referred to their fellows as, and that of "magister" reserved for the president-elect, and generally the identical nature of the names of the corresponding officials in both guilds and brotherhoods (I have made the same observation concerning the *maistores* and *keryces* of the Byzantine guilds).

Returning to the honorary patrons, we later see German emperors like Maximilian received as an honorary journeyman and master (*Briefmaurer*) of the *Bauhütte,* and the brother of Louis XIV did not scorn association with the *cayenne* of the milliner journeymen of the *devoir* of Master Jacques. But it was especially in England starting in the sixteenth century that curious high lords and intellectuals sought to be admitted as "free and accepted masons,"[14] as either honorary

14. The term *freemason,* whose German form (*Freie Maurer*) we find in the fifteenth century, appeared in England as early as 1375 in a document (from the London masons' guild) as "ffremassons," then in 1396 on a list of craftsmen employed on the construction site for Exeter cathedral. The term *lodge* (*ye loge* or *ye luge*) also appears in the fourteenth century (archives of York cathedral).

masters or journeymen into the operative lodges. For example, the book of protocols updated daily since 1598 of the lodge Mary's Chapel of Edinburgh mentions that on June 8, 1600, John Boswell of Auchinleck was received as a "non-operative brother." He was followed later by the Lords Eglinton and Cassilis. In 1618, the Earl of Pembroke was even president of a lodge; in 1663 the Earl of Saint Albans was elected honorary grand master of the Lodge of Saint Paul with Sir Christopher Wren (who would be the architect of the new Saint Paul's Cathedral) as first overseer. Tradition maintains that Charles II took part in the meetings of this lodge under the name of "Brother Rowley." It is certain that Christopher Wren's royal friend used the trowel of the former (now this lodge's grand master) when placing the first stone of the new cathedral. The trowel still exists along with two mahogany candlesticks—gifts of Christopher Wren—in the lodge treasury. This lodge first known as the "Old Lodge of Saint Paul" successively took on the names "Goose and Gridiron" (this is the name it held in 1717, taken from the name of the tavern where its meetings were held), the "Queen's Arms," and finally—since 1768—"Antiquity."

On October 16, 1646, we find in the journal of Elias Ashmole, the famous humanist, scholar, and Rosicrucian (founder of the Ashmolean Library), the entry: "I was made a Free Mason" (in Warrington).

The creation of a "society of philosophers" devoted in complete mental tranquility to the study of the sciences was in the air and naturally appeared more easily in the Protestant countries, in other words, Germany and England. Sir Francis Bacon imagined this "college" in his *New Atlantis;* the Rosicrucians went further by claiming its existence to the readers of their earliest manifestos.

This was precisely the era of the extraordinary vogue of these Rosicrucian ideas, first expressed publicly by the Württemburg pastor Johann Valentin Andreae (the *Fama Fraternitas* of 1614 and the *Confessio*), a disciple of Paracelsus and friend of Comenius; as well as in the works of Michael Maier (1568–1622), the personal physician of Rudolph II. (These include *Arcana Arcanissima, Themis Aurea, Hoc*

Est de Legibus Fraternitas Roseae Crucis, and especially the *Atalanta Fugiens* with the bewitching engravings by Johann Theodor de Bry.[15]) His personal contact with English Hermeticists like Robert Fludd, and the interest displayed in his ideas by Elias Ashmole (to whom is dedicated the English translation of the *Themis Aurea: The Laws of the Fraternity of the Rosie Crosse,* 1656) especially contributed to spread the influence of this movement in England.

The *collegium lucis* from which the Rosicrucians claimed to receive their continuing directives, despite the efforts by many thinkers and scholars to join it or locate it, remained completely inaccessible.[16] The English adherents of the Rosicrucian movement, the majority of whom were part of the operative masonic lodges of London or Scotland as honorary members (free and accepted masons), decided, while retaining the rites, symbols, and denominations of these lodges, even their

15. The list of books by Michael Maier is quite long. In other pamphlets on the brothers of the Rose-Cross (*Apolegeticus; Silentium Post Clamores*), he emphasized their "law of silence" that did not allow them to respond to the "cries of indignation" aroused by the famous manifestos. In his *Septimana Philosophica* Maier features Solomon, the queen of Sheba, and Hiram; in his *Symbola Aureae Mensae,* as in the *Silentium Post Clamores,* he mentions a "college of philosophers" existing since the most remote times, which was connected to the Egyptian, Eleusinian, Orphic, and Pythagorean Mysteries.

16. However, proof for one of these "esoteric" colleges is provided in a letter sent from Lyon to Agrippa von Nettesheim by the physician Landolphe (circa 1510). The document recommends to the famous master of the occult sciences a "diligent seeker," the bearer of this letter, and suggests he be tested with an eye to possible initiation into a certain society (*si in nostra velit jurare capitula nostro sodalicio adscitum face*).

For more on the existence and role of these secret societies of philosophers and alchemists affiliated with artists guilds and stone-carver lodges, see G. F. Hartlaub, *Giorgiones Geheimnis: Ein kunstgeschichtlicher Beitrag zur Mystik der Renaissance* (Munich: Allgemeine Verlagsanstalt, 1925).

Concerning the fateful recurrence of November 10 (1618, 1619, 1620) in the life of Descartes, Gustave Cohen, in a fine article in the *Nouvelles Littéraires,* writes: "One would even be tempted to imagine the three grades of a secret initiation, especially when one thinks of the curiosity that Descartes admitted he had about these mysterious Rosicrucians with whom he spent time in Ulm at the Faulhaber home and whom he continued to frequent in Holland in the person of his best friend Hogelande." It was in the third dream that he had on the night of November 10, 1619, that Descartes saw an old man point out the "Yes and No" of Pythagoras in an open book.

meeting sites (in London toward the beginning of the eighteenth century these were taverns; the Old Lodge of Saint Paul as we have seen bore the colorful name of "Goose and Gridiron"; the other operative lodges of London, of which thirteen peers of England were members at this time as "accepted masons" had similar names like "The Apple Tree," "The Crown," and "The Roman and Grape"), to completely remove their still theoretically professional purpose and simply keep the social, philosophical, and "speculative" nature that had become their motivation. The idea of construction remains, but the builders of temples had become the builders of systems in the social sphere.

This was why the Grand Lodge of England was founded on June 24, 1717, which gathered under its suzerainty the four principal "operative" lodges of London. This date marks the beginning of "speculative freemasonry," which we now simply call Freemasonry.[17] And we should note

17. The transition of operative freemasonry to speculative Freemasonry was much less abrupt and definite than generally believed. Here are some clues that illustrate this position:

The Mason's Company (an operative guild) of London, which already existed officially in 1376, obtained its coat of arms (naturally bearing a plumb line, compass, and so on) in 1472, and had its privileges confirmed by Charles II in 1677, included—since 1620—a semi-autonomous lodge of honorary "accepted masons" (its rolls for 1620 and 1621 were discovered and published in 1894).

Elias Ashmole, cited earlier as having been initiated in 1646 as a "Free Mason" of the lodge of Warrington, explicitly mentions in his journal that when he was invited to London's "Mason's Hall" on March 11, 1682, for a ceremony, he saw six "gentlemen" being accepted into the fraternity of freemasons of a certain lodge. He also noted that only three of them belonged to the "Mason's Company," but that the master and two wardens of the company attended the ceremony as members of the lodge.

The Scottish Code of 1599 offers us official proof of the existence of three Scottish mother lodges in this year: "First and Principal Judge" of Edinburgh; "Second Judge" of Kilwinning; "Third Judge" of Stirling. The entry rolls of these mother lodges and many Scottish subsidiaries, fragments of which have been found, prove that both accepted honorary patrons as well as operative master masons. In this way the roll of the lodge of Aberdeen for 1670 shows forty-nine brothers "all with their marks recorded," except for two, of which only ten were operatives; the others were "patrons" such as the earls of Finlater, Errol, and Dumferline, Lord Forbes, and so on. And the old lodge of Melrose did not become affiliated with the Grand Lodge of London (which is to say, became speculative) until 1891! Some English lodges continued to remain operative even after joining the Grand Lodge, their rolls mentioning the reception of technical apprentices until 1754.

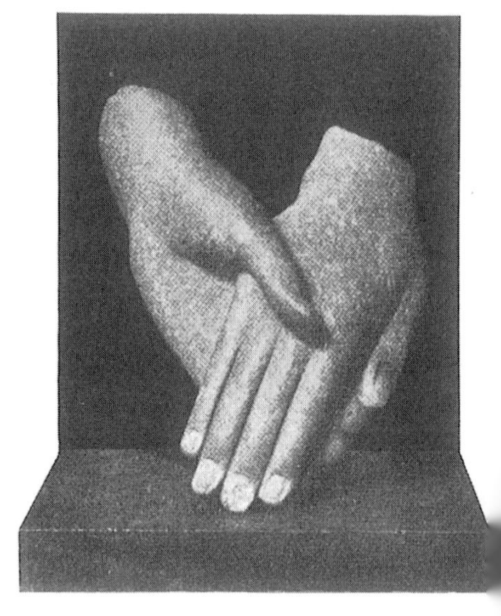

Plate 56. The mystic handshake: *left,* the goddess Hathor and Seti I (Florence, photo Alinari); *right,* fragment from el-Amarna (Berlin, photo Grantz)

that it was Saint John's Day that was chosen as the date for this recasting or transformation of operative freemasonry into speculative Freemasonry.

The first grand master was Anthony Sayer. In 1723, the Reverend John Anderson who, as an accepted mason of the former operative lodges, had devoted special study to all their archives regarding their traditions, and who was one of the initiators of this transformation, wrote the Masonic "constitutions" intended to serve as a ritual code for the "lodges" that, in imitation of this Grand Lodge of London (it took the Duke of Wharton as grand master in 1723), were successively founded all over. The former York lodge was set up as an autonomous speculative lodge in 1721. The first French lodge ("Louis d'Argent") was established in 1725 by the Jacobite exile Lord Derwentwater in Paris on the rue Boucherie. In Germany the first lodge was that of Hamburg (1733), the lodge of Boston was the first in the Untied States (1730), and the first in India was that of Calcutta (1730). The dates the subsidiary lodges of the Grand Lodge of London were established are: for France, 1732; Holland, 1735; Switzerland, 1740; Italy, 1763; and so on. By 1776, 480 lodges were already in existence. They were no longer satisfied with the purely social and ceremonial side of their ritual, but—in compliance with the program of occult societies of seekers and reformers hoped for by Francis Bacon, Micahel Maier, John Amos Comenius,[18] and Robert

18. John Amos Comenius (1592–1670) is claimed by modern Rosicrucians as one of their high apostles. Born in Moravia, it was in England in 1641 that he met the Rosicrucians Hartlieb, Durseus, and so on, and founded in Patak, Hungary, the first Pansophist College in 1650. Here is the program of the "Pansophia" (the name reveals its ties with an ancient Gnosis that had never died), outlined at this time in a letter to Durseus:

Collection of all existing scientific material in an encyclopedia (this idea was realized in France a century later);
Study of all languages;
Special study of mechanics, chemistry, and philosophy;
Study of the "mysteries" of a prophetic nature;
Study of the universal language of "magic";
Organization of a college and society that could achieve all this.

Material from this program outlined by Comenius was used in 1717 at the time of the founding of the Grand Lodge of England. The Pansophist College was recently

Fludd, whose initial plan was now fully realized—suddenly developed not only philosophical orientations but politically active tendencies. The tradition of absolute secrecy and mutual aid between "brother"[19] Freemasons in any and all circumstances, also inherited from the operative lodges, granted them great potential for action in this new field of activity. The public authorities and Church quickly grew alarmed and in the Catholic countries raged against Freemasonry (this term henceforth only designates speculative or political Freemasonry, with the operative lodges, deserted by their patrons and even the architects for the "philosophical lodges," fading completely into the shadows).

Just as in the thirteenth century, sometime before Philip the Fair's power play, a pope had promulgated the bull *De insolentia Templariorum,* so before Louis XV gave the command to close the Lodge of Paris Pope Clement XII (1738) had cast anathema on Freemasonry and ordered a vindication of Masonry published in Dublin burned (1739). Benedict XIV renewed this anathema in 1751.

For the society that had just formed or reformed was clearly this

---

(cont. from p. 325) reconstructed on "modern" foundations, and serves as the nucleus of a significant faction of the German Rosicrucians.

19. The tenacious transmission of the idea-force of fraternity and the very word *brother* may now be clearly apparent to the reader. In the Pythagorean church by the Porta Maggiore, the heavenly brothers or twins, the Dioscuri, hold each other's hands with their fingers entwined, the left hand is held in the right hand and the right hand is held in the left hand (in a funerary fresco at the Louvre, Hathor, goddess of love, suzerain of the Cainite temple of the Sinai, holds the hand of Seti I in the same manner; in plate 56 there is an illustration of a variation from the Turin Museum). This gripping of the hand has been handed down, like the sign of secrecy (hand held horizontally beneath the chin with the thumb held up to form a right angle) made by the effigy of an anonymous master at Saint Stephen's in Vienna, like the "sign of the master" (hand raised with two fingers held vertically touching the forehead), like the uncovering of the knee (the "golden thigh" of the god shown to Abaris, the unclad neophyte placing his left foot on a ram's hide during the lustral aspersion at the beginning of the Lesser Mysteries, Christ with his left knee uncovered on the tympanum of Saint Stephen's), like the gold apron of Ramses, and like the beribboned thyrsus of the villa of the "mysteries" in Pompeii (staff of the wandering tribe [*rekabim*] of the Cainite smiths, ritual cane of the participants in the Jewish Passover—symbolism of the journey—the beribboned canes of the *compagnons*), and so forth.

Plate 57. Square and plumb line on a funerary mosaic in Pompeii (photo Brogi)

time the "political" society, the "great *hetairia*" that was based on secrecy, above human laws, and outside the sole religious law accepted by Rome.

Agrippa von Nettesheim called the Creator "the eternal Master Builder," the Freemasons venerated the "Great Architect of the Universe," and this great architect is closer to the god of Pythagoras and Plato than that of the Roman Catholic Church. Passed down by the Kabbalah, alchemy, and the Rosicrucians, the "inexpressible" Tetragrammaton of the ancients and other "words of power" appeared in the lodges on the diplomas and insignias of the "brothers" of the new covenant. In the symbolism of the entrance examinations inherited by the medieval stone carvers from the *collegia* of antiquity there still lived that spirit of the initiatory ceremonies that had inspired them; and in the new "dens of philosophy" it is the voice of the *hierokeryx* (great herald) of Eleusis that through the ages forewarns the new mystes of the severity of the tests, the terrible penalties that await whoever betrays the sworn secret. Reappearing in the very lodge where the "brethren" gather are the cavern of Metapontum, Plato's symbolic cave, the underground chamber of Maat, and perhaps also the large crypt of the immemorial sanctuary in the "Malachite Mountains."[20] It is, finally, the eternal Pythagorean and Mediterranean pentagram that,

---

20. The temple of Serabit, which had first been hollowed out in the mountain, where Cainite smiths and miners and Egyptian engineers of the Sinai worshipped the "Goddess of the Green Stone." It already existed as a complete temple with colonnades and more during the Twelfth Dynasty (1900 BCE). Cf. Robert Eisler, *Die Kenitischen Weihinschriften.* Maat, goddess of truth, "Mistress of the Chamber of Judgment," where she is accompanied by Thoth. She is the most important of the Egyptian goddesses as she also represents order, the law of the world. The Chamber of Judgment, or purification, with its walls immersed in the water of a rectangular canal, is mentioned in the ancient initiatory text attributed to Thoth-Hermes and mistakenly called the *Book of the Dead.* We know of Marsham Adams's theory that identified the Great Pyramid as the "House of Mysteries" mentioned in this text. He supports his claim of the existence of a "chamber of Maat," bordered by water at a certain depth beneath the pyramid, on a passage from Herodotus. A large "chamber of Maat" similar to those described by this preeminent "Hermetic" text (the oldest hieroglyphic version known, in Saqqara, dates from the Fifth Dynasty) was found in Abydos. Its dimensions have the Fibonacci-based approximations 100 and 60 (in Egyptian cubits) of the sides of a Φ rectangle (golden section).

having become the "blazing star," and already present on the writs of the last English operative houses of the seventeenth century, takes the place of honor on the bulk of the Freemasonic documents of the eighteenth century and flashes above the throne of the master of the lodge or the altar, with the enigmatic *G* always at the center.

Despite the austere technical and geometrical symbolism of the accessories—compass, square, triangle, plumb line, pillars, cubic stone—the Church was certain it could recognize here an old enemy; and while the Masons claimed to be revering the first letter of the word *geometry* in the mysterious *G*,[21] Rome read it as the initial of Gnosticism.

Marsham Adams's hypothesis is not without interest. Here is the last passage from the last chapter of the *Book of the Dead* (whose true title is *The Book of Going Forth by Day*): "This book is the greatest of the mysteries. May no (profane) eye be allowed to see it. Its name is the 'Book of the Master of the Hidden Dwelling.'" Turin Papyrus.

Among the symbols the initiate should mention during his journey toward the light is the star "Sopdet." This is Sirius, which will become the five-pointed star of Hathor-Aphrodite, then the pentagram of Hermes, and finally the pentacle of Gothic-era magic.

21. In a Freemasonry manual of the Adonhiramite Rite, printed in 1786, the neophyte at his initiation into the grade of journeyman was questioned on the meaning of the letter *G*. He answers "geometry." The "great word" for the seventh degree is "geometers."

The respect for geometry as the preeminent science and for Pythagoras as the great initiator of the "royal science" has come, moreover, from both the reverence held for the genius of Pythagoras in the guild traditions (*Bauhütte* and guilds were only passing on here the spirit of the builders' colleges of antiquity) and the prestige he held in Rosicrucian ideology. Robert Fludd stated that the Rosicrucians held, among other secrets, that of the use and composition of the "wheel of Pythagoras" to "give its number" to all things, even God.

On the frontispiece of Anderson's *Constitutions* we find the classic diagram of the "sacred" triangle of Pythagoras (3-4-5), the one that permits drawing the right angle ("angle of equity") by means of a surveyor's chain, with the eureka of the master beneath it. Recently I saw in the window of a small London antique store an earthenware jug from Liverpool, dated 1845, showing on one side a frigate at full sail, a really clichéd image, with the name of the sailor who had been given this jug by his brothers, and on the other the square and pentagram with the central *G* with a Masonic poem ending in these lines:

For stamp'd upon the mason's mind
Are Unity and Love.

The rigorous persistence of these Pythagorean symbols, the concise quintessence of the *Ieros Logos* suddenly terminating in doggerel, resonates like a tenacious, strange echo in the Louis-Philippe-era romanticism of the object.

As we know, Freemasonry picked up the glove tossed at it by the Church: the role it played in the preparation of the French Revolution is common knowledge.

Already in an anti-Masonic pamphlet of 1747 (*Les Francs Maçons écrasés*), "subversive" ideas of liberty, fraternity, and equality were assumed to be leanings of the Freemasons. In a clever pamphlet from 1744 (*La Franc Maçonne*) their ideal is presented as being a universal democratic republic with Reason as its queen, and an assembly of sages as its council.

On October 22, 1773 (the year the Jesuits were expelled), the Grand Orient of France was founded, whose watchwords are specifically "liberty" and "equality." This lodge had a large part in the writing of the Declaration of the Rights of Man. The astronomer Lalande, Benjamin Franklin, first envoy from the United States,[22] the Marquis de Lafayette, and so on, were members of the lodge the Nine Sisters, but the most influential lodge during the Revolution itself was that of the Amis Réunis (Friends Banded Together).

It is curious to see that one of the ideological components of this antimonarchical activity pursued by French Freemasonry was the spirit of revenge against the Capets and the papacy, to avenge the dissolution of the Templars and the execution of the grand master Jacques de Molay. Moreover, one of the high grades of the Scottish Rite still includes as a plan of action vengeance against the successors of the king and the pope.[23]

---

22. George Washington was a high official of American Freemasonry; there is a painting of him wearing a Masonic necklace and the letter *G* above his head. Goethe was received into the Amalia Lodge in Weimar in 1780.

23. The fifth category of members (including the 19th, 20th, 23rd, 24th, 25th, 26th, 27th, and 29th degrees) of the classic Scottish Rite (which has 30 "visible" degrees, from the 4th to the 33rd, in addition to the 3 classic Johanite degrees—the Scottish rites were organized in Charenton in 1801, and reformed at the Congress of Lausanne in 1875) is connected by its titles to the Templar Order. The degrees of the 6th category (22nd and 28th degrees) are Hermetic and Kabbalistic, of Rosicrucian lineage and titling; the 33rd degree's first password is "de Molay." Viollet-le-Duc points out a Templar's tombstone in Saint John's Chapel in Cuac'h, Brittany, that depicts the Masonic square. We know that Dante (*Purgatorio* and *Paradiso*) exhibited his sympathy for the Templars by excoriating Philip the Fair and Clement V.

Plate 58. Christ the "Initiate" (left knee uncovered), Saint Stephen's Cathedral, Vienna, north choir (photo B. Reiffenstein)

By a strange coincidence, it was in the tower of the Temple, last remnant of the French residence of Jacques de Molay, that Louis XIV and Marie Antoinette were incarcerated before mounting the scaffold.

The first English and German speculative lodges (Grand Lodge of England and its subsidiaries) at first had only three degrees. These first three degrees (entered apprentice mason, fellow-craft mason or journeyman, and master) constituted the classic Johannite Masonry (*Johannes Brüderschaft*), which came down in full from the builders' lodges. These lodges also had, as we have seen, the three degrees of apprentice, journeyman, and master. The name of Johannite Masonry obviously derives from the mysterious role played by the apostle John and Saint John the Baptist in the traditions of the guilds and especially operative masonry. June 24, Saint John's Day, was the ritual day for the inauguration of the *Bauhütte* conferences and congress. It was also the day reserved by the Templars for the most important ceremonies of this order whose patron saint was John. I should also mention here the mysterious kingdom of Prester John, popular with medieval chroniclers, and the persistent survival of the fire ritual associated with Saint John's feast day (summer solstice).

It was also on Saint John's Day in 1717 that the Grand Lodge of England was founded, and the Gospel of John[24] was often found on the altar of Masonic "temples."

(cont. from p. 330) A fairly questionable tradition attributes to the king of Scotland, Robert Bruce, the founding of an order of knighthood with a secret ritual. It so happens that 1314, the year of Jacques de Molay's execution, was the year Bruce freed his country from English domination.

I have already mentioned that the English Rosicrucian Robert Fludd (1574–1637) and Elias Ashmole were also accepted masons. This strange couplet is found in a poem published in Edinburgh in 1638 ("Muses Threnodie") by Henry Adamson:

For we brethren of the Rosie Crosse

We have the Mason's Word and second sight.

The "lost word" plays a large role in the verbal symbolism of the speculative Masonic lodges.

24. I noted earlier the link between the work of the Beloved Apostle and Gnosticism. Masonic exegetes believed that the Essenes (who had a meeting hall adjacent to the

To finish this overview of Freemasonry's role in history we should note it was supremely important during the nineteenth century. The political activity of the new "fraternity" based on secrecy, mutual aid, and absolute obedience to sometimes-unknown leaders made it a force with a radically international character that transcended national boundaries. Paradoxically, it became a powerful instrument in nationalist struggles for freeing and reconstructing oppressed groups (the Carbonari, Garibaldi's supporters, and so forth). This was the romantic period of Freemasonry.

In France its political activity has been almost fully in the open since the advent of the Third Republic; the battle against Rome has taken on a very bitter tone, and the freethinking, deist Grand Orient has become violently anticlerical and, almost as if by accident, atheistic.

So, embarrassed by the "Great Architect of the Universe," it has removed any reference to him (in France) on its diplomas and documents.

It would be the naturalistic, utilitarian, "homaistic" period of the end of the nineteenth century that would be marked by the laws against the congregations and, once again, the expulsion of the Jesuits.

The treaties concluded at the end of the Great War both enshrined the triumph of nationalist principles, such as desired by romantic-era

---

temple of Jerusalem and, as we saw, were spiritually akin to the Therapeutae and other neo-Pythagoreans of Egypt) constituted a stage of ancient Freemasonry and claimed that Saint John the Baptist was an Essene.

The Templars had their initial headquarters on the grounds of the temple of Solomon, which reappears frequently (particularly the pillars Jachin and Boaz mentioned earlier in connection with the *Bauhütte*) in the symbolism of the lodges, as does the legendary Adon-Hiram, who had given his name to a branch of Freemasonry in the eighteenth century (the Adonhiramite Rite). The same Adon-Hiram (the "son of the widow") also figures consistently in the traditions and rituals of the French *compagnonnage,* as well as the idea of vengeance and reparation for the murder of the just, which has the triple face of Master Hiram, the mysterious Master Jacques, and the proud grand master of the Temple.

In Mesopotamia, the sect of the "Christians of Saint John" still exists. They are also called Nazarenes or Mandaeans, and place Saint John the Baptist above Jesus.

Freemasonry, and the foundation of a truly international body (League of Nations) that brings to mind, all things considered, the assembly or college of sages as dreamt of by the cosmopolitanism of the seventeenth- and eighteenth-century Rosicrucians and Freemasons.

It has resulted in a kind of period of détente, introspection, and regrouping of ideas and goals. Even in France, the utilitarian, earthly, if not to say down-to-earth atmosphere of the lodges colors what could be described as the "initiatory renewal" of Freemasonry.

A new breath of air is passing through—or rather a breath of very old air. The rites and symbols, a little crumpled, and secularized as well, are being scrutinized anew, as are the slowly spelled-out passwords that have come from so long ago. This Great Architect of the Universe—is it merely a dusty, pedantic, allegorical secular formulation? The dust has been blown away, allowing the Great Geometer, the Great Ordering One of Pythagoras and Plato, to reappear. The square? Is it a lay symbol or the angle of equity, of truth used by Maat to recognize the righteous in the Chamber of Judgment? The acacia? Is it the tree of Hiram or that of Osiris? The blazing star? We have acknowledged it for a long time.

I am making no claim for doing more in this chapter than summarily marking off some important trails, but what we can now consider as a given, is that this modern universal "fraternity," which has a political influence whose power even the profane can measure, is connected by a bizarre lineage—a series of grafts, cuttings, and symbols that appear in its ritual, its practice of secrecy, and its geometrical ideology—through the medieval builders' guilds, craftsmen colleges, and initiatory groups of antiquity, back to the other great political "society" we saw perish in the flames of Metapontum.

The phoenix has withstood the test of fire.

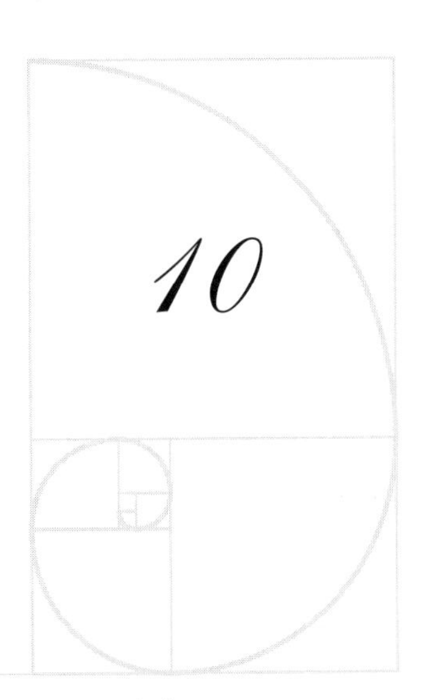

Modern Science and the Return to Pythagoras

Perhaps the oddest thing about modern science is its return to Pythagoreanism.

Bertrand Russell

I already quoted this quip by Bertrand Russell in my previous book when illustrating the triumph of the two methods of exploring consciousness, set theory and group theory (with the calculation of invariants as a branch of the latter), thanks to which the law of number allows—starting from several abstract symbols—the deduction, on the one hand, of the theory of numbers, the theory of functions, analytical mathematics, and logic, as branches of one same "logistics." On the other hand, it makes possible not only geometry, and all possible geometries (through examination of transformation groups and sets of 2, 3, or n dimensions), but also (through Einstein's latest synthesis[1] that applies to the set of four dimensions representing the physical universe the principle of least action or Hamilton principle) the laws of gravity and electromagnetics, with classical mechanism and dynamism as particular cases.

It is the triumph of abstraction, of the mathematical symbol become not only the condensed image of a cycle of abstract operation but the very key to the physical universe and its accessible realities, as effective

1. *The Unified Field Theory,* 1929.

as the "words of power" of the old magic stories. It is most especially the triumph of what Oswald Spengler rightly calls "the last and final notion of Western mathematics, the amplification and spiritualization of the theory of functions in group theory."

Descartes had already sought to realize the synthesis of science through "concatenation," the chaining of symbols. After his first refinement of the algebraic tool expounded by Fibonacci, Pacioli, and Viète, it is interesting to note that in his "illumination" of November 10, 1619, it was the "Yes and No" of Pythagoras shown to him in his third dream by the unknown old man who indicated to him, after the wind gusts and the shower of sparks (his two preceding dreams), the path of knowledge.[2]

Through the operation of our latest mathematical symbols we have drawn out an image of the physical world in which structure alone matters, a philosophy of pure form, form and rhythm, or at least periodicity. For in this world of physical phenomena (what used to be called the world or the "material" plane), we shall see later, in accordance with the words of Nicomachus (cf. part 1): knowledge can only encompass relationships and structures, and that number, not substance, is the sole eternal reality.

But let us first revisit the latest developments in mathematics.

The paradoxical subtlety of Georg Cantor's theory of "transfinite numbers" (the basis of set theory) had already jolted certain mathematicians and the controversies between the proponents of finitism and the proponents of infinitism on the logical possibility of a "present

---

2. We know that during the year 1620, between his two "illuminations" of November 10, 1619, and November 10, 1620, Descartes, a volunteer officer in the Duke of Bavaria's army, spent several months in Ulm, where he became a friend of the Rosicrucian mathematician Faulhaber. It was at this time that, picking up the Pythagorean theory of punctual or figurate numbers at the point where Nicomachus and Theon of Smyrna had left off, Descartes wrote his *Progymnasmata de Solidorum Elementis,* in which he analyzes by means of the intermediary of the pyramidal numbers already studied by Theon, the "solid" numbers contained in the five regular polyhedrons and in the nine semi-regular polyhedrons. He called "weights" these "polyhedric" numbers obtained by establishing the law of the formation of "gnomons" or solid differences that illustrate the "homothetic growth" of the polyhedrons.

infinity," a mathematically realizable infinity (in thought) that was no longer simply a never-attainable boundary like the exasperating ∞ of classical algebra. Cantor uncoils, manipulates, numbers, and lays out in tiered processions (conceptually) "realized" infinities of different orders. At first glance it gives one the impression of a hallucinatory fantasy rather than a serious discipline worthy of finding a place in the classical temple of Mathesis. The audacity of the concepts and the symbology in which the Hebrew aleph of the *Zohar* and tarot becomes the insignia of the cardinal numbers, and the Gnostic omega that of the transfinite ordinal numbers, brings to mind some Kabbalistic flight of fancy, a *sefirotic* pyramid and white-magic tower, a golem of symbols growing at a terrifying speed, of some disciple of Rabbi Loew forgotten on the slopes of the Hradschin.

And yet all this phantasmagoria, in which the finite is only a particular case of the most modest of transfinite numbers (aleph zero, the "countable" infinity), in which the reflection, the endlessly analogical recurrence, find their brightest mathematical image in an infinite development of the concept of the monad and the decad, allowing the flood of numbers and rhythms to gush forth, has won and held its position. The theory of functions, the continuum hypothesis, and so on, have adapted it as their framework and can no longer imagine themselves without it.

Hardly had the creations of this alarming magician of the transfinite been incorporated into our mathematics and logic as a sumptuous armature of set theory, when another Kabbalist, a tamer and charmer of symbols, using, as I noted earlier, group theory, led us in three transcendent leaps from one paradox to another, to the aforementioned ultra-Pythagorean synthesis of the physical universe into idea-numbers.

Here again people shout it is sleight of hand and trompe l'oeil. Couldn't we have fallen victim to an illusion like the famous "rope trick" that makes a Hindu crowd see a child climbing into the sky on a rope the fakir had tossed into the air? But no: the child is still climbing the ever-lengthening rigging of symbols. The second magician appears to have successfully pulled off his trick, and if, as we have seen, the entire scientific substratum of the Hebrew Kabbalah, Tetragrammaton and *sefirot* included, was once, during the glory days of Alexandria, borrowed from Pythagoreanism, we could say the Semitic genius in the persons of Cantor and Einstein has repaid in sparkling fashion what they borrowed from Greek science.[3]

I noted earlier that form and rhythm alone count in what remains of the ancient material substratum of the world. Chemistry initially brought us back to the unity of matter, thus to the essential aspect of the "alchemical" notion. We should recall in passing how useful the studies of the alchemists were for modern science. If they seem too preoccupied with magic for our taste, conforming with the blend of Gnostic, neo-Pythagorean, and Neoplatonic ideas inherited from Pymander's Alexandria at a time when Hermeticism served as a liaison between effervescent metaphysics and the even then highly advanced technique of Egyptian "chemistry" (alloys, enamels, balms, and perfumes), if they were overly devoted to the pursuit (still for our taste) of the great spiritual work and the generation of the "new man" instead of simply sticking to the quest for the materia prima or *quinte essence,* we should not forget that the ever-mobile fringe that forms the hinterland of the unknown at the borders of science is constantly shifting. What is mysterious and occult today (and I am thinking here of this or that chapter of the "paranormal") becomes the standard science of tomorrow.

If the materia prima has finally been found, we have also learned that all the so-called material and solid bodies, including our bodies as living beings, are in reality, because of the immense distances between the molecules that make up their apparent framework, in a gaseous state (the sole relatively "solid" bodies known in the universe are three

3. An amusing coincidence: in the symbolic expression $G\mu\nu = O$ that provides an abridgement of Einstein's equations, we find a reminder of the mysterious *G* that has intrigued us in the preceding chapters. The decad also appears, in the form of the ten coefficients of curvature that cancel out in "empty" space and Einstein's ten potentials (the $G\mu\nu$ that supply the mechanical and gravitational properties of space in general).

recently discovered stars like the white dwarf or "companion of Sirius" in which matter has been compacted to a density 60,000 times greater than that of water; a ton of this matter could be held in a matchbox, the nuclei of its atoms being so close to each other so as to largely remove the electron orbits and orbiting electrons themselves).[4] Furthermore, we know now these molecules and atoms, which were still thought to be indivisible only forty years ago, are little "solar" systems, practically empty zones in which, in their turn, at immense distances, astronomically relative with regard to their size,[5] the ultimate particles of "substance" (but no longer matter as they have lost its sole material quality: constant mass) are gravitating around each other, particles of

4. The whole atom (including its "planetary" electrons) has a diameter on the order of 10 millionths of a millimeter; its nucleus (stable neutrons and protons) has a diameter that is 10,000 times smaller.

5. Here are some colorful comparisons borrowed from Sir Oliver Lodge or Sir Arthur Eddington:

The size of an electron in comparison to that of an atom (which is a simple spatial zone that contains the orbits of the electrons forming the atom of the given element) is like that of a grain of wheat in comparison to Albert Hall (London's largest concert hall), or like that of a mosquito in comparison to the size of a cathedral. In other words, all the electrons of a "heavy" atom together take up much less than a billionth part of the atom's volume, which primarily consists of empty space. As atoms themselves are—even in solid matter—just as few and far between, we can see how this pseudo-matter is maintained in reality. The electrons and protons that form its apparent substratum would, if pressed together, occupy a relative space that was truly infinitely tiny (all the electrons and protons forming a human body pressed together at the density of the atomic nuclei would occupy a space the size of a grain of wheat, at the density of the matter in "the companion of Sirius" [Sirius B], the volume of a hazelnut).

The "atomic weight" of an element is equal to the number of protons (particles of positive electricity) concentrated in the central nucleus of the atom. The "atomic number," which varies from 1 to 92 on the scale of simple bodies, marks its place in Mendeleev's table, and determines its chemical properties, is equal to the number of "orbiting" electrons freely gravitating around the central nucleus. When this number is lower than that of the protons, a number of additional neutrons equal to the difference is combined in the nucleus with its protons so that the total electrical charge of the atom is null, or neutral, rather, each (negative) electron neutralizing a (positive) proton. As the pseudo-mass of protons is 1,850 times greater than that of the electron, it is clearly the number of protons (atomic weight) that determines the pseudo-mass, the pseudo-weight of the given atom. Conversely, the volume of the proton is 1,850 times smaller than that of the electron.

pure—positive or negative—electricity (protons or electrons).

The ninety-two elements [in 1931], the ancient "simple bodies," are distinguished from each other by their "atomic number," the number of orbiting electrons that gravitate around the nucleus that is itself composed of protons and neutrons, in other words, particles of pure electricity. The number of protons and electrons in a normal atom are equal because their opposing but equal electrical charges neutralize each other.

The simplest element, which is the submultiple of all the others as guessed by a number of dreamers like August Strindberg, is the hydrogen atom composed of a single electron gravitating around a central proton (Oliver Lodge compared this system to a penny moving around another penny in an orbit with a diameter of one thousand). Between this atom and the densest one—the uranium atom that has 82 orbiting electrons—the 90 other elements are grouped on Mendeleev's periodic table in "octaves," series of 8 elements each having as its first or "fundamental" term an inert gas, the lightest gas being helium, first term of the first chemical octave.

Moreover, periodicity, numerical rhythm,[6] comes into play in all

---

(cont. from p. 339) Examples: the hydrogen atom contains a central proton and an orbiting electron: atomic weight 1, atomic number 1. The helium atom has a nucleus composed of two protons and two neutrons and two orbiting electrons: atomic weight 4, atomic number 2. The carbon atom has a nucleus composed of six protons and six neutrons, and six orbiting electrons: atomic weight 12, atomic number 6. The nitrogen atom has seven protons and seven neutrons in the nucleus, and seven orbiting electrons: atomic weight 14, atomic number 7. We know that different atomic weights can correspond to the same atomic number as is the case for isotopes, due to the presence of neutrons in the nucleus that do not cause any variation in the charge of the nucleus but do in its mass.

6. This is a good time to recall that there are two kinds of rhythm, the one with strictly regulated periodicity, either because identical phrases repeat uniformly (constant, "symmetrical" rhythm in the static sense of the term) or because their range diminishes by becoming softer following an exponential or sinusoidal law as the energy, the initial living force, is absorbed; and the other with a dynamic pulsation in which the periodicity is no longer regulated, given, so to speak, in advance, but forms an elastic pulsation that can undulate, making itself more amplified and more intense in unanticipated, symmetrical spurts as if under the effect of an influx of energy, an external impulse. This latter case is that of "living" physiological or mental rhythms (including music and poetry).

physical phenomena through the wavelike nature of all the physical manifestations perceptible to our senses, whether they are phenomena of sound, light, electromagnetism, and so forth. We know that since scientists have succeeded in producing nuclear fission, eight new, artificially produced elements have been added to Mendeleev's table, numbered 93 to 100.

I mentioned the scale of sound waves in our considerations of musical rhythm (part 1). These waves can be transmitted by air, water, or any other material environment.[7]

The other waves known in physics are transmitted through the hypothetical ether, which, while it has momentarily vanished in time as the "milieu" of absolute, detectable "reference" (independent of the movements of "material" bodies) as a consequence of the negative results of the Michelson-Morley experiment and the concurrence of this result with the theory of relativity, nonetheless remains, until we learn otherwise, as the environment and agent of propagation for cathodic, luminous, and electromagnetic waves. They can be classified as the vibrations of an immense keyboard, staggered in fifty octaves from the long Hertzian waves to the ultrashort waves of X-rays (0.00000001–0.000005 millimeters), by way successively of wireless waves, infrared calorific rays [now known as infrared radiation], the light spectrum (vibrations on the order of 500 trillion a second), and the ultraviolet rays. Some octaves of this keyboard have yet to be explored (between the Hertzian waves and infrared radiation, and between ultraviolet rays and X-rays).

Let's return to the small planetary world formed by the atom: electrons, particles of pure electricity, which is to say completely immaterial (their sole "material" quality, pseudo-mass or inertia, is only the apparent result of the effect of a purely electrical property[8] or the self-induction of these particles on measuring instruments), were long considered to be

7. We have seen that the frequency of sounds perceptible to the human ear varies from 32 (C or low C of a 32-foot organ pipe) to 33,000 vibrations per second (B_{10} or the piccolo of the organ emits 31,249 vibrations). The A of the usual tuning fork (850 vibrations per second) is located in the fifth octave.

8. A liquid propelled at great speed can produce the apparent physical effect of an

whirls, spheres, or tori (rings) of ether, which acquired their electromagnetic properties (self-induction and so on) from their extremely rapid gyrations, orbital translation, and perhaps vibration.

But the most recent studies seeking to explain certain contradictions between these hypotheses and quantum theory, and that have achieved their goal with the new "wave mechanics," appear to have relegated the quantity of pure electricity, even in the form of ether rings, to joining the material atom and mass in the attic of suspect epiphenomena. In Schrödinger's and de Broglie's wave mechanics, the electron is not even a concrete transmitter of waves or energy, but a simple geometrical point, or rather an ideal zone where waves from a new environment meet and overlap (the "sub-ether" as Eddington calls it to denote that this environment is on a more remote plane than Maxwell's ether, that of electromagnetic and light waves). Following the interpretations of this new theory offered by Heisenberg and Dirac, we have reached what appears to be the final possible degree of abstraction: these pseudo-electrons, "singularities," sites of the greatest wave density in the sub-ether. When attempts are made to examine them more closely, these sites even lose their identity as precise geometrical locations and simply become the probability for the singularity to be found within a certain zone. As a last resort, they are not only depicted symbolically but actually by a purely mathematical entity, or "matrix," an infinite two-dimensional grid composed of columns and rows of "pure numbers" whose elements are the coordinates of position and speed, or at least the synthetic unfolding, based on these two coordinates, of the "harmonic components" defining all the possible vibrations of the system.

I recommend to all who wish to enter the arcana of the new subatomic

---

(cont. from p. 341) extremely hard object (a jet of water that cannot be cut by a saber); all the apparent "solidity" of matter that in reality is an endlessly restrained gas is only due to the enormous speeds at which its molecules are oscillating. Purely electromagnetic phenomena produce effects of "hardness" that are even more paradoxical, for example, the Thomson effect, thanks to which a copper axe "bounces back" without being able to enter a fairly powerful electromagnetic field. And the purely immaterial electric spark produces extremely disruptive and explosive mechanical effects.

physics a bit more deeply to read the fine book by Eddington, *The Nature of the Physical World*.[9] These few lines should suffice to provide a glimpse of the fact that we have managed to replace matter with pure electricity through active quanta, which in turn have been replaced by Heisenberg's grids or matrices in which pure numbers sparkle as the final and only noumena, realizing a concept of matter as strictly Pythagorean as the most crystalline abstractions of the number mysticism of Nicomachus of Gerasa. Furthermore, to go back from the infinitely small to the infinitely large (it is still the physical universe that concerns us here), it is strange to see that even the limited spherical cosmos of the Pythagoreans has reappeared—as a hypothesis initially—as a possible consequence of the "curvature" of our non-euclidean four-dimensional world, then as a probability based on a beginning of experimental proof: a movement toward the red spectrum of the light rays coming from the most remote spiral nebulas,[10] once these rays had completed a "world tour."

Hubble's calculation gives 84 billion light years for the radius of the curvature of the universe, thus its diameter in thousandths would produce a figure on the order of 10^{24} (it is fairly odd to find the exponent 24 that appears in the most varied constants of physics, both in the infinitely small and infinitely large).[11]

I should make explicit, moreover, that according to relativity theories

9. Cambridge University Press, 1928

10. Eddington, ibid.

11. In this theory our world is not a simple (three-dimensional) sphere bordered by nothingness; it is a "hypersphere," a volume of a consistent curvature, which curves into space (four-dimensional space-time). The sole analogy that can make this comprehensible is the surface of a sphere in our classic space (three-dimensional euclidean space). This surface, for beings of no density who share its same curvature and are thereby compelled to slide over it without breaking contact, is a two-dimensional space (two coordinates are enough to determine the position of one point). But in reality this space is curved into a three-dimensional space. In similar fashion, three-dimensional beings (us) living "on" the hypersphere will have the natural impression of living "inside" an infinite three-dimensional volume. It is only as a result of certain scientific observations that they will be able to see that this volume is not infinite but boundless because it is curved (like the surface of the pure sphere for its inhabitants who have no density).

the size of the universe is not constant, but varies in the opposite direction with regard to the quantity of condensed matter in the universe. We can add to this the theory of the expanding universe, according to which the galaxies move away from the observer with speeds proportional to their distances.

We have thereby rediscovered a "macrocosm" of finite radius, limited (in its four-dimensional expanse), that is more similar to those of Plato and Herrad of Landsberg than to the infinite cosmos of Laplace's celestial mechanics and Lucretian chemistry of 1900.

To recapitulate: on one side, matter, the sole tangible reality accepted during the era of materialist determinism, has become not only an "epiphenomenon"—to replace the clever expression invented by orthodox materialists or "mechanists" precisely to simultaneously label, explain, and spirit away consciousness—but even, since the acceptance of de Broglie and Schrödinger's theory, an epi-epiphenomenon, an epiphenomenon of the second degree, with respect to the forms and states of equilibrium governed by the principle of least action or Hamilton principle. I would like to beg your indulgence for a brief digression here

---

(cont. from p. 343) Every point on a circumference is equidistant from a center of inner symmetry; the same thing is true of every point on the surface of a sphere. Similarly, every point of the hypersphere (our universe) is equidistant (the just-cited radius, R = 100 to 200 million light years) from a center of symmetry *that does not form part of our physical universe* while being enveloped by it.

I should add that the formula $2\pi^2R^3$ provides the volume of hyperspace (the three-dimensional finite volume, not the four-dimensional hyper-volume that has this volume as a border).

Our galactic world is only one island (spiral nebula, "island universes" as Eddington put it) of the many in this finite universe that holds several million. The galactic world (in the form of a rounded disk) has a diameter of around 100,000 light years. The closest spiral nebula is 850,000 light years from us. These are Eddington's figures; other authors have reduced the Milky Way's diameter to 20,000 light years, and the distance to the Magellan Nebula to 100,000 light years. On the other hand, the Andromeda Nebula would be 1 million light years away. Finally, an "island" has been recently discovered, a star system some 30 million light years away with a diameter of 2 million light years. I am only providing these obviously provisional figures to give an idea of the general proportions of the universe.

in which I will attempt to elucidate this concept of the epiphenomenon that we will revisit in the next chapter.

*
* *

Determinist mechanics did indeed coin the ingenious expression of "epiphenomenal consciousness" toward the end of the nineteenth century to summarize the materialist theory of consciousness. This means that the phenomena observed in organized (living) matter are in no way different (according to this theory) from those taking place in inorganic matter, and these biological and psychological phenomena can be explained by the physicochemical laws that govern inorganic matter (we saw in the first part of this book that this premise is not quite accurate from a strictly mathematical perspective). This means that consciousness (the discerning awareness of higher beings, the elementary instincts of others) is only an appearance, a mask over the phenomenon, in other words, an "epiphenomenon." This means more precisely that this apparent consciousness is only the contour, the reflection in our plane of observation, the summation, the arbitrary junction, the phosphorescent track that gives an impression of continuity of physicochemical reactions (attractions, repulsions, electrical transfers, energy transformations, and so on) taking place according to the normal physicochemical laws in certain physicochemical systems called "living organisms" (this being a diluted variation of the short formula with mathematical pretentions: "the soul is only the integral of the body's physicochemical reactions").

By trying to find other examples of pseudo-phenomena that would only be "epiphenomena," we can see that the concept analyzed logically always implies a deceived observer (or several), the victim of an illusion or appearance, who is victimized as such in the following fashion:

1. Everything takes place "as if" (the epiphenomenon took place, and was real);
2. Everything takes place practically (within certain limits) "as if," and so forth;
3. But the reality is something else: the cause of the perceived or

> observed processes is the true phenomenon masked by the epiphenomenon, which, despite 1 and 2, does not occur and only exists in appearance.

The observer (the deceived victim of the illusion) is therefore more essential to the concept of the epiphenomenon than to the interrelated ones of phenomenon and noumenon (the thing in itself); but we can see that the collocation to consciousness of this quality (epiphenomenon) leads to an immediate misinterpretation in that the observer who is the victim of this illusion would specifically here be that—illusory—consciousness.

John Stuart Mill, despite the integral determinism that the science of his time imposed, at least as a table of reference, on all impartial minds, had already collided against the amusing absurdity of a consciousness that would only be a series of physicochemical reactions and yet "could know itself as a series." As noted by Joseph Needham in *Science, Religion, and Reality* (London: Sheldon Press, 1925): "Mind, therefore, and all mental processes cannot possibly receive explanation or description in physico-chemical terms, for that would amount to explaining something by an instrument itself the product of the thing explained."

Epiphenomenon and phenomenon are only the projections in the plane of an observing perception—or a perceiving observer—of a noumenon belonging to the world of "things in themselves." The materialists forget that the physicochemical phenomena that appear to be scientifically proven to them always assume conscious observers experimenting in their planes of respective perception and comparing their experiences; it is the epiphenomenon "consciousness" that would confer whatever rigorous qualities that scientific reality (science of phenomena) might have.

The contradiction inherent to the epiphenomenal theory of consciousness thus creates a vicious circle similar to the pseudo-paradox of Epimenides: "Epimenides said: 'all Cretans are liars'; but, Epimenides is a Cretan, therefore he is lying; but then Cretans are not liars, therefore he has not lied, and so on."

Here the cycle would be:

> X (the mechanist philosopher) says the consciousnesses of so-called living organisms are epiphenomena (are appearances that result from the interplay of physicochemical forces). It so happens that X is a member of the class of "so-called living organisms," therefore (if his proposition is correct) he does not exist as a real conscious entity. His mental activity as a being issuing judgments is an illusion that an omniscient observer (Laplace's "universal brain") could explain by the simple play of physicochemical reactions in a material zone formally called X. But then X's opinion about his own reality or that of other similar systems is of no value, and the fact that he appears to have produced a reasoned opinion is an illusion. But if X's opinion is valueless, his initial proposition falls, and living organisms are perhaps endowed with real consciousness capable of producing valid judgments, and so on.

The triple sophistry of the pseudo-paradox of the Cretans is obvious because Epimenides's initial assertion, whether he was a Cretan or not, only has the value of an individual statement from which nothing can logically ensue, and the two therefores have no determinant force. Epimenides could be part of a minority of Cretan non-liars or a minority of Cretan liars (and who had lied in particular in this case), or of a majority or even totality of Cretan liars, and had spoken the truth on this day. The puerile nature of the sophism sticks out immediately. In the case of the materialist philosopher, on the contrary, the vicious circle is real, and will only accept as a solution the falsity of X's statement, or the paradoxical hypothesis that X would be an exception in the class of pseudo-living organisms with epiphenomenal consciousness, an exception capable of producing valid judgments.

But, and this is the culmination of this long digression, if the term *epiphenomenon* can in no way be applied to consciousness (I am saying awareness, not personality) for which it had been coined by materialism, it can now be admirably applied to matter, or, if one wishes, to its former characteristic quality, a former that is not so long ago—"material mass."

For, everything seems as if matter was endowed with a characteristic quality that could be defined and measured as the conventional quantity called "mass" in pure mechanics; everything takes place within certain limits as if the appearance in question was a reality, but the "phenomenon" giving birth to the effect described above (mechanical mass) is the self-induction (electrical inertia) of the charges of pure electricity that make up intra-atomic particles (electrons and protons).

In 1903, Gustave Le Bon's analytical mind glimpsed this paradoxical "dematerialization" of matter and presented it in a book that hardly seems to have aged (*The Evolution of Matter*). The speech given by Max Planck when he assumed his duties as rector of Berlin University (October 1913) was the first official consecration of the new regime (Planck was also the first to recognize the importance of the principle of least action as the law governing all energy transformations in inorganic systems).

*
* *

As we have seen, electrons and self-induction are in turn in the process of becoming epiphenomena, and the doors of the pseudo-material "noumenon" or ancient hyle or materia prima are the cabalistic grids or the matrices of Heisenberg's pure numbers.

The Hamilton principle therefore remains the great victor of the transformations and revolutions of physics over the course of the last thirty years. It was through it, as Weyl predicted, that Einstein succeeded at his final synthesis; it was also this, as the principle of stationary action, that allowed Heisenberg to tame and apportion to the alveoli of their "transfinite" hive the hosts of probability that teem beneath the "epiphenomenal" mask of the electron. As I explained in my *Esthétique des proportions* and revisited in part 1 of this book, this principle of least action, which under its individual or statistical form entirely governs *the becoming of all inorganic systems,* be they as large as a galactic world or contained inside the active sphere of an atom, can no longer be strictly applied when the system contains life (we could also say that life acts, or can act, from an energetic standpoint, like a force "outside" the system, with the word *outside* being precisely defined geometrically as acting in a dimension perpendicular to the other dimensions of the system). It is therefore a veritable mathematical criterion of life's transcendence, a "test" for distinguishing between inorganic and organic systems. This criterion has not been discredited by the latest developments of intra-electronic explorations by means of wave mechanics. But these studies have reduced to rubble, inside the purely inorganic system that forms the electron itself, a principle that, while in the case of life accepts the possibility of bending (insertion of choice, of Bergsonian indeterminacy: life "acting as an outside force" as noted above), in the case of a "material" or energetic inorganic system (lacking living centers) appears intangible, and not only in the domains of mathematical physics. For it was imposed as a principle of causality, as an a priori law of logical understanding; the new sensational fact would be that inside the atom the elementary processes whose resultants are "quantum" phenomena *would escape determinism*. There is a probability for such an elementary process to be produced, but as these are only probabilities, an entirely different process or even the absence of any process would be possible. This is to say that the law of causality would no longer apply in the "sub-ether," or, at least, no longer apply except as averages, as in psychological or sociological statistics.

I confess that here (this concerns what takes place inside any inorganic atom, to the exclusion of living phenomena), I am not tempted to join the scalp dance around the causality principle. While I accept that even in classical dynamics it was possible to imagine functions such that the body in motion describing the corresponding curves could at one point, like the mathematical equivalent of Buridan's ass, swing between several equally probable courses to conceive an infinite number of forces that, proportionate to the terms of an oscillating series like this, refuse to produce a determined resultant, I grant myself the right to doubt the actual presence of this anarchic indetermination in the world of subatomic energies, and am instead inclined to accept the interpretation of Leslie Walker, outlined as follows:

> If the substance of the universe, as quantum phenomena seem to indicate, is not infinitely divisible, a diagram whose positions are assigned by means of points and whose speeds are attributed to point-instants is too refined (overly precise) to adapt to the granular structure of the world in which we live. It should come as no surprise, then, that in Quantum Theory, the more we try to rigorously (punctually) fix the position, the speed becomes indeterminate, and vice versa.[12]

It therefore is a question here of one of those antinomies that stems from how, with our continuous, irreversible consciousness—and with the help of a logical tool capable of realizing this ideal continuity as far as the absolute, which is reflected in it—we strive to interpret a discontinuous physical universe.

* * *

Nothing definitive remains of the physical universe except for structures and diagrams, in Hamiltonian balance for the entire inert macrocosm (the *natura naturata* of the Kabbalists, of cubed and hexagonal symmetries), but still betraying the temporary breath of life in the forms and designs emanating out of the biosphere in which symmetry, the tendency toward entropic leveling, has been—through being exempt from the principle of least action—shaken and shaped into rhythms (often ones governed by the number 5) and subject to this asymmetrical growth pulsation for which the golden section is the dominant "proportion."

And all this, including the "material" substance of our own bodies that are so fine, so "gaseous," and so "transparent" despite their apparent opacity, is resolved in the final instance into a finite Platonic macrocosm, in ideally mathematical "structures," in these "matrices" of pure numbers that obey the old saying from the *Ieros Logos:* "Things are only the appearances of number."

12. "The Physical World," *Journal of Philosophical Studies,* July 1929.

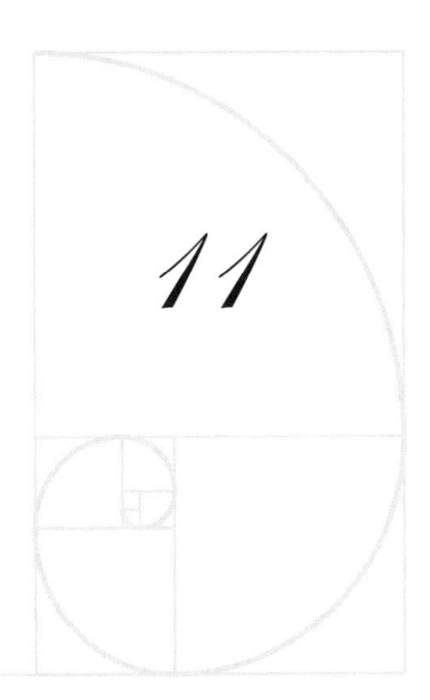

11

Life Force, Rhythm, and Duration

Non fui, quod eram, nunc sum, dum morior!

Gnostic motto

Behind the contorted, deceptive sparkling of the patterns—matrices of numbers, grids of probability—of this physical universe reduced to being the epiphenomenon of an epiphenomenon, which undulates eternally in accordance with the principle of least action to travel from the least probable to the most probable states, to perfect equilibrium, to the symmetrical nirvana of entropic death, appears, escaping from the inexorable law, acting like a force from an "outside" dimension, the sole directly given reality: the spirit, the breath, the "pneuma" of life.

And if we examine this physical cosmos, this macrocosm of forms of Einstein-Weyl and of de Sitter become finite, spherical, and even, if you like, geocentric[1] like that of the Pythagoreans, where beyond the Milky Way gathered the souls turned into spirits who had completed the cycles of incarnations through metempsychosis, we can see that this

1. Any point of the hypersphere can be chosen as the symmetrical center in the curved volume that makes up both the substance accessible to our understanding and its boundary. In order to grasp our relationship to the hypersphere, we must always refer to beings "without density" who can glide along the surface of an ordinary sphere. If these beings were geometers or astronomers, they might one fine day discover that their "world" was three- and not two-dimensional, then demonstrate the existence of the third dimension. This is what has happened to us with the fourth dimension.

extra-dimensional "élan vital" leaves morphological traces of its passage through the world of pseudo-matter that even differ from the simply mathematical perspective of the configurations of crystalline balance imposed on inorganic systems by the principle of least action.[2] These geometrical types and these "pulsing" rhythms that are *never* found in inorganic matter, which appear like the signature stamp, the sign of life's reactions (growth and reproduction), which cause the Hamiltonian geodesic tapestry to sway as if blown on from the outside, are, as has been noted on several occasions, pentagonal symmetry and the pulsations, the spirals of growth akin to this same pentagonal (sometimes decagonal) symmetry through the intermediary of the twofold additive series of ideal homothetic growth (Φ series), or its asymptotic approximation in whole numbers (Fibonacci or F series), both series having as their directing number the algebraic invariant $\Phi = \frac{\sqrt{5}+1}{2} = 1.618...$, the golden number or ratio of the golden section.[3] This is especially present in plants, marine organisms, and the human body.

It is therefore through a brilliant intuition that the Pythagoreans

2. Vladimir Vernadsky in *The Biosphere* observes that Gibbs's law of phases only applies to living matter.

"Living organisms, energy-transforming mechanisms, are formations of a particular kind clearly distinct from all other atomic, ionic, or molecular systems, building matter from the earth's crust outside of the biosphere while at the same time being part of the matter of the biosphere. . . . Their energetic character, as manifested in their multiplication, cannot be compared geochemically to the inert structures that build inorganic matter. . . .

"The existence of a fundamental difference (that appears immutable) between living and inorganic matter can be considered an axiom that may, at some time, be fully established."

It should be noted that Vernadsky was not a vitalist philosopher but a chemist and biologist, a member of the Science Academy of Leningrad.

3. Let me remind the reader that the F series, the development of the tenth kind of proportion established by the neo-Pythagoreans, is the series: 1, 1, 2, 3, 5, 8, 13, 21, 34, 55, 89, 144, ... , in which each term is the sum of the two previous terms. The ratio between two consecutive terms tends toward that of the golden section, Φ, and as the strictly geometric series at the rate of Φ ($1, \Phi, \Phi^2, \Phi^3, \ldots, \Phi^n$) also plays this property in the twofold additive series (because $\Phi^n = \Phi^{n-1} + \Phi^{n-2}$), we could say that this Φ series is the model, the ideal paradigm of organic growth, for which the F series is the practical approximation (discontinuous, with whole numbers) realized in nature.

chose the pentagram as the condensed symbol of life, and which their descendants, the neo-Pythagoreans, Gnostics, and Kabbalists, made the emblem of man as microcosm.

Even if life leaves rhythmic traces of its passage through the world of forms, and souvenirs of its own "law of number," it seems that, conversely, certain periodicities of an astronomical nature have an undeniable effect on living beings, without even concerning ourselves here with the great cycles—the "Great Year" of the Pythagoreans, which is $25{,}920 = 72 \times 360$ years in duration (this is the approximation of the cycle of the precession of the ecliptic poles), the Drayson cycle (31,756 years, in which, in addition to the precession of the line of the poles, the annual movement of the great ecliptical axis is taken into account)—that are supposed to lead to a perfect recurrence of the great "seasons," which have the glacial periods as their winters and floods as their springs, thereby determining the migrations and extinctions of races and the fate of civilizations. The role of the seasons and the succession of day and night in life's rhythms seem sufficiently obvious to forgo further examination of this here. We also have the recurrences and more complex actions due to solar flares (eleven-year cycle), then a meteorological cycle of around seventy years due to the combination of the previous cycle with the recurrence of the respective positions of the Sun, Earth, and Moon.

Even without adopting the theories of Fliess[4] and Svoboda on the decisive influence of certain dates and certain cycles of days in the life of every human being, it is impossible to even barely scientifically study the influence of the lunar cycle of twenty-eight days on women,[5]

4. The date of conception with respect to the full moon is the starting point for these computations that use twenty-three-day phases for men and twenty-eight-day phases for women, combined with the seven-year cycle noted by the Pythagoreans. This system claims to be as rigorous in its "statistical" predictions as that of the planetary correspondences in astrology.

5. The influence of the lunar cycle on the female physiological metabolism (menstruation) and its psychological consequences has been corroborated by a long series of observations (particularly in Danish hospitals). It was established that the dates of menstruation oscillated around that of the full moon for the vast majority of women with normal periods, much like the "law of large numbers." For each individual,

reproduction, and love in general without taking into consideration the stock of truth in folk beliefs on this subject.

Many are familiar with the truly "astronomical" regularity that every year in a certain zone of the South Pacific on the eve of the full moon preceding the austral summer solstice sees millions of large sea worms rising to the surface and covering the sea with their intertwined forms. This worm is the *Palola viridis* or "palolo," as it is called by the natives of Fiji and Samoa, who are quite fond of it and take advantage of this nuptial solstice to fill their canoes with this living manna. The Romans had taken note at an earlier time of the lunar cycle of a certain edible spiny sea creature of the eastern Mediterranean that regularly produced its spawn at the time of the full moon. The sea thimble jellyfish (*Linuche unguiculata*) of Haiti always released its eggs at eight o'clock in the morning, and so on.

It appears undeniable that, while not in a scientifically predictable way for individual fates as astrologers would like[6] and as, within certain limits, some supporters of Fliess believe, the "constellations," which is to say, the reciprocal positions of the stars at a given moment, and the "rhythms" governing these positions, have a very substantial effect on Earth's life cycle, and even more exactly on the passions and destinies of human beings[7] and nations.

(cont. from p. 353) the influence of the moon combines with the state of atmospheric electricity and the momentary electrical "tonus" of the given individual. Hence the discrepancies, which can sometimes be several days.

6. I did not include astrology in this study because this discipline, despite its scientific context and the considerable role it played in antiquity and the Middle Ages, is not part of the specifically Pythagorean domain but originated in Chaldea. The incorporation of time into a spatial dimension, which results from a certain interpretation of Einstein's ideas (see later) has naturally given birth to a new theoretical explanation for cases of clairvoyance.

Some verifications of the *Centuries* by Nostradamus (published during the rule of Henry II) are odd; for example, the prediction of the execution of the "Great Montmorency" (under Louis XIII), which gives the name of his executioner (Cleyrepeyne), and the announcement of a new era (Republican calendar) in 1792.

7. Gustave Cohen has pointed out the role played (for three consecutive years) by the date of November 10 in the life of Descartes. The mathematician Cardan claimed to have similar "illuminations" every year on April 8.

And the selection of Hecate, the dark lunar hypostasis of the Thracian Artemis as the goddess presiding over spells and enchantments, was quite justifiable.

It could be said that on our globe at least, a living "macrocosm,"[8] a "biosphere,"[9] actively comprises all living things, including the pseudo-material substance of their bodies (a substance that as we saw earlier is extremely fine and gaseous, pure undulating forms) and the rhythms created from its life force, its "pneuma." This living macrocosm, also considered as a single organism, a collective personality with communicating minds, or as the juxtaposition of independent organisms, appears to be, from the perspective of life and reproduction, influenced by the planetary and stellar rhythms mentioned above. Leaving aside the question of the origin of life (why and how life appeared on our planet),[10] we can now see that to the theory of evolution through successive and increasingly "successful" adaptations to external conditions (environment, climate, food)

---

8. Of the *natura naturanda* type, a decadic macrocosm as opposed to the inert, hexadic macrocosm that was the *natura naturata*, the hyle of the Neoplatonists.

9. "An awesome power is at work every moment on our planet, a power comparable by the magnitude of its efforts to those being studied by the physicists. This power is life. In four and a half days, one bacterium can produce a progeny of individuals that could be expressed by a number with *thirty-six* zeros. Their volume would be equal to that of the world's oceans and their weight at least of the same order of size. In five years, one paramecium can produce a mass of protoplasm that would be 10,000 times the size of our globe. The great strength of nature, visible here, has shaped our planet. All the free oxygen in our atmosphere is produced through the work of green plants. And the entire atmosphere travels back and forth through living beings several times. We could describe the atmosphere as an appendix of living nature. This same strength created the continents. Enormous geological layers are made up solely of the remnants of skeletons and shells, or again what remains of vast forests." André Mayer.

"Living organisms have never been produced by inert matter. Through its life, its death, and its decomposition, the organism incessantly returns its atoms to the biosphere and takes them back, but living matter penetrated with life always draws its genesis from life itself. . . . There is no chemical force on the Earth's surface more immutable and in its final consequences more powerful than living organisms taken in their totality." V. Vernadsky, *The Biosphere.*

10. The most plausible hypothesis on "how" life appeared on our planet remains the theory of Svante Arrhenius: germs, infinitely tiny spores from interstellar space that under pressure by cosmic radiation entered our atmosphere. This would not exclude

of the initial "quantity of life" (scattered spores or the ancestral "great sponge," *Bathybius,* the concrete "biosphere" encircling the world's oceans with its gelatinous armature) has been added—explaining and justifying it—the Bergsonian notion of creative evolution, the élan vital or life force, a vertical shower, a continuous series of fireworks from which fall like cinders corporeal forms momentarily illuminated by the eternal flame. In the first part we examined prosodic and musical rhythms as emanations and reflections of phrases suitable for this breath, this life force, and noted that they introduce into musical or lyrical expression modulations in time or rather *duration.* This notion of psychological duration as a model from which we forged after the fact a concept of scientific or absolute time really flowing in an irreversible direction, independent of observing and perceiving consciousness (just like we forged the concept of geometric continuity in space as a reflection of the continuity of our own experience, it too is the result of the pressure and tension associated with the surging operation of our life force) is also specifically Bergsonian. It is life that would introduce duration[11] into our universe (the complete macrocosm this time, thought, life, and pseudo-matter, not only the Hamiltonian world of inorganic shadows).

(cont. from p. 355) the successive arrival of living germs; thus an apparently spontaneous generation could seem to appear before our eyes. However, I do not have the impression that the trials of spontaneous pseudo-generation in vitro are close to succeeding.

11. The most penetrating analysis of prosodic and musical rhythms as reflections of the flow of living time was made by Ernst Levy in a speech to the First Congress of Rhythm (Geneva, 1926). He drew a distinction between "meter" and "rhythm" and related them to two different physiological functions. He believed that meter corresponded to the normal frequency of the human heart (80 beats per minute) and rhythm to respiration, which provided the prototype of rhythmic process (a slow, deep breath lasts around 12 seconds, which corresponds to 4 measures of 4 beats at 3 seconds per measure). Levy also drew a distinction between two modalities of rhythm as such: the normal primary, iambic form (downbeat on the exhalation) and the trochaic rhythm (crusis, the downbeat falls on the inhalation), with this latter corresponding to an exceptional mental and physiological state. "Rhythm is essentially nonmetrical; its regularization is a development of meter's influence. The identifying feature of the rhythmic phrase is, on the one hand, an inner tension leading toward a culminating point followed by a break, and on the other hand, a possibility of limited material duration."

In chapter 7 we saw that Plato and the Pythagoreans believed in a kind of universal soul, Panpsyche, in which were combined on one side men, animals, and all living creatures, miniscule as they might be, and on the other, the "daimons," the supermen or genies, the gods, and lastly, God.

The Bergsonian theory of the élan vital and the modern ideas about collective personalities are far from contradicting this hypothesis of collective souls, or even a universal soul, from which all "individual" souls are issued as emanations, destined to perhaps rejoin it one day, after having traveled and exhausted what the Gnostics called the cycles of earthly generation and the play of pseudo-material forms.

*
* *

This question of souls or collective personalities, however, is worth lingering over; it will first enable us to make a logical distinction between the notions of personality and consciousness. Consciousness (not moral conscience, but the feeling of existing, of being, "awareness," that can develop into self-observing thought, consciousness of having consciousness) does not necessarily involve the feeling of individual personality, although there is no personality without consciousness. We could define personality (the feeling of having a personality) as the feeling of being isolated from the rest of the universe, and the feeling of being "someone" with his own past (the memory of his past), accompanied by the awareness of its duration.[12]

We can try to classify the various degrees of consciousness of living organisms according to their level of evolution, from the vague vegetative sensation we could assume to a mushroom to the "I think therefore

12. While we saw in the previous chapter that the label of epiphenomenon cannot be applied to consciousness without contradiction (for which it was invented), the same does not hold true for the perhaps temporary personality attached to consciousness during a defined life. We can without contradiction in this instance assume that the feeling of being a unity, a real, continuous individuality distinct from the rest of the world and other similar individuals, which is normally felt by superior organized beings, does not correspond to an absolute reality, and that the personality is "epiphenomenal" in relation to consciousness. It is an illusion of perceiving, discerning consciousness (development of simple "awareness" or vegetative consciousness); this is the Buddhist theory.

I am" of *Homo sapiens.* Even in the individual who is unaware of his temporary personality, we can assume this rudimentary world system: "there are irritating states and moments," and "there are pleasant states." These diverse degrees of awareness can be encountered in the same individual at various phases in his history. The zero in this gradation would not be dreamless sleep but the case of the suspended animation of certain dried organisms (*Rotifera*) that once moistened can be restored to life after years of apparent death. Personality, consciousness, and the subconscious are obviously suspended during this period of dessication.

But one can also, and this is what concerns us now, consider the organisms made up of several individuals connected physiologically, and seek to determine at just what point the plurality of individuals forms a plurality of awareness, with or without global consciousness. This is the question of collective personalities.

A typical example of a colonial organism raising this question is the case of the Siphonophorae, made up of a certain number of individual zooids, each of which is a medusa or polyp that, if detached from the others, can reproduce through budding and rebuild a colony. But the zooids have different forms and functions in the colony. Division of labor exists here, and the individuals are divided into engines (polyp-sculls), hunters (casting lasso filaments with a paralyzing fluid), eaters (they swallow and digest for the whole community), and reproducers (male and female). This is somewhat similar to the republic of termites except that the individuals are all physiologically connected, which explicitly allows some of them to eat for everyone. But the colony functions, swims, and finds its way as a single individual.

Marine organisms (especially the colonial medusas of which the Siphonophorae are a variety) offer many examples of this type in which, on the one hand, the component individuals or polyps behave as if they had a separate "conscious" life and, on the other, the resulting colony itself acts like a single animal with its own consciousness and personality.

The question of collective personality posed this way leads to a classification of this kind:

INDIVIDUALS OF THE FIRST ORDER

A. Protozoan (unicellular organisms)
One Cell. One Consciousness

B. Metazoan non-colonial (multicellular organisms); this consists of the higher animals and man.

A large number of cells. A consciousness (and a personality).

Here is the place of an intermediary class, B', in which the individual, such as the hookworm or tapeworm, is made up of a head or scolex containing a principal nervous mass and a series of identical segments called proglottids, each containing a complete system of reproductive organs and motor organs. One segment detached from the rest of the animal will continue to live independently and can become a complete animal.

INDIVIDUALS OF THE SECOND ORDER

C. Colonial organisms formed of identical individuals (with no other differentiation except sexually, if called for), combined and connected physiologically. There is no division of labor. This class includes:

1. Colonies of unicellular organisms, like certain colonial diatoms who, while reproducing endlessly through simple segmentation, remain linked in chains. The purely vegetative "psychological bond" between colony members is quite loose here, and generally one cannot speak of this as cooperation or association in the strict sense. But in some colonial protozoans like the *Volvox globator,* the mobile colony acts as a whole, with the colony members being contained inside a commonly secreted, transparent, organic membrane.

2. First, colonies of identical multicellular organisms, such as social Anthozoa, which include the "madrepores" or corals, strictly speaking. All the polyps of a coral colony are connected by a physiological serum circulating between their alveoli and even by a fleshy membrane

covering the limestone exoskeleton (the coral as such) inside of which the alveoli open. Second, colonial Ascidians, whose individuals (a kind of sea cucumber of the *Tunicata* genus) making up the floating colony are identical to the solitary members of this species, except instead of each possessing a sac or membrane—the envelope that gives the genus its name—they are all enveloped in a common membrane, which is often cylindrical. As the respiratory orifices of all the individuals (ascidiozooids) are arranged on the periphery of the membrane, the excretory orifices open in the center in a cloaca hole (of the common sac). *Botryllus violacens, Pyrosoma elegans,* and so forth, are examples.

D. Colonial organisms made up of physiologically connected individuals as in C, but with morphological differentiation (division of labor) in accordance with the duties assigned the individual. The Siphonophorae mentioned above fall into this category; they belong to the family of colonial medusas (a group of Hydrozoa) who offer the most interesting examples of mobile colonies of individuals with a pronounced morphological differentiation. Each individual acts in accordance with its defined function as if provided with an independent consciousness, and the colony also behaves as if, above its component individualities, it possessed its own individuality and overlying awareness, corresponding to the overall system considered as an isolated animal. The component polyps generally possess as a common organ, the "bell" of the super-medusa, which can be used as an organ of rapid propulsion beneath the waves, as an air-filled floater on the surface, and like a veil in certain varieties.

Another interesting case of a colonial animal in which the component individuals of the colony considered as a whole exhibit overlapping "personalities" is the colonial freshwater Bryozoa called the *Cristatella mucedo.* The component "polypides" (on a higher level of physiological development than the Hydrozoa polyps described earlier) stand up like the trees of a little forest on a common muscular "sole," like a slug, which allows the members of the colony to communicate (the colony is made up of two kinds of individuals: eaters and reproducers), and permits the whole entity to move about by creeping exactly like a slug.[13]

INDIVIDUALS OF THE THIRD ORDER

E. Social or colonial organisms with morphological differentiation of their constituent individuals based on their duties (as in the preceding

---

13. Other kinds of organisms forming colonies with functional differentiation of the constituent individuals include:

Bougainvillea fruticosa: a colony of hydroid zoophytes with a shared attachment stem ("stolon"). Dimorphism: fixed eater polyps in the form of hydras and reproductive polyps that detach themselves in order to live apart from the swimming medusas, during which time they feed themselves.

In *Clavatella prolifera* once the reproductive polyps have become detached they no longer swim but crawl and have an intermediary form between hydra and medusa. In *Tubularia indivisa,* the reproductive polyps grouped at the base of each feeding polyp are degenerate medusas that do not detach but produce their eggs on site.

In Millepora, a colony of coral-form hydrozoa that, like the preceding examples, are members of the Coelenterata phylum, we have trimorphic individuals. These individuals are "gastrozooids" (mouth and stomach) and fixed "dactylozooids," (mouthless polyps with a tentacular crown), and mobile reproductive medusoids that emerge from medusoid buds (blastostyles). These medusoids have no mouth or tentacles but only their swimming organ (bell), their battery of aggressive filaments with a paralyzing solution, and a large quantity of eggs. Once they have swum a certain distance and laid their eggs, they die as they are unable to feed themselves.

Bryzoa also offer colonial organisms with a fixed or crawling stem, and can be dimorphic or trimorphic. For example, *Bugula turbinate;* the skeleton of the colony is made up of cornea-like capsules arranged in branches, each capsule contains a complete polypide with tentacles, stomach, and an egg-producing ovicell. Another polypide stands outside each capsule. This "avicularium" looks like a kind of bird's beak or crab claw on a muscled stem. Its beak catches any organisms passing within range and holds them until they decompose, meanwhile the currents created by the tentacles of the primary polypides pull the decomposed particles into the feeding holes. In *Bugula bicornis,* [*Camptoplites bicornis* is now the accepted name—*Trans.*] each capsule is flanked by two avicularia on a very long stem; in *Flustra foliacea* each of the main polypides has the reproductive organs of both sexes, and their embryos emerge from the ovicell and swim about for several hours before attaching and giving birth to new colonies by budding. In other trimorphic types, the main polypide is not only flanked by an avicularium or bird beak, but also be a long individual "sweeper" (vibraculum) that drives intruders away and cleans its neighbors.

class), but the individuals are not connected physiologically (meaning in a "visible" fashion).

This is the case with the social insects, Hymenoptera (bees and ants) and Orthoptera (grasshoppers and crickets).

The case of termites is most interesting; trimorphism is present: sexed individuals, workers, and soldiers (in some cases there are two types, normal warriors with large mandibles and "machine-gunners" or rather "flamethrowers" who shoot out jets of corrosive fluid).

Close study of the social life of these extraordinary insects (for example, Maeterlinck's *Life of Termites*) is enough to show how natural the transition from the previous class (D, colonial organisms in which the individuals communicate physiologically) was; one could even say it was unconscious.

What Maeterlinck called the soul of the hive or the soul of the ant-hill seems to govern the functioning of these insect communities with a formidable reliability of guidance. The collective instinct here appears conscious, implacably conscious: "everything takes place as if a community consciousness and even personality existed."

This takes us by a completely natural gradation to a second kind of collective organism of the third order in which, as in the previous case, the component individuals are not "visibly" connected and in whom the morphological differentiation does not appear, only the functional specialization:

F. Human society organized in the form of a clan or national state.

Here, too, the transition is natural and appears to take place by degrees depending on flexibility in a direction that is favorable to perpetuation of the species. There is no essential difference from the viewpoint of the psychic bond between the individuals who make up the groups, and among the individuals of the second and third order—C, D, E, and F.

Because types E and F exhibit a unity of purpose to an outside observer and global consciousness that is as "visible" as the colonies of types C and D, it is valid, as a hypothesis, to believe in the actual existence of this soul or collective global awareness (overlying the consciousness of the individuals, by supposing that even though the physiological bond has been undone or removed among the component individuals, "psychic" communication may still exist. The phenomena of telepathy, collective reactions, group psychoses, and so forth, make this a plausible hypothesis.

In any case, "everything takes place as if" these global consciousnesses of societies or states existed, and we say: "France is suffering . . . , France hopes . . . , France will not accept . . . , France remembers . . ." as if the conscious, global personality "France" existed in addition to the personalities of French citizens. Those who do not believe in the existence of these type-E and type-F global consciousnesses (it can never be proven, as only entities of the same order could perceive them by first perceiving themselves) can in complete tranquility apply the label of epiphenomenon. It is yet another case where this statement does not imply a contradiction.

Before leaving this question of integrations, mergers, and divisions of consciousness or personality, I would like to cite a few examples illustrating the impact of experimental biology on a problem that is purely metaphysical in appearance. We find cases of individuals dividing or merging not only in protozoa and paramecia (who have alternating phases of reproduction through segmentation and reproduction through sexual coupling, the two individuals that couple merge together in such a way as to form a new rejuvenated individual), but even in multicellular organisms.

For example, many worms, starfish, and sea anemones, cut in half, can reform into two distinct individuals, and certain parasitical worms in the liver also experience alternating reproduction, one phase consists of the sexual union of two adults (as in the protozoa paramecium) who merge and become one single individual. In both cases we are at liberty to ask what is the link between the consciousnesses and memories of new individuals formed directly by sectioning or merger, and their "parents."

The personalities of individuals produced from the same "stock" can continue to communicate in space and duration alike. The chain, the continuity in duration is that of individuals produced by successive "ovulations" of one same ancestor; the perfect chain in space would be the whole of colonial individuals communicating physiologically. The two chains coexist in an ideal coral colony that nothing could disrupt. The "global consciousness" of this colony is renewed by minute fractions (death of elementary polyps, appearance of new polyps through budding), and the continuity would be so perfect in space and duration that this "global consciousness" would have the impression of always being the same, although at the end of a certain cycle, all its "elementary consciousnesses" would have been replaced by others.

In the final category of living collectivity mentioned above, that of human society, whose most organic type would appear to be the nation or the national state, the rejection of morphological specialization permits the human being—by moving away from the path of instinct in order to reserve for himself according to the Bergsonian ideal "the largest sum of indeterminacy possible"—to dominate matter and the elements by thought and scientific invention, and to extend his range with tool and machine, truly making himself the king of creation.

No point will be served by placing any additional emphasis on the connections between this problem of collective consciousness, the Bergsonian theory of the élan vital (life force), and the Panpsyche of the Pythagoreans and Platonists. But before leaving the domain of biology, I would like to cite two more cases of "metamorphosis" in the insect kingdom that strangely evoke the Pythagorean and Orphic ideas on the successive rebirths of the soul as symbolized in the tests of the mystes at Eleusis, for example. The first is that of a locust that the Chinese call "the seventeen-year cicada" and whose larval stage underground does in fact last seventeen years, to culminate in a year of "intense life" like a cricket. Hence the symbolic jade cicadas sometimes found in ancient Chinese tombs. The other case, which is even more bizarre, of a cycle of transformations of the same "personality," is provided by what happens in the four utterly different stages in both form and kind of life—before emerging from its final slumber as a chrysalis to live the normal life—of a scarab.

If we were to consider the whole of humanity as a collective personality that has detached itself downriver from Panpsyche to attempt its proud adventure, we could say that it appears to be torn between an insect ideal (the communistic termite mound)[14] and an ideal of a constructive demiurge (Caesar, Napoléon).

We have thus recorded the intervention by a third magician in the most recent evolution of European thought. This intervention is no longer in the domain of the abstract sciences or mathematical physics, but in that of metaphysics. Bergson is also in some way a "Kabbalist" in the noble sense of the word, by his ancestral perspicacity, his razor-sharp intelligence, the paradoxical abysses his thought stands alongside without slipping into, and lastly by the strict scientific basis (pure mathematics and physicochemistry) of his intellectual training.

It was in fact through mathematics and biology that he recognized the inadequacies of the mechanistic explanation for life and consciousness. He also shows (and this is another way of expressing the mathematical criterion of the transcendency of life formulated several times over the course of this book) that life introduces, inserts, in determinism's physicochemical or pseudo-material world, indeterminacy and choice. This élan vital is a push "from behind" like that of the wind in the sails of a ship running downwind or broad reach, in a direction that is not that of "scientific" time (which does not have one) but that of duration. We could also say that the push, from another perspective (growth), manifests from "the inside outward," or rather (which amounts to the same thing, mathematically) comes from an external, transcendent dimension; this is why in the physical world of four dimensions accepted as a consequence of Einstein-Weyl's theories, one of the four dimensions (it does not matter which), corresponding to the former "scientific time," is assigned the imaginary coefficient $\sqrt{-1}$. It is

14. We should note that the "worker" (the soldier as well as the nursemaid) of the communist insects is asexual (or sexually atrophied, rather). Among humans, the sole harmonious communistic societies are monastic orders where the sexual issue has been eliminated.

parallel to the "psychological duration" of the "outside" observer, with it too being transcendent in relation to the ordinary space represented by the three other axes.

This life, riding high with destination unknown, has become again, through Bergsonian indeterminacy, a magnificent adventure, the "creative evolution." We see crossroads and bifurcations passing by on the unknown road; for example, the mysterious side road of the social insects—bees, ants, termites—in whom individual passion and intelligence have abdicated before collective instinct. It would have been sufficient for these insects to have grown a bit larger in size (like the wasps in H. G. Wells's novel who ate "the food of the gods") for them to have acquired, in addition to their almost infallible social instinct, a spark of organizational "thought" (like the giant ants in a novella by Arthur Conan Doyle that took possession of a town on the upper Amazon and were held in respect by the troops and navies sent to dislodge them), which would ensure, in a merciless war between the communist insects and man, the latter's defeat and extermination, or being reduced to going to ground in inaccessible retreats, shrunk to a small number of ferocious, hunted individuals like the giant gorillas of Mount Kivu.

But the human family appears to have escaped this war of extermination on the part of the social Hymenoptera or Orthoptera; it has greater fear of the infinitely small bacilli or bacteria than it does of ants or termites.

This human branching off, which has kept, according to Bergson's phrase, a maximum of living indeterminacy, which has refused to become morphologically specialized in a determined domain, allowing itself to grow claws, telescopes, wings, or fins, through the inventive power of its thought has managed to forge, to "build" the material organs that, as discretionary extensions of the body, have assured humanity domination and material kingship over the planet.[15] Having

15. Let's add (so as not to forget one of the characteristic elements of human society): who was it that imagined transforming and accumulating in metal or paper "signs," the "quanta" of potential, the manual or intellectual labor of its members.

avoided biological cul-de-sacs, it is to be hoped we shall also avoid the sociological impasses or "dead ends" like ossification inside a purely material well-being, or the return to the ideal of the termite mound advocated by the communists. It appears to be holding back for itself the possibility of all unforeseen circumstances as culmination, or even as a stage, a race of "supermen," of daimons, intermediary spirits between man and the deity, like those glimpsed by the ancients.

And, by holding the question of "god-men" like Christ and Buddha in reserve, we can count in this category of "geniuses" Pythagoras, Leonardo da Vinci, Shakespeare, and Goethe, as well as Alexander, Caesar, and Napoléon.

Lord Abernon's memoirs include a personal conversation he had with General von Kluck on the respective merits of the Allied and German generals during the war of 1914–18. It is common knowledge that German soldiers (and this applies equally to the general staff and individuals) who happily acknowledged the merits of Marshal Joffre and liked to say that they lost the war in September 1914, avoid by an odd, automatic "censorship" resulting from an as yet unhealed psychic trauma discussing the qualities of Foch and are unable to "see" the second battle of the Marne, the symbolic battle for France in 1918, as a strategic action; they see it as a series of unfortunate accidents, due to an internal defeatist attitude, to the blockade, to the "hypocrisy" of President Wilson, or, more exactly, to chance.

When forced to be more explicit because of a direct question of the English ambassador, von Kluck responded: "It is difficult to judge Marshal Foch, because he has something of the 'daimon' in him."[16]

I have said that the Bergsonian notion of life's "adventure," of

---

16. In the same order of things, a quip by Count Keyserling in his *Travel Diary of a Philosopher:*

"As always, astonishment is followed by sorrow. It is tragic to see one's intelligence outstrip its capacity for realization. Why am I not a god? Simply because I lack the physical power; it is the amount of energy available, nothing else, that distinguishes the metaphysician from the deity. If I had sufficient means, my ideas would become physical forms of their own volition and, while my thoughts grieved, worlds followed upon worlds."

psychological duration, of the different psychological durations as the sole periods of time provided with direction, agrees perfectly with the framework of Einstein and Weyl's four- or five-dimensional world, in which purely conventional, "scientific" time is any one of its axes equipped with the "imaginary" coefficient of transcendence $\sqrt{-1}$, because its direction has been made to coincide with that of the "duration" of the observer.

There would not in this case be rhythms of the universe except those rhythms of periodicities perceived by the variety of consciousnesses, either within themselves or outside. The periodicities that have not been seen or experiencd as unfolding sequences marked off in the duration of a consciousness would not exist (as rhythms) because they would contain neither time nor duration, only a spatial pseudo-periodicity, provided once and for all, as in a finished architectural piece or in the harmonic text of a melody notated on paper. And again, the architectural piece, the sonata, contains traces of living rhythms "lived" at least once and coinciding at that time with a living duration, or the reflection of a duration, one that can always be resurrected as rhythmic experience for a new consciousness; whereas the pseudo-rhythm death, that of stars, for example, would have elapsed once and for all—for the reason that natural phenomena, periodic or not, have neither future nor direction, no before or after, if perceiving forms of consciousness are removed from the universe.[17]

Or again, we can only artificially make these apparent periodicities coincide, in a period of time that does not exist, with the direction of what is to come and the psychological duration of the observers.

However, the phases of these astronomical "dead rhythms,"

---

17. This extreme opinion, an absolute negation of time in itself, or rather of an irreversible direction in physical time, has not been adopted by everyone. Some accept an "entropic future" independent of consciousness and life. This is the law of the growth of entropy (and the deterioration of energy—a form of the principle of least action) that provides them an irreversible direction. But they also admit that in fairly small regions, or those containing living systems, the direction of this physical future or natural time can be reversed, and that after complete deterioration (the thermal balance that is the

stripped of any cause and living impulse, can coincide so well with living rhythms that it shows that more is at work here than an artificial parallel or clever manipulation of the coordinate axes. The rhythms of the seasons, the lunar rhythms are, as noted earlier, more than simply perceived periods of time with respect to the minds of living beings: they have a direct influence on their metabolisms, their growth and reproduction cycles, and all the feelings connected to them. We therefore suddenly find ourselves, having combined these aspects of time in the theory of relativity and in Bergson's concept of duration, which appear to fit together so well, confronted by the following contradiction: (*a*) inasmuch as time, as a continuous flow endowed with direction, *does not exist* outside of its representations in the thought of conscious observers, there are no rhythms in time, there are no "natural" periodicities that exist objectively (outside the periodicities perceived by living minds with awareness, these perceptions become blended with other psychological currents deriving from the "becoming" of a consciousness in the flow and the direction of its own duration); but, on the other hand, (*b*) the not only physiological but psychological processes of living beings, the cycles of their individual and collective evolution, appear to be at least influenced if not regulated by natural rhythms, located outside any form of duration, any living pulsation, for example, astronomical time periods, although these, as we recently noted, have no reality.[18] The contradiction is quite serious, and perhaps may be the sole paradox in the audacious

fatal culmination of a physical system left to itself, without "outside" impulses) the irreversible direction of the future should vanish. ("Time will have lost its arrow . . . will have become pure extension. . . . Time will be eternity.")

18. A rhythmic living duration, a periodicity reflecting a living cycle, can, without any antinomy, influence another living cycle inasmuch as the psychological durations *exist* separately, or globally as future, as irreversible succession and directed flow, and have the right to have or transmit rhythms. It is the conventional scientific time of classic mechanics and astronomy that no longer has the right to engender periodic successions and especially influence living rhythms because, once turned upside down by the theories of Bergson and Einstein, they no longer correspond to any concrete reality.

skyscraper of today's mathematical metaphysics. I see only two ways to get out of it: the first, quick and clear-cut, states that by the light of the new cosmogony and new physics, and independent of Bergsonian epistemology and its "psychological duration," it is not only time that has been tossed overboard (this time, without light and without the consistency of the speed of its propagation, would never have meant anything), but space as well, and our paradox thus finds itself swept away with the attempt, which no longer has any sense, to construct an objective, physical universe independent of thought.

I would like to cite in this regard, a razor-sharp passage from an article by H. Wildon Carr:

> The concept of a finite universe thus entails new concepts for time and space. Instead of considering them as the preliminary conditions for mass and radiant energy, we must now envision them as created by mass and radiant energy. . . . We must rid ourselves of the idea of an absolute time and space, independent of the movement of masses. . . . It is possible that it is not thought that belongs to nature (to the natural order of things, to the physical universe), but nature that belongs to thought (which would be a derivative or consequence of it). . . . The question we must finally ask ourselves is this: "If there were no conscious thought, would there be a universe?". . . What is impressive in the new Cosmogony, is not the reality but the *unity* of the physical world. Without (observant) thought, not only would this unity have no meaning, but reality itself would not find its place in the world.[19]

Eddington suggests in the same way that not only the unity of the physical world but the permanence of the so-called laws of nature (conservation of energy and so on) arise from the structure of our minds, which draws from these laws only truisms resulting from conventional

19. "Some Reflections on the New Cosmogony," *Journal of Philosophical Studies,* July 1929.

definitions. Hence the observation by S. Alexander: "If all this aspect of balance and permanence formulated in the laws of nature (Eddington's 'identical laws') only arises from a tendency to permanency that resides in the mind, there is something practically miraculous in this presumed capacity of the mind to create the notion of permanence this way without being pushed to do so by the experience of external things" (*Journal of Philosophical Studies,* July 1930).

The other solution, one less radical and less anthropocentric, introduces God as a physicomathematical hypothesis. Astronomical rhythms would be the reflections and traces of a creative act or a transcendent thought and no longer be the "dead waves"[20] of Hamilton's balanced-wave models. Once they have been part of an irreversible duration (like the material forms of life or the traces of life, flowers, shells) they could be attuned to living durations, or at least perceived as real periodicities by them, and even influence them. Astronomical rhythms, which, if the world were stripped of conscious life, would no longer exist anymore in "time" than the spiral fossil of an ammonite shell, are relived as rhythms by the simple presence of life, and are capable of inserting themselves as rhythms into psychological durations—like the record on the gramophone turns back into rhythm and pulsation for whoever puts it on the player and listens to it.

For the hypothesis of the deity, of a divine or at least transcendent consciousness, which is not necessary for Bergson's vitalism from the creation perspective[21] (once the élan vital has been observed and accepted as experimental fact, all creative evolution unfolds from it without any need for an initial conscious creator: supermen, geniuses, gods, or God

20. Moreover, even the apparently "dead" waves of the entropic future could be the reactions determined by an impulse, a rupture of the balance that is so "unnatural" that it could also be attributed to an act of God.

21. It could paradoxically be said that the sole philosophical doctrine that cannot do without an aware God as creator and regulator is materialist determinism, for which living organisms are only admirable physicochemical mechanisms, automatons that have been put together so well it demands the search for their "author," if not more exactly a supervisor who puts the machines back together. The same contradiction can be found in sociology where it is the strictest proponents of materialist determinism (Marxists) who are seeking to redirect or break the course of capitalist evolution.

can appear as final results of the process) is, on the other hand, highly tempting when it comes to cutting short the difficulty resulting from the problem of the interrelations of "psychological durations" experienced by all living organisms. Could they be paralleled by distinct streams like the "particular times" of observers endowed with different speeds in the theory of relativity, or do they blend together (these durations lived by all participants in the adventure of life), despite their discrepancies and turbulence, like trickles of water into a large river? Are they exactly the reflections and emanations of one same transcendent "Duration"—that of the global awareness that we cannot attain directly any more than the rudimentary irritability or awareness of one of our body's cells can realize the unitary consciousness that transcends it?

In any case, as the apparent continuity of a geometrical space, as the apparent flow of an astronomical time heading in an irreversible direction, rhythm—this rhythm that we have seen as "number" playing such a major role in prosodic and musical expression—is introduced into the domain of sensation by the essential quality, the conscious impression, typical of collective or individual life, by the sensation of "duration in the process of being lived"; or rather, in contrast to these two illusions of time and continuous space, it is it (rhythm), the direct emanation of life—either as life-reproduction or life-thought—that permits the human being, alone of all the living organisms we know, to see himself, to express himself in his awareness of having awareness, and his ability to transcend and exceed himself.[22]

Rhythm-*gamos* and rhythm-*logos* culminate in knowledge.

And the rhythmic harmony of the cosmic symphony finally poses our final dilemma: Is this harmony the emanation and rhythm of God, who is living the adventure with us as "world soul," of which we are truly the

22. "The most exquisite pleasure of men, beneath the infinite variety of things, is the grasp of simple mathematical ratios. Made into sensation, it is art; made into concepts, it is science; nebulae that are resolvable or not from one same universe, aesthetics.

"Thus rhythm is everywhere. And in these disorderly spaces that surround us, the sole friends of man, the sole that offer reassurance in clear words, are rhythms." P. Severin, *Essai sur les rythmes toniques du français.*

reflections and "microcosmic" correspondences? Or could all this drama, so marvelously arranged, only be taking place in our minds; are harmony and unity only within us, with the entire world outside of consciousness being unknowable or even imaginary? This is Eddington's melancholy parable of the strange footprints on the shifting sands of scientific knowledge in which, as eternal Robinson Crusoes, we only find our solitary traces.

At least as Robinson, we have company on this desert island, a company that does not appear to be the illusory flights of fancy of just one of us, the me that writes or the you that reads, unless we are all simply images passing through the nonchalant daydream of a god who would not deign to create anything.

However, love, the dart of the genius of the race, or the quest for the "kindred soul," gives us the impression that even in this world with its epiphenomenal décor, this protean cosmos in which, all materiality having evaporated, we have been abruptly urged to make our way from the materialist factory to a farandole of pure spirits (or at least pure desires to live clad in waves), that voices other than simple echoes of our own are answering us.

> Life is both quotidian and cosmic. . . . Awareness of this awesome existence hardly reaches us except through human love. For a moment it gives us an exultation that puts us in touch with the universal, before we tumble back into opacity and everyday life. . . . The regret for this short and paradisiacal state is almost the only thing that explains sorrow spread like a mask of ashes over human faces.[23]

So, even if the hour never sounds for the divine banquet in which, like the epopt of Eleusis, we would be welcomed by the smile of the beloved, waiting for us on the other shore, we can still say that the sole feeling that gives us an impression of reality in this world, a spark of absolute reality, is love.

23. Francis de Miomandre, *Nouvelles Littéraires,* December 14, 1929.

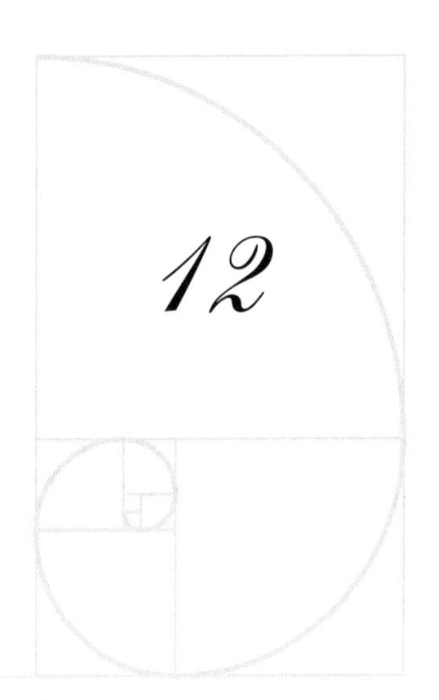

12

The Phoenix of Metapontum and the Duel of the Magicians

The wind is rising! . . . We must try to live!

Paul Valéry, "The Cemetery by the Sea"

Two and one-half millennia have passed since the catastrophe at Metapontum that cut so deeply into the roots of the tree of Pythagoreanism. The very stump appears to have been burned, the torch of the master overturned and trampled underfoot, with its sparks swept away into the night of antiquity. And yet, we have been able to glimpse that this flame, never truly extinguished, has been passed on from one torch to the next like that of the legendary racers of Olympus.

Like the luminous serpent of the mystai during the great night of Eleusis, we see emerging from the past and coming toward us not a single procession but various chains of torchbearers who went their separate ways from the altar of the main home and wandered in remote concentric spirals that are once more drawing near to the human mind.

Science first: the law of number triumphed there through the chain of Pythagoras, Archytas, Plato, the Alexandrians,[1] and then

1. If it was from Egypt that Pythagoras and the first Greek geometers learned respect for number and geometrical form, it was also in Egypt, in the city founded by Alexander, that the definitive "corpus" of Greek mathematics by Euclid, Eratosthenes, Diophantus, and so on was formed, the sole exception being Archimedes, who lived on the island formerly ruled by the "society."

Nicomachus, Leonardo of Pisa (Fibonacci), Luca Pacioli, Descartes, Hamilton, Cantor, Einstein.

We have seen, thanks to Einstein's final dissertation and the studies of Planck, Schrödinger, de Broglie, and Heisenberg, that the complete synthesis of the physical universe in number-ideas appears to have been achieved, with the help of one same instrument, group theory, and of one lone law, Hamilton's principle (of least action);[2] the "substratum" of the matter we have reached consists of points-numbers, intersections of immaterial waves, discontinuous sparkles in the countless pigeonholes of abstract "matrices" in which Archytas, Plato, and Nicomachus would have no trouble recognizing their number-ideas and their "figurate" grids.

Thanks to the characteristic property of transfinite series in which the whole can be equivalent to the part, reflecting it in its entirety, we have found in the hallucinatory series and processions of alephs and omegas of the Kabbalist of infinity, Cantor, the uninterrupted generation of numbers by the decad dreamt of by Nicomachus (to be exact, the very image of the "Great Continuity," itself a reflection of the flow of our internal duration),[3] and also the endless reflection of the same in the other, of the whole in the similar part, the abstract model of the great law of analogy.

In other domains of knowledge, the transmission of the torches occurs in underground passageways such as tunnels of initiation. The procession vanished for centuries; symbols and passwords continued to be passed down, sometimes in piety and sometimes mechanically,

2. Eddington's criticism has shown that the law of energy conservation and the law of gravity are only truisms, identities unfolding from the way in which the conventional elements included therein are defined. The principle of least action stands out as the sole "law of control" accessible in physics systems. It remains intact in the intra-atomic domain: "At all times every electron occupies or tries to occupy the position that requires the smallest expenditure of energy possible"—the Lennard-Jones potential.

3. These transfinite domains of Cantor paradoxically permit the figuration by whole series of not only the Archimedean geometrical continuity that appears to respond in fact to the flow of our consciousness, but even "non-Archimedean" continuities of higher density whose counterpart in the world of phenomena we have yet to meet.

by sentinels who no longer grasped their meaning; among these signs the magic pentacles grimacing like mysterious neon signs at the dark crossroads where the tunnels intersected, or on the pediment of ideological ruins crouching in the night; then all at once by the light of an intuition, of an explanation, an entire monument is illuminated. Like the moistened rotifer, the symbol has come back to life and resumed its quasi-magical form, and transforms, condenses, and shows the *analogy,* or even the rediscovered identity.

In my *Esthétique des proportions,* and again here, I have sketched out how the transmission of the geometrical designs of the Mediterranean builders took place, as well as their doctrine of the harmony between the whole and its parts, based precisely in the directing notion of analogy, and on its extensions: symmetry and eurhythmy. Building was the great ambition of Western civilization; dreams or a life of harmony were not enough. It was forced to build and organize desperately, heedlessly, not only ideas, systems, clear and well-proportioned syntheses like Alexandrian geometry or Roman law, but monuments, cities, empires; the Pharaohs raised their pyramids, Rome drew its paved roads and built its aqueducts, the Middle Ages covered Europe with its cathedrals; designs and techniques wended their way almost esoterically, like the abstract science that inspired them; architects and craftsmen religiously passed down procedures and rules whose profound reasons they had sometimes forgotten. They would be suddenly brought back into the light by the "daimons" who recognized the signs—Alberti, Pacioli, Leonardo, Dürer—but then the light vanished, artists forgot and lost the science of proportion, misplacing the precious legacy, loyal respect for royal geometry. The seventeenth-century architects translated Vitruvius without understanding him; the pulsing, "dynamic" symmetry of Plato, the *commodulatio* of Vitruvius, the *concinnitas* of Alberti became the static, mechanical symmetry, the death-in-equilibrium (repetition of identical elements on either side of an axis or "plane of symmetry") that we still know under this name. But the signs did not cease to be transmitted through other tunnels by dreamers who only half comprehended the symbolism, which they sometimes used to sound the troubled ocean of their desires.

However, unknowingly incarnating the white gods of Atlantis heralded by Toltec and Carib legends, the conquistadors landed in the New World, completing the great loop.

And as our fervor to build, our taste for exploration, our adventures in mathematics and science, and the synthesizing spirit have not abandoned us in the intellectual domain, it is through engineering science that our old Mediterranean notion of the cosmos—not merely harmonious and endowed with souls, but the source of a harmony that is defined and perceived geometrically, whose Organizer is both musician and geometer, and whose proportions, the relationship between the whole and its parts, are grasped as mathematical ratios and as musical chords, where the same word [*rapport*] means "relationship," "ratio," "reason," and "logos"—has been gradually coming back to life, regaining its effectiveness, compelled by our need to tame, harness, and organize the forces of nature. It is through our ideological journey "in quest of the absolute," rediscovered through the space and time dynamic, that—in accordance with Plato's prediction—we have been given dominance over the world. It is finally through the agency of the engineer that the architect has been returned to our midst, guiding his rediscovery of beauty in the truth of necessary forms.

It was initially in the New World that Western civilization rediscovered this beauty of pure volumes; the American engineer is the heir to the bridge builders, the geometer-builders who collaborated on the lighthouse of Alexandria, the aqueduct of Segovia, the Coliseum, and the Hagia Sophia.[4]

And little by little, thanks to the studies of Zeysing, Cook,

4. In the presence of the magnificent sequoias of California, which had already inspired the prophetic powers of Walt Whitman ("Song of the Redwood-Tree"), Count Keyserling also glimpsed in 1913 that the American coast of the Pacific would one day be the new Hellas of European civilization: "It is in America where we will complete our development, if we are to do it anywhere." *Travel Dairy of a Philosopher.*

Hambidge, Lund, and Mössel, the old Mediterranean geometric symbols passed down in the esoteric shadows have, despite the distrust of the "a-geometers," become again the "regulating designs" consciously used.

As we have seen, Zeysing rediscovered the golden section in its aspect of linear continuous proportion during the middle of the nineteenth century. As the ideal plane proportion (in the form of the rectangle governed by module Φ), it is especially highlighted by Fechner's experiments some twenty years later.[5] This golden-section rectangle—whose characteristic property as Thiersch, Timerding, and so on, explained was its ability to automatically suggest an "analogical" partition recurring in similar diminishing forms to the observer, in other words, to procure, at the same time as the insertion of the infinite within the finite, the impression of "unity in variety" (which it did in the simplest way possible: satisfying the hedonistic principle of the least effort, the minimal wasting of nervous energy)—established itself among others in commercial and industrial practice, independent of any aesthetic considerations.[6] Hambidge gave it the starring role among his "dynamic" rectangles that finally allowed us to grasp how the ancients were able to manipulate proportions that were irrational or "potentially commensurable," which is to say, when the surfaces formed part of an architectural or decorative design, its method of decomposing areas not only permitted verification of the proportions of the ancient drawings, but the creation of new harmonic compositions. This "Φ rectangle" has

5. In these tests, Fechner had a large number of subjects select the rectangle they preferred out of six rectangle types whose proportions were staggered from $\frac{3}{1}$ to $\frac{1}{1}$ (square); the vast majority chose the rectangle of the golden section, $\frac{1.618...}{1}$, followed by the rectangles $\frac{3}{2}$ (1.5, fairly close to $\sqrt{2} = 1.414...$) and $\frac{2}{1}$, then the square.

6. The customers of large German banks were quite surprised when in 1928 they were given checkbooks in a new format (they were shorter in length). This measure had been taken to introduce rectangular checks in what was called "DIN-Format," which had been imposed earlier by the "Committee for the Standardization of Forms" of the League of German Engineers for the bulk of rectangular forms used in industrial manufacturing. When the press raised questions about this mysterious "DIN-Format" or rectangular standard, they were told that this standard was the rectangle of the golden section, $\frac{1.618...}{1}$.

made a triumphant return in architecture and has done so in the recent blueprints of the most famous apostle and representative of the new tendencies, Le Corbusier.[7]

Here is an interesting detail: It was the first Benedictine abbeys (Monte Cassino) that saved the texts of Vitruvius and Boethius. It was in the Benedictine abbey of Beuron that a religious aesthetic of strictly Pythagorean tendencies was born (around 1870). Not being able to expatiate further here on the subject of Beuron, I will permit myself to cite some statements its founder (Father Desiderius Lenz) shared with a Dutch painter who was also a member of Saint Benedict's order:

> For many years I had meditated on nature and its ever-changing appearances until I came to the conclusion that simply copying nature would never lead to the artworks with the quality of the works of antiquity. The works of the early Christians and Byzantines, as well as those of Giotto, had shown me that geometry and geometrical division were the principal factors in the execution of their works, but I did not find the conscious and reasoned use of

---

7. I am citing Le Corbusier for his blueprints for the "Mundaneum" or World Study Center, where coordination of artistic and scientific studies could be implemented, planned for Geneva.

"The Mundaneum has been conceived as a rectangular city. The ratio between the width and depth of the rectangle is provided by the 'golden section,' thereby assuring a great unity and ensuring harmonious proportion will rule. . . . The departments of the Mundaneum are quite diverse. Each building is an independent entity that is more or less enclosed within walls that . . . open, however, onto useful sites, particularly to the right of the two main axes whose crossing determined the top of the pyramid of the World Museum. Both of these axes are established by the golden sections of the sides of the general surrounding wall. . . . In this way . . . the rhythm (of the World Center) is organized on the 'golden section,' the measure that has established the harmony of so many works throughout history." Plates 59 and 60 printed here are from a special issue of *Architecture vivante* (Morancé) on regulating designs and depict the blueprint of the Mundaneum and that of a villa in Garches in which Le Corbusier also applied the golden section by using the diagonal as the element for establishing proportion.

More recently, Le Corbusier integrated the golden section into his famous Modulor. Two scales of proportion were used here, both derived from the proportions of the human body.

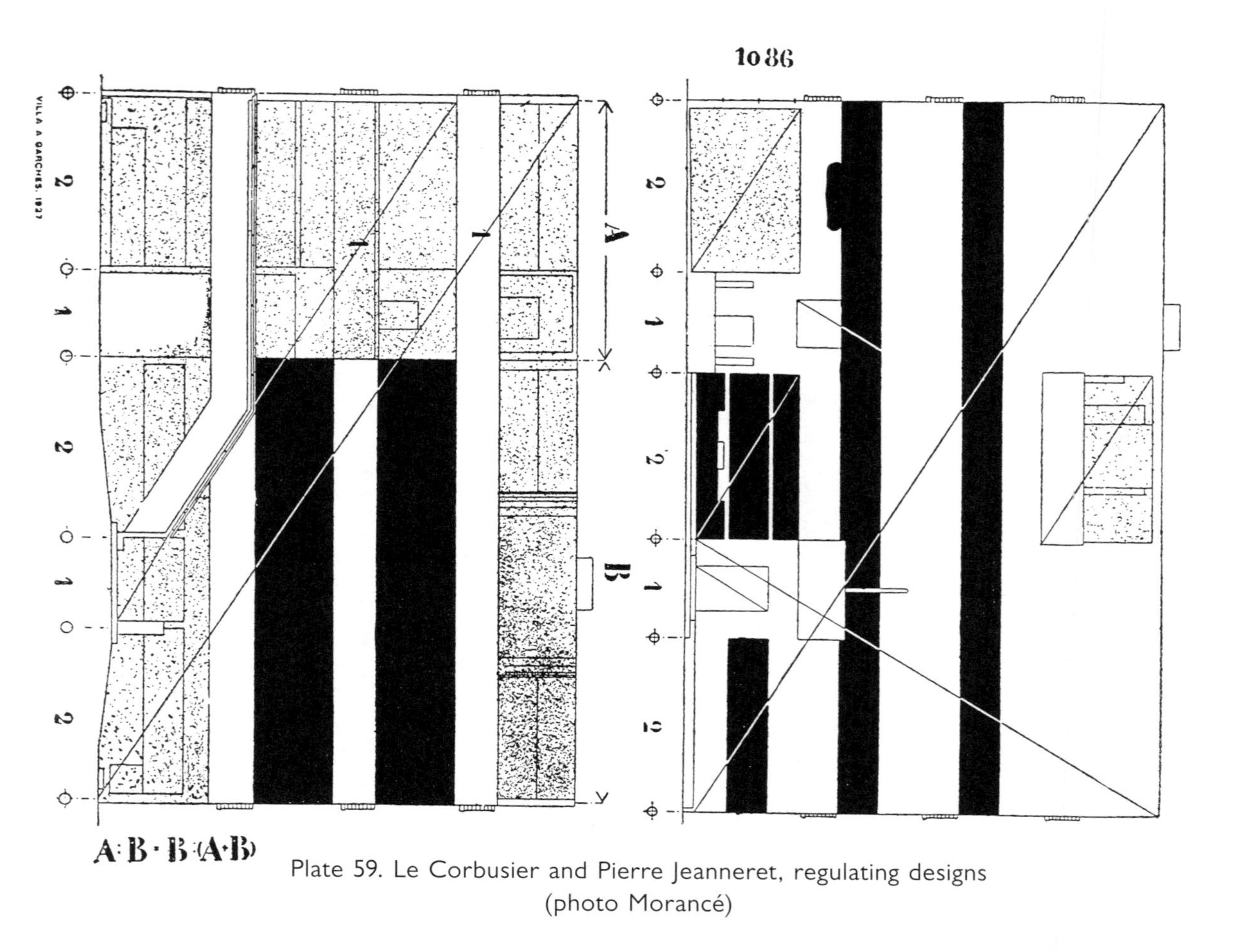

Plate 59. Le Corbusier and Pierre Jeanneret, regulating designs
(photo Morancé)

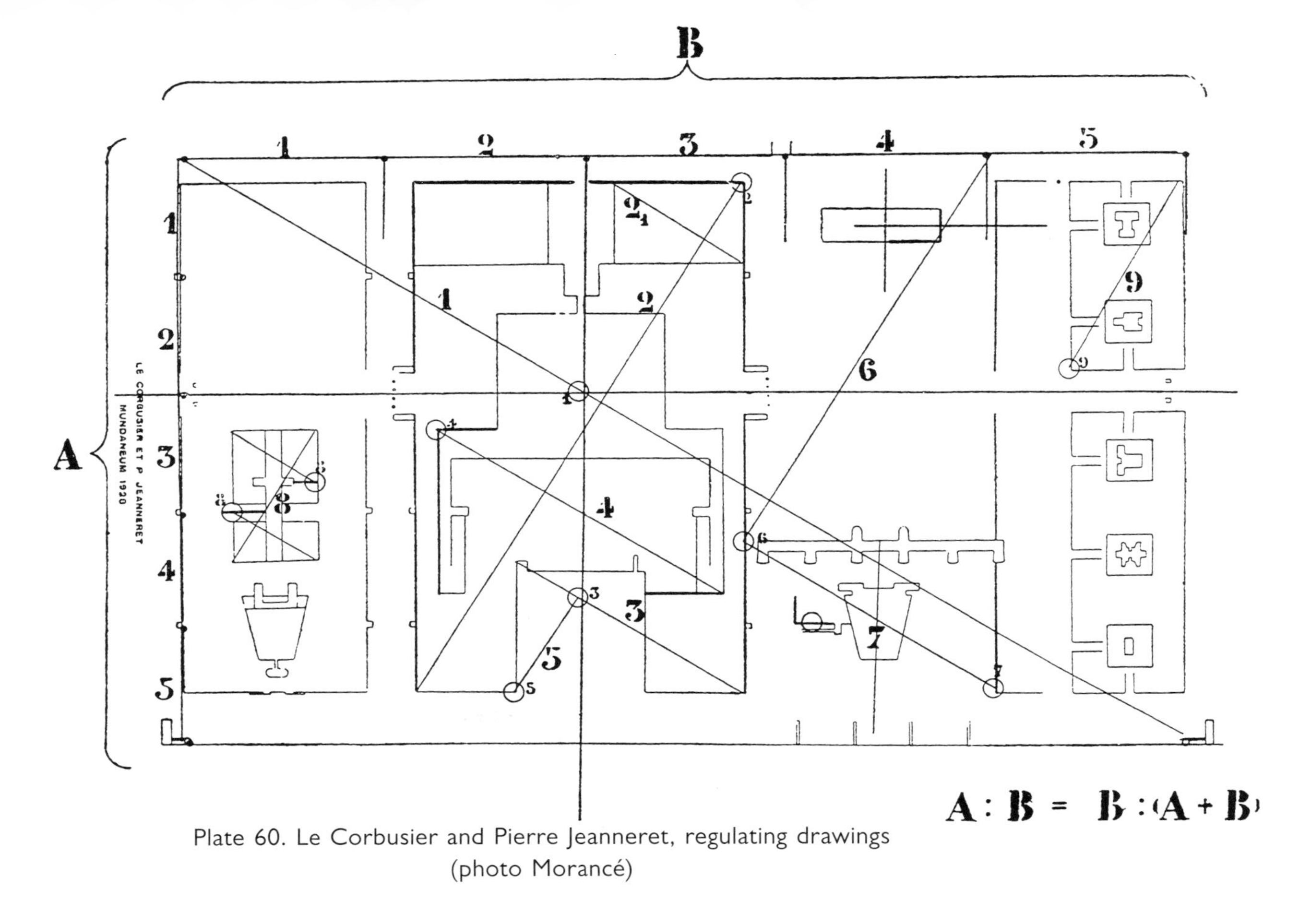

Plate 60. Le Corbusier and Pierre Jeanneret, regulating drawings
(photo Morancé)

> these essential means in them. . . . For the Greek masters of antiquity had employed exact laws in their measures and divisions. What were these laws? A long study of plant structure and the paintings on Greek vases helped my progress on the road of knowledge. . . . But it was through studying the great work of Lepsius in Egyptian temples . . . that my innate desire for number, balance, and order found complete satisfaction for the first time. It was here I found the religious sentiment such as I understand it. . . . And two factors seemed to predominate here to me: first, logic, a merciless criticism into the very depths of life's necessities; next, the notion of balance and harmony of the dimensions. The domain of music revealed this perception of the harmony of magnitudes to me. And then all at once I clearly saw that just as music is founded on melody and harmony, the plastic arts are based on the ratios of numbers. The mysterious force of certain simple ratios, both rational and geometrical (irrational) speaks to us in classical temples and statues. The secret of their beauty is here. . . . Number is, in fact, divine. (William Verkade, *Die Unruhe zu Gott*)

Today the star-shaped symbol of "divine proportion," the pentagram, has shaken off the dust of the magic grimoires, and with its two amplifications to three dimensions (the two star dodecahedrons) and the two "Platonic" polyhedrons dear to Pacioli (dodecahedron and icosahedon) has regained its place in theoretical and applied aesthetics.[8]

By meditating again on the "science of space," on proportions, as did Plato, Alberti, Piero della Francesca, our generation may perhaps reconnect the golden chain, not only in architecture but in all the arts of form.[9]

---

8. The dodecahedron, the icosahedron, and the two star polyhedrons (star dodecahedrons of the first and second type) are again, as during the early Renaissance, employed as hanging lights. And, a major Parisian perfumer gave the bottle of his Essence Rare the shape of the icosahedron.

9. As examples of "Cartesian" studies in this direction, I am reproducing several paintings by D. Wiener in plates 61, 62, and 63.

Plate 61. Harmonic variations on the dodecahedron
(D. Wiener)

For simplicity, the frank adaptation to purpose that has given the geometrical spirit back to architects has also led the decorators back toward harmony. Le Corbusier and the "Esprit Nouveau" group have contributed powerfully in France to greatly sweeping away "styles" as well as the anemic residues of that "modern style" of 1900, which was so ignorant of life's geometry. The progress of taste, of the sense of proportion in the last ten years is striking. Let us hope that it returns with the confidence it had among those who were once the great masters of form on the banks of the Nile.

Equally essential to the artist as knowledge of pure geometry is knowledge of what I call the geometry of life, the special study of the human body, based on the "dynamic" analyses (Hambidge), of which the ancient Pythogorean symbol, the emblem of the microcosm, is also the condensed summary. It is not only a question of the empirical measures of aestheticians seeking to restore by means of the golden section the canons of Polykleitos, Phidias, or Leonardo; it is men of science, naturalists, botanists, and biologists who know nothing of neo-Pythagorean aesthetics—a domain utterly foreign to them—who have in biological morphology scientifically made explicit and explained the presence of the golden section, of the Fibonacci series and its related pentagonal symmetries, and who, by proving mathematically[10] that these forms and proportions cannot, conversely, appear in crystalline configurations as such, which are devoid of life (those in which hexagonal and square symmetries, especially the former, rule unchallenged), have justified the extraordinary intuition of Plato and the Pythagoreans on the subject of the roles meted out to the different numbers in the world of forms.

---

10. Cf. *Esthétique des proportions,* and chapter 2.

We can crudely sum up why square and hexagonal symmetries share the *inorganic* crystalline grids (homogenous assemblages of points in space) where pentagonal symmetry *never* appears by recalling that it is not possible to "compactly" divide the plane into pentagonal sections, whereas its division into squares, hexagons, or triangles is easy by the simple fact that the vertices of the square, hexagon, and triangle are submultiples of 360°, which is not the case for the angle of 108°, the angle at the vertex of the pentagon.

This is why the great German botanist and naturalist R. Francé was compelled to acknowledge in the golden section a special character of algebraic invariant connected with the functions of life.

> Plato who, as we know, incorporated Pythagorean doctrines, wrapped the theory of the proportions of the human body in mystery. . . . And this knowledge found refuge through the centuries in the Hermetic and magic tradition, for it now appears to us more and more clearly that the secret so well guarded by the lodges of medieval masons and stone carvers, was nothing other than the canon alluded to by Vitruvius, the secret Pythagorean spell expressed in the magic pentagram, the symbol that condensed the ratios of the "golden section." . . . The poor brains of the Hermetic dreamers, alchemists, and magicians had forgotten this meaning of the pentagram.[11] They were only haunted by respect for the great idea exhibited in this sign that remained their magic pentacle. It is in fact one of the world's great mysteries. Even by ridding ourselves of all illusion and sticking to the terrain of strict reason, we are compelled to admit that this proportion (the golden section) is one of the great constants of nature. (*Harmonie in der Natur,* Stuttgart: Kosmos, 1926)

Francé himself stumbled on the Φ ratio of the golden section in a biochemical study of the phosphate content in an ideal compost.

By calculating the optimal proportions for a certain kind of electromagnet, Roland Boris (a brilliant marine engineer) produced the formula $\frac{d}{D} = \frac{1}{\Phi} = 0.618...$ (*d* and *D* represent, respectively, the diameters of the soft iron core and the exciter coil).

---

11. Not all. Agrippa von Nettesheim (the most scientific of the magicians, whose *De Occulta Philosophia* was practiced by Descartes) appears perfectly aware of this "secret," as he cites without attribution entire phrases of Vitruvius on the correlations between the proportions of the temple, if not the work of art in general, and those of the human body, and published the famous engraving of the human body or microcosm framed within the pentagram.

Plate 62. Harmonic variations on the icosahedron
(D. Wiener)

The ratio Φ (or rather $\sqrt{\Phi}$) also appears in calculations related to some kinds of airplane wings.

In *Esthétique des proportions,* I listed the main flowers with pentamerous symmetry. They form, moreover (especially as calyxes with five petals), the vast majority of flowers (we could say that the lily was the sole "major" flower with "crystalline" hexagonal symmetry)—for example, all the flowers of fruit trees and plants with edible berries. The immense calyx of the *Victoria regia* [now *Victoria amazonica*], a giant water lily, is pentamerous like all of the genus *Lotus* and all the genus *Rosa*. But the largest "organic" pentagram found in nature is the immense starry pseudo-flower (one yard in diameter) of the parasitical *Rafflesia arnoldii* (corpse lily), whose purplish-blue sticky bosom that smells like rotting flesh attracts and swallows "carrion-eating" insects in the jungles of Sumatra.

As an attempt to relaunch this philosophy of musical harmony by employing this group theory that is the triumph of Western analytical thought, I have already cited (in chapter 5) Pius Servien's essay in which study of all possible structures for the "invariant" nuclei (leitmotifs or musical ideas) is pursued in tandem with that of expression, thereby requiring examination of the transformation groups possible for each idea.[12]

In a recent short work (*Paul Valéry,* Editions de la Nouvelle Revue

---

12. Here are a few conclusions from Servien's *Introduction à une connaissance scientifique des faits musicaux* (Introduction to a Scientific Understanding of Musical Elements) (Paris: A. Blanchard, 1929): "It seems that the value of the invariant, the leitmotif (the 'depth' of 'musical thought') would be measured by the more or less large quantity of thematic alterations to which it lends itself, in other words, its musical generality, the number of musical phenomena for which it is the recognizable type. The most general and deepest leitmotifs are found in the category of those called the melodic leitmotifs . . . of the interval-pitch type. [This brings us back to the motif of Bergson's indeterminacy as a characteristic of life.] The most interesting subgroup is that of the transformations inside the octave."

The table of the scales (from 462 = 66 × 7 possible "modes") appears . . . as the complete table of the most important transformations of a theme: those that change the intervals (without changing the dynamics, for this would attack the invariant—the leitmotif itself).

Plate 63. Lilies (D. Wiener)

Critique), Pierre Guégen perceives in Valéry's poetic work a tendency to persistently apply, in the domain of the sensations and ideas of one same individual, this quest for all possible transformations and substitutions in order to reach the core and the common invariant (this is the basic principle of group theory). The purpose behind this play of operations and substitutions is to touch the individual, "the metaphysical enigma that each soul is for itself," the mysterious subject of the "intellectual comedy." This veritable intellectual comedy, which is both mathematical and sensual at the same time, and represented, for example, by "The Young Fate," is summed up by Guéguen as "the play of substitutions in one consciousness during the period of one night."

This observation is all the more exact as it is explicitly illustrated by Valéry himself in the symbolic analysis of the cinematic transformations and movements of Athikté, the dancer of *Dance and the Soul,* and in his direct definition of the soul as "the invariant accepted by the most general group of our transformations."

In chapter 11 I sketched out this new spiritualist monism founded on the élan vital that appears to be the sole reality emerging out of the ghostly residue of ex-matter, now that instead of the universe-factory of the turn of the century and its collection of mechanical "models," gyroscopes, gear wheels, and flywheels that were swept away by the exorcism of several abstract mathematical symbols, all we have left are Heisenberg's grids in which numbers dance and sparkle as in the compartments of the Kabbalah's magic squares. Traveling through this

---

(cont. from p. 387) "And in this set of the possible tempered scales, the maximum asymmetry and precision is provided by the pentatonic and (especially) the heptatonic diatonic scales. This latter, thanks to its extreme asymmetry, permits by means of its sounds the closure of the most varied combinations of intervals."

(This is the Pythagorean scale. In it we find, with its seven "modes," the number of the world soul from the *Timaeus.* It is interesting to note that the scales with 3, 4, and 6 intervals exclude themselves by excess of symmetry, as well as all those with an even number of intervals. It is worth recalling that in the proportions of life and "eurhythmic" designs, the square and hexagonal symmetries give way for the same reason to the pentagonal system and its asymmetrical pulsations.)

arachnid "maya" contained in a limited macrocosm like the ancient "sphere of the fixed stars" are conscious entities coming from a fourth or fifth dimension, individual or collective psyches in continuous creation, perhaps arising from a transcendent Panpsyche.

It is obvious that the new non-euclidean cosmos, a theater of creative evolution in which none of the Pythagorean-Platonic ideas would be out of place (even the incarnations among us of "daimons" or demigods, even the encounters with the "true life" from the other side of the "hypersphere of the fixed stars," in the light of the transcendent oneness, of kindred souls lost on the cusp of the ages), where life's rockets lift this pseudo-material[13] substratum and start it rhythmically quivering, a substratum reduced (for those terrified by pure numbers) to grids of overlying waves (waves of electromagnetic ether transmitted by the great fifty-octave keyboard, which connect the peaks, nodes, singularity points, Schrödinger-de Broglie's sub-etheric waves), and it all appears excessively well ordered, and also from this perspective (this new macrocosm) would meet the requirements of the most demanding Pythagorean. And because it does so to such an extent, moreover, this suspect rhythmic perfection has placed us on the horns of the old dilemma: Does harmony emanate from a consciously "music-loving" Great Ordering One, or does the entire universe only exist as dream in the harmonious imagination of *Homo sapiens* (and why would it be so harmonious, then)?

The second hypothesis sparked the paradoxical metaphysical commentaries of the astronomer and physicist Eddington cited in the preceding chapter. The first inspired the pure logician and mathematician Whitehead to later develop an "organic" theory of the world that, as well as the philosophy of the harmony of life, which Rignano[14]

---

13. "This wrinkling of space-time is not merely due to the presence of matter, but actually is matter." S. Alexander. "Apparently, whatever electrical charges are, they are not substance; so in the final analysis substance is non-existent: it is an illusion!" G. A. de Moudray, *Journal of Philosophical Studies,* July 1930.

14. Publisher of *Scientia.* We can also compare the organic concept of the work of art (cf. chapter 4) to this organic theory of the world.

discovered by way of biology, would have appeared perfectly obvious to the Alexandrians of the second century. Here are the thoughts of today's mathematician-philosophers that could be easily found in Plotinus or Proclus, or at least are precisely suggestive of the way Plotinus or Proclus would have analyzed the theory of relativity. R. D. Carmichael, on the way Einstein discovered the laws of gravitation by posing the ideal conditions of invariance a priori, wonders if "the structure of the universe is not 'one' with the structure of human thought," and concludes: "The ideal imagined by abstract thought was thereby achieved experimentally, confirming that there is a profound solidarity between the laws of human thought and the laws of external nature."[15] Eddington, meanwhile, at the end of his final book *The Nature of the Physical World,* writes straight out:

> The idea of a universal mind or Logos would be, I think, a fairly plausible inference from the present state of scientific theory; at least it is in harmony with it. But if so, all that our inquiry justifies us in asserting is a purely colorless pantheon. Science cannot tell whether the world-spirit is good or evil, and its halting argument for the existence of a God might equally well be turned into an argument for the existence of a devil.

This observation would have delighted Simon Magus and the other fathers of Gnosticism.

In any case, the directing impulses that philosophy can detect in these life rhythms in which we are both actors and spectators are the desire to live, to reproduce, indeed, for us men, to perpetuate and "go beyond" ourselves, all without possessing any "official" information on the final purpose of the vertiginous odyssey.

15. R. D. Carmichael, in the article, "La Théorie de la relativité et son aspect esthétique dans la covariance des lois de la nature," *Scientia.*

* * *

Christianity frowns at this evocation of a "Great Adventure" the deity would share with us, or of which it would be the final result instead of his being the author and administrator of the "mystery." The Church is likely grateful to the third of the Hebraic magi mentioned earlier, Bergson, for having "deboned" materialism even before matter was completely exorcised by the physicists as pure electricity, space curvature, or matrix-skeletons of probabilities, but this unexpectedly resurrected Neoplatonism with all the charm of youth, this "world soul" that it would easily tolerate in the reveries of Gothic abbesses (as it more recently tolerated Beuron's neo-Pythagorean aesthetic) it could only condemn in Bergson as the serious awakening of the old enemy it had so much trouble taking down before. And it is not simply one head of the serpent that appears to be rising back up, but two, or three. I have already mentioned the connection between operative masonry that sought to build harmoniously in conformity with the traditions of the geometrical fervor passed down since the time when the royal art of architecture was handed down as a secret from one initiate to the next, and speculative Freemasonry that wishes, like the great "fraternity" of linen-clad philosophers once did, to harmonize the rhythm of human society with the great cosmic harmony, to build also (and for the earthly world alone) on the basis of unquestioning obedience to the Master, of unreserved mutual aid among brothers, of absolute secrecy.

As we have seen, these connecting processions have sometimes become commingled with those that through the ages transmitted Pythagorean geometrical esotericism in rose windows, pentacles, ritual signs, symbols, and directing designs.

Was Pythagoreanism reborn, not only as a metaphysical and aesthetic discipline, but as the body of a complete doctrine with its own dogma and morality? Because, while Bergsonism has not yet outlined its morality, the Pythagoreans did have one based on harmony and love. To break harmony was a transgression, and sinning against love was a mortal sin. It was necessary to love all creatures: like Saint

Francis, Pythagoras treated animals as brothers, and their legends reveal astounding parallels in this regard.[16]

The old Pythagoreanism also required the renunciation of material ambitions, wealth, the joys of the senses, and even the integral communism we find among the Therapeutae of Lake Mareotis and again in the monastic orders. The moral precepts of this ancient Pythagoreanism and the early Church are obviously close. Those of the Pythagoreans are broader and less demanding in theory (for the penalties of crimes against divine harmony); those of the Church could be more accommodating in practice. In any case, the punishment of the evil and the rewarding of the just in the next life already formed part of Pythagorean dogma (Plato admits his belief in this subject in his *Seventh Letter*), and Isidore Levy even finds in a descent into hell starring Pythagoras (in the *Abaris* of Heraclides Ponticus) the exact parallel of the parable of the rich and the poor in the Gospel of John. We have already discussed the role played by physical purity and morality in the life of the Pythagorean "brothers," and we looked at the examination of conscience and other practices that we might have thought were essentially "Christian."[17]

Many of these convergences have been noted and recorded, but it took Isidore Levy in his two recent books[18] (mentioned in chapter 7) to connect all these coincidences or borrowings, and come to the fairly sensational conclusion that Christ's morality and teaching were derived directly from Pythagoreanism, which in the form of its Alexandrian renewal attracted, as we saw (chapters 7 and 8), the intellectuals of Egypt and Syria—including Palestine.

This would explain not only the strange antipathy displayed by

16. Méautis, in *Recherches sur le Pythagorisme,* has indicated these correlations: the sea-bear in Daunia and the wolf of Gubbio; the fish that Pythagoras bought from the fisherman and threw back into the water and the carp a fisherman gave to Saint Francis, who threw it back into Lake Rieti.

17. Excommunication, for example.

18. *Recherches sur les sources de la légende de Pythagore*; *La Légende de Pythagore de Grèce en Palestine.*

Jesus (in the Gospels) against the Hebrew spirit in the strict sense (in its orthodox form of harshness, severe intolerance, and attachment to the letter of the law), but also the immediate success of this religion founded on love, which echoed that of the vague pantheism of then stylish neo-Pythagorean and Neoplatonic philosophies.[19]

This theory, which the Church has welcomed to this point with a somewhat mocking indifference, but without anxiety, and whose exegetic value it accepts, affects in no way the question of the person of Jesus Christ: man, superman, or God. He obviously drew the bulk of his teachings from a complex of preexisting dogmas but gave them, by means of their association with the drama of his life and death, a new life. If up to the present day, we were lazily accustomed to believing that the traditions from which he had drawn were Hebrew traditions, we can also clearly admit after examining Levy's arguments that the sole Hebrew element of the Gospels (in addition to the context of their Galilean charm) is the messianic notion. All the rest can actually or potentially be found in the stock of Pythagorean ideas and leanings, such as they were grafted and developed in Egypt and Syria. From an ideological and emotional standpoint, Christianity would not be a Semitic religion but a Greco-Egyptian religion, the Greek contribution being Pythagoreanism. Jesus's personal contribution and gift to the world was charity.

Examining this statement more closely, we should recall that Pythagoreanism itself, although representing (especially if you

---

19. I cannot put this any better than to repeat Isidore Levy's conclusion: "This is what explains the mysterious fact of Christianity's triumph. How is it possible that a doctrine crafted in Judea at the end of a most singular religious evolution should be found to supply a food suitable for the spiritual needs of Greco-Roman society? The answer is easy for anyone who has recognized the lineage that connects to Platonic-influenced Hellenism though the Judaism of Alexandria, then Judea, and the Gospel. The essential aspect of the religion that came out of Palestine during the reign of the Caesars had been introduced into Jerusalem a century earlier. The Gospel . . . seduced the ancient world, because it brought it, imprinted with the most invasive of exotic charms, a product of Greek thought, heir to a remote Indo-European past."

include, as I do in this study, Plato as the indisputable spokesperson of the Master of Samos, the only one whose voice has explicitly reached us) the most remarkable and powerful products of Greek thought, also shows two foreign components of different origin: First, an Egyptian contribution represented by everything that Pythagoras himself had borrowed from Egyptian thought; the reading of Herodotus, Plato, and Plutarch can leave no doubt as to the importance of these Egyptian elements, which I can describe this way: the quasi-sacred character attributed to geometry and then to the perfection of form, the importance of secrecy (esotericism), the magic value of the Word (and sometimes the word), the magic value of the sign, then the symbol, the magic value of the rite, and rhythm, and so forth;[20] and second, a contribution that I can immediately call "Hyperborean" from the very name bestowed on it by Herodotus and Heraclides Ponticus.

Without going into detail on the complex question of "Nordic" influence on Greek thought and religion in general, we should recall that when Pythagoras was identified in the legend with a god, it was not with Horus or Thoth, but with the Hyperborean Apollo, the one whose mysterious temple north of the Ister [Danube] floats in the clouds around which soar the sacred swans.[21] It is the Valhalla we shall see again in the Scandinavian sagas, and our Hyperborean Apollo is the handsome Balder, god of eloquence and love (Porphyry cites "the handsome Pythagoras" among the geniuses who received the hierophant in the sanctuary after the final ordeal of the last initiation).

The legendary disciples of Pythagoras have always shared a mysterious bond with the north, like Abaris (because he was a priest, specifically, of the Hyperborean temple), like Zamolxis of Getae or Dacia,

20. Concerning the obligation of Pythagoreans, like Egyptians, to wear only linen, Herodotus says: "This conforms with Orphic ceremonies, which are called Bacchic, and also the same as the Egyptian and Pythagorean. . . . In fact, it is forbidden to bury anyone who has taken part in these ceremonies in a woolen garment."

21. The swan is specifically assigned to the Hyperborean Apollo in Hellenic legends.

who returned to his northern land after being a slave and disciple of the Master of Samos. Following his return he became a legislator, pontiff, and king.[22]

The Hyperborean connections of Artemis-Diana, the sister-goddess of Apollo, are well known. She also has a Nordic sanctuary (generally identified with the temple in Tauris made famous by the legend of Iphigenia). Herodotus mentions the two Hyperborean virgins Opis and Arge who came to Delos with the goddess and were buried behind her temple. Lastly, we know that Orphic idea and rites, which held a great influence over Pythagoreanism, as it did on all the esoteric cults affiliated with the great center of Eleusis, were always associated with the Thracians,[23] a people who were always "Nordic" in relation to Greece as such, and related to the Dacians, Getes, Scythians, and so on.

We can therefore speak with certainty of a Nordic component in Pythagoreanism, but it is difficult to discern the subtlety other than through elimination or analogy. By comparing the scant clues we have in hand to the nature of the Celtic-Nordic influences that appeared in other eras (cf. part 1), we can approximately sum up this factor as follows: intense and refined idealism, particularly the quasi-mystical notion of friendship, the bond between chosen "brothers" (we can find it again in its "feudal" incarnation in the equally Celtic-Nordic notion of chivalry, and finally in the attitude of the Romantic spirit); the extension of this fraternal feeling to all that lives (animals and plants); the possibility of amplifying this "cosmic tenderness" to the point of intoxicating, tumultuous, and orgiastic ecstasy; and finally, we have the possible culmination of all these enthusiasms into a transcendent form of divine love.

Let's now take a look at what remains as a specifically Greek

22. Then a demigod. Josephus compared the priests of Getae who subsisted on milk and honey to the Essenes. One of the rare ancient rings bearing the Pythagorean pentagram was found in Poiana, on the Sirit (Romania).

23. Herm. Callimachus calls Pythagoras "θρακῶν δόξος μιμούμενος."

element after this breaking down of the Pythagorean system of ideas and tendencies.

The spirit of synthesis, and clarity in the synthesis;
The realization in artwork of perfect formal beauty;
Development and completion of geometry as the ideal model of an overview founded on the axioms and on the series of unassailable logical (axiomatic) deductions;[24]
In parallel: establishment of the theory of "numbers," the entire universe being "governed" or "arranged" in accordance with numbers;
Concepts of proportion and rhythm derived from the aforementioned disciplines (theory of forms and theory of numbers) and applied to the quest for beauty;
Theory of musical harmony;
Harmonic conception of the cosmos.

We can see the major part played by Pythagoras himself in the crafting of the specifically "Mediterranean" part of the Pythagorean-Platonic concept of the world.

With Levy's theory serving as the starting point for this short analytical recapitulation—to wit: Christianity was a religion with a Pythagorean ideological basis, in other words, it was Greco-Egyptian—we have yet to mention the purely Egyptian infiltrations, which in addition to the Egyptian elements of the old Pythagoreanism, were introduced into early Christianity by virtue of its kinship and contact with the neo-Pythagorean schools, which then in their full flower again

24. "What we owe to Greece is perhaps what has most profoundly distinguished us from the rest of humanity. We owe it our mental discipline, the extraordinary example of perfection in all the orders. . . . Greek geometry was that incorruptible model, a model offered to all knowledge seeking perfection, as well as an incomparable model of the most typical qualities of the European intellect . . . a European spirit of which America is a formidable creation." Paul Valéry, "The European" (1924).

plunged their roots deep into Egyptian soil. This is what we could sum up as the "magical" component of Christianity, the science of ritual and incantatory technique.[25] We saw in chapter 8 how the Church, while incorporating this incantatory technique, refused to adopt "a passion for magic" and initiatory recruitment, and deliberately expelled Gnosticism from its bosom. This Gnostic religion comprised all the semi-Christian, semi-pagan sects focused on setting the pure crystal of the new gospel of love not only in the Pythagorean mysticism of numbers, but in the talismanic magic of letters and words (which through adding their combinations gave birth to the Kabbalah), the musical pantheism of the Neoplatonists, and Iranian dualism,[26] all under the suspect patronage of a feminine Holy Spirit, Thoth-Hermes, Isis, and Abraxas.

The Ophites or Naassenes (one group of whom were known as Cainites) were the best-known Gnostic sect and worshipped snakes as the emblem of knowledge. And among the now-resuscitated Hydra heads present at the rendezvous is that of the Gnostic serpent, because the posterity of Simon Magus and his Helen-Ennoia formed another line of our secret processions: Manicheans, Paulicians, Bogomils, Waldensians, Albigensians, Kabbalists, magicians, alchemists, Rosicrucians—all representing the phosphorescent progression of those who, despite the castigations of Rome and the severity of its secular arm, did not wish to give up the pride of knowledge acquired individually, and who tirelessly sought, by applying the old symbols to new experiences, answers to the eternal questions: why, and how.

But in these issues in which life and soul are the unknowns and the

25. It is possible that the bulk of the elements borrowed by Christianity from Hebrew ritual were also of Egyptian origin. Here again, the Cainite sanctuary in the Sinai, the worship of the goddess with the turquoise eyes, and of the thundering Yahweh of the desert tribes could have played a role. Robert Eisler reminds us that Moses was the son-in-law of the high priest of Midian, another colony of Semite miners southeast of the Sinai. It is odd that the name "Cainites" was also borne by a sect of Luciferian Gnostics at the start of the Christian era.

26. The equal division of the influences and responsibilities between two divine, conscious principles, Good and Evil.

actors, it is the reproductive cycles, including the act itself that alone visibly transmits the flame of life, that offer (especially the gestation-birth cycle) the most striking symbols of the experiences, rebirths, and journeys of the soul and intelligence in the real or allegorical pilgrimages attributed to them. So we have seen an insinuating eroticism gradually color all the speculations of this lineage of seekers and dreamers. While the symbols for reproduction (the ear of wheat transformed into a golden phallus during the "Mystery of the Seal," the voyage into the bosom of Persephone during the "Mystery of the Circle"[27]) remained chaste for the participants in the Eleusinian Mysteries, either because of the gravity of the ritual in which they were participating or because of the high degree of the participants' initiation (these were third- and fourth-degree initiates, epopts and *holoclere* initiates), the spiritual descendants of the Gnostics, on the other hand, allowed themselves to be captured by the murky seduction of these images. Incense, perfume, and songs ceased being cathartic and henceforth served not to put their thought in harmony with the universal soul, but their desire in tune with universal desire. Like Persephone, they succumbed to the juice of the pomegranate offered by the King of Darkness, transformed into the sparkling scales of the serpent, and no longer considered evading the circle of generations. Here again, Hyperborean sentimentality will color with a subtle softness the generative heat of the Mediterranean ram. The frozen mysticism of the pure number lends itself to all kinds of resonances, and it is the child of Great Pan's flute that will, little by little, around the Hermetic androgyne, the athanor—crucible of the Great Work identified with the womb of the Great Feminine—and the

27. In the Petelia inscription, already mentioned in chapter 7, the soul of the deceased evokes this part of his initiation: "I am buried in the breast of the Mistress (Despoina), the queen of the underground world." Persephone is no longer the soul here as in the first grades, but the "Great Goddess."

We know that during the initiation into the fourth degree (the third if we do not include the initiation into the Lesser Mysteries, or Purification), or *holoclere* initiation (the Mystery of the Circle), the candidate was symbolically (or actually, at least for the duration of the ceremony), "emasculated" by a certain brew.

five-petalled rose of love led alchemists, Rosicrucians, and pansophists into the "garden of earthly delights" in search of the "wild strawberry."

It is always the same evolution. Just as the Egyptian crux ansata, the "ankh" of life, after having become identified with the chrismon (the Chi-Rho cross topped with a circle) as seen on the funerary portraits of Antinoe,[28] openly became the planetary sign of Venus and the alchemical symbol for copper (sacred to the goddess of love[29] because of their common association with the island of Cyprus), the proud Artemis became the Cottyto-Bendis of the Thracian orgies, and all the spirits, principles, and symbols associated with fertility entered into the *rotas* of the pangeneration whose planetary sirens became maenads.

Already in the time of Alexandrian Gnosticism, the Christian sect of "Stratotian" Gnostics had invented the "collective" rite in which ceremonies opened with a meal that women attended naked and "were offered to one and all."[30] This collectivism was revived around 1872 and, based on the interpretations of certain passages from Saint John and on Rosicrucian ideology, sought to realize the "universal (androgynous) man" through the collective or common marriage of couples: "Determined by the reciprocal aspiration of two human halves in search of each other to form the individual androgyne and the human androgyne. The individual androgyne is produced when the two halves of

---

28. Cf. the handsome portrait of Lady Krispina at the Musée Guimet.

29. Again the goddess with the turquoise eyes, the lady of the sanctuary in the "Malachite Mountains" (turquoise, copper, and malachite are combined symbolically here as they are chemically): "Luciferian" Hathor-Ishtar-Aphrodite. The star with five rays, Set or Sothis (Sirius), sign associated with Hathor, then Thoth-Hermes, and its Pythagorean variation (pentagram of harmony) will reappear as the pentagram of Venus Genetrix, then finally as the Mason's blazing star.

Gérard de Nerval, as shown by the story of Solomon and Balkis he inserted into his *Voyage au Orient* (Journey to the East), appears to have suspected the role of the Cainite smiths and miners of the Sinai in the development of guild and masonic traditions and legends. Among other things, he mentions the "Child of the Widow" ruling over the "Sons of Fire" in an underground city.

30. Sixtus of Siena, "Bibliothecs Sancta," cited in the *Revue Internationale des Sociétés secrètes,* supplemental issue of June 1, 1928.

the individual body, the man and the woman, following the harmonic marriage, are no longer a mystery to one another, and together form a perfect union."

Then:

> As soon as the perfect union of the two individual halves has been achieved, the man and woman can, indeed should, each wed a second, then third partner, or more if the perfection of the couples hastens their time of unification. . . . The consequences of these direct or indirect communions, of this divine chain that is occultly completed in humanity through marriage . . . is the creation of the true man, who is neither you nor me, but the collective individual, enjoying awareness of himself and living a unitary life. When this time comes, humanity will no longer be a fictional being, a being of reason. It will be a positive being, with its own life, thought, love, will, and potential all in proportion to the number of individuals of which it is constituted. When, through harmonic marriages, everyone will be unified with everyone, then humanity will find itself formed in its entirety and everyone will feel alive in all because at this moment, all will be one. On this day will be the final realization of the mystic gospel and the triumph of the Consoling One on earth. . . . The desire is for good to replace evil, love to replace hate . . . something that cannot happen until all men form a single family of brothers.[31]

This ingenious method for hastening the return to the collective soul, to prepare the incarnation of the soul of humanity by creating clans, psychic madrepores, through affinity and erotic welding, which combines the androgynes of the *Symposium* in a concrete realization of the *solve et coagula* of the Hermetic androgyne, also garnered the favorable opinion of the "fathers" of Gnosticism in preference to the "political" method illustrated by the "League of Nations."

---

31. Ibid.

Plate 64. Ancient nude: Bronze in the style of Polykleitos (Atalanta?) (Providence Museum)

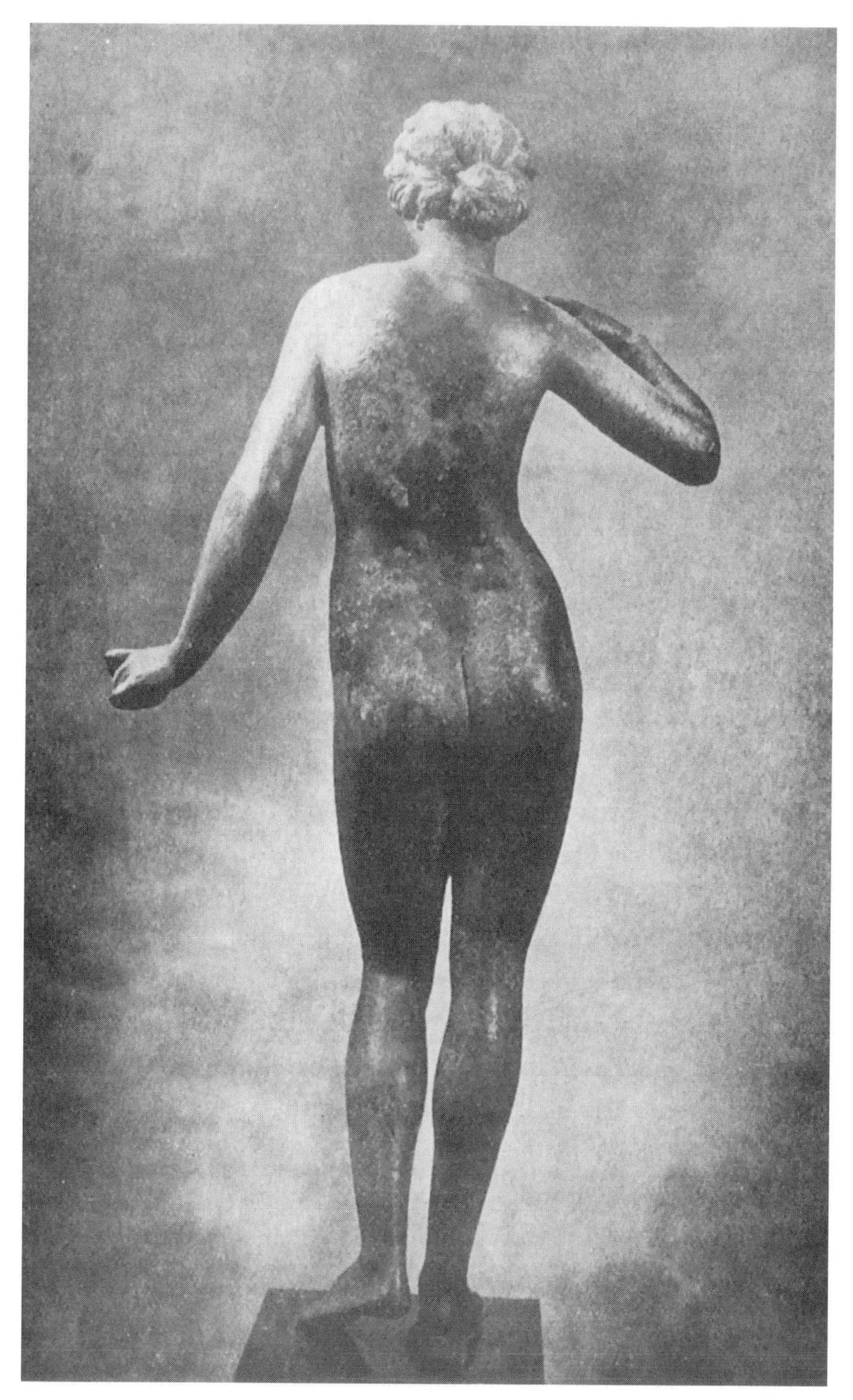

Plate 65. Another view of plate 64

Furthermore, Gnosticism itself, raising like a challenge its name formerly condemned to anathema, audaciously renewing its half-Christian, half-pagan traditions, connecting its new priests to the lineage of Simon and Valentinus, has been resurrected for some twenty years and has provided its first "pontiffs" with a successor in the person of Valentinus II, "patriarch of the Universal Gnostic Church." As once the "extremist" wing (Cainite) of its Alexandrian elder did, the church of today uses the dualist principle to imagine two almost equally powerful gods, a god of love and light, and a wrathful, jealous god whom they identify with the harsh god of the Church and the Old Testament, who through a secular misunderstanding usurped the place of the other god in Christian theology. As the Ophites once did, today's Gnostics are inserting invocations to the forces of nature into Christian hymns and incantations in which the ancient gods of Egypt and Syria appear in the transparent disguises of Greek or Hebrew alliterations.

Parallel to philosophical neovitalism and to ritual neo-Gnosticism, a new desire has manifested independently, a desire for actual daily communion with the currents, forces, and beauties of nature, joined with a fervent, aesthetic worship of the human body. Here it has been the Nordic peoples who have rediscovered the initial Mediterranean enthusiasm toward beauty and physical harmony. Walt Whitman, the lover of humanity, has taken an intoxicating look into the titanic crucible inside of which a new race is being crafted, and with the freeing of pent-up sensuality accumulated by generations of Puritans has sung, veritably possessed by pagan Eros, the body of man and the body of woman with an almost cosmic fervor: "I sing the body electric."[32] In the Germanic countries, the nudist movement initiated before the First World War by several scattered eccentrics has succeeded in turning over laws, practices, and prejudices, and audaciously introduced into everyday life, for a growing number of adepts of both sexes and all social classes, the shared

32. In his preface to *Leaves of Grass* he says: "Of physiology from top to toe I sing, / Not physiognomy alone nor brain alone, is worthy for the Muse, I say the Form / complete is worthier far, / The Female equally with the Male I sing."

Plate 66. The new Nordic naturism (nudism)
(photo Riebicke)

performance of sports in that state of complete nudity that antiquity described as "magical" during the ritual of the Mysteries.[33]

We could say in summary that at this historical juncture Western civilization is witnessing the embodiment of the great technical dreams of the "bridge builders" and Mediterranean inventors who, on the one hand, can finally see the flight of the great dove of Archytas and the birdman of Leonardo and, on the other, see dissolving into pure numbers the material carcass of the world, where the echoes of Eleusis again become one with those of the *Ieros Logos* and the *Timaeus.* All the initiates and the descendants of initiates are comparing their symbols and passwords, striving to recognize one another and regroup almost right out in the open.

Freemasonry, even in France, shakes off its ideological torpor, shakes off its comfortable lethargy and its prosperous backroom dealing, in a "neo-initiatory" phase, to rediscover that its symbols are the same as those of the Roman *liberi muratores,* the stone carvers, the ancient alchemists, Kabbalists, and Rosicrucians, and that the grand master, the "Old Man of the Ages" who taught us the joy of establishing proportion and building in conformity with the paradigms that earlier inspired the "Great Ordering One," was the Master of Samos,[34] whose seal has become its blazing star bearing at its center the *G* of the royal science. This was actually the Kabbalistic symbol for the Pythagorean decad (the letter yod, the first letter of the Hebrew name of God, is also the sign for the number ten).

---

33. In each degree of initiation, at Eleusis, for example, the candidate should at a prescribed moment strip off his clothes and remain naked during a phase of the ceremony. This was one reason that women's initiations took place at a separate ceremony. But as figurants or priestesses (*hierophantides*), women attended the male ceremonies. We also know that the initiates of the first four degrees only saw the statues veiled. Only the high officials were allowed to contemplate their nakedness.

34. "It can be stated with conviction that the masons of the first centuries of Christianity remained, in a more or less Christian way, connected to the ancient Pythagorean doctrine like the alchemists to ancient Hermeticism. Pythagoras's role in Masonic legends is moreover so fundamental that it can even withstand the way his name is twisted by English workers. One of the old manuscripts, in fact, speaks of a certain Peter Gower . . . building contractor." Armand Bédarride, *Symbolisme,* June 1927.

But Germanic pansophists, Anglo-Saxon Rosicrucians and neo-Templars, and French Gnostics, having exchanged their signs and resumed contact, are not attached so strongly to the geometric meaning of their symbols (this is of greater interest to mathematicians and the creators of forms) as they are to the continuity of the traditions for which they are witnesses and the cryptic messages they form. For them, the picaresque Mephistopheles of the Faust cycle has turned back into Hermes Trismegistus, the Hermes with the serpent. In their final "word," the "lost word," which is not revealed until the final initiation, they find again the Tetragrammaton of the Kabbalah, which is also an avatar of the pure tetractys of Pythagoras. For others it is from the "unspeakable" name beneath the severe mask of YAHVEH (from "He who is") that emerges the Yod Heh Vuv Heh, the invocation to the two engendering principles, the breath of fire and the womb it fertilizes, the spear and the cup, the serpent and the cave; and the echo of the round dance reflects the ancient EVOHE [the cry of Bacchic frenzy].

For Great Pan is not dead. The cycle begins again eternally. Like the Eleusinian ear of wheat, the serpent of knowledge intermittently transform into the phallus. From the adoration of the naked virgin who represents wisdom and love in the modern Gnostic ritual[35] to the

35. "The celebrants (of the Johannite Gnostic mass as practiced by the neo-Gnostics whose bishop-primate lives in Lyon but has adherents throughout central Europe, Russia, and America), are six in number: (1) the high priest dressed in white and carrying the holy spear; (2) the high priestess who, depending on the ritual, should be "*virgo immaculata*"; her garment is white, blue, and gold and she wears a sword at her side hung from a red belt; she carries a paten on which sits a holy wafer. . . . During this time (the first invocation by the high priest in front of the altar) the high priestess is undressed girded only by her crown and belt. She stands erect before the altar and utters the naturist hymn: "Love me: here is the most supreme of all things." (Then, after a second invocation by the high priest) the high priestess, still standing, steps in saying: "There is no other law but this: Do what thou wilt and love under will."

She wraps a robe around her shoulders and body then sits in the center of the altar. The high priest then shouts: "Io, Io, Io, Iao Sabaoth, Kyrie Abraxas, Kyrie Mithras, Kyrie Phallus, Io Pan, Io Pan, Io Pan, Iao Khaire Phallus, Khaire Pamphage, Khaire Pangenetor, Hagios, Hagios, Iao!"

adoration of the feminine principle, then the organ, there is only a step.

If Pan has resurrected, he regained his footing in Europe after an odd trek through the Caribbean Sea and American continent, and like the virgins of Chartres, Saragossa, and Montserrat, he is black. It is the bass note of a black Great Pan who, by way of panpipes, now chants the round dance of the Corybantes. It is the black magician, Hor, son of the Negress, who has once more returned proclaiming his syncopated incantation. As Saint Peter was once opposed to Simon Magus, the virtuous Rabbi Loew to the saturnine John Dee, and Abbot Trithemius to Doctor Faustus,[36] now facing each other as antagonistic and twin couples are the heads of giant black and white *nagas,* towering over the free-for-all.

(cont. from p. 407) "The high priestess holds the paten with the holy wafer in her right hand and the chalice and the wine in her left. The high priest points the tip of the spear, which she kisses three times. Then holding the spear between her arms and thighs, she presses it against her chest. The high priest kneels before her, extends his arms around her hips and kisses her thighs nine times. He remains in this position while the dean recites the opening prayer and the organ softly plays."

The service ends with the two forms of communion. These details have been taken from an article by A. Dalmas on the Universal Gnostic Church that appeared in a supplemental issue of the *Revue Internationale des Sociétés secrètes,* February 1, 1928.

We find a recollection in this ritual of a phase of the *holoclere* initiation of Eleusis that was also adopted by the Ophites (Gnostic Naassenes). The following passage from the *Philsophumena* (*Refutation of All Heresies*) alludes to this rite of "regeneration by a virginal breath":

"For this is the gate of heaven and this the house of God . . . into which will enter no man who is not purified, in mind or in flesh, but it is kept for the pneumatics only (the third degree that for the Ophites corresponded to the epopt and to the *holoclere* initiation), who having reached there, must cast aside their garments and become bridegrooms emasculated by the virginal breath. For this is the virgin who bears in her womb, and conceives and gives birth to a son who is neither of mind nor of body, and is blessed down through the centuries. . . . For the Savior has said that straight and narrow is the gate that leads to life." Victor Magnien, *Les Mystères d'Éleusis.*

36. Here is another of those curious cases of convergence or recurrence like those we examined in part 1. The Faust legend was inspired by both the life of a German magician who really lived around 1500 and by a Gnostic romance (*Les Reconaissances de Clément*) from the second or third century CE in which a certain Faustus associates temporarily with Simon Magus. Cf. Oscar Cullmann, *Le Problème littéraire et historique du roman pseudo-clémentin* (Paris: Félix Alcan, 1930).

Facing the white, triple-crowned miter is the proud serpent of Hermes, the royal uraeus of knowledge and desire, which fecundates and slays tirelessly.

The duel of the magicians has started over, with the war cries A.M.D.G. [*Ad majorem Dei glorian*] of the Jesuits on the one side and the Freemasons' A.L.G.D.G.A.D.U. [To the glory of the Great Architect of the universe] on the other. Both sides, like the geometers, moreover, who staying outside the melee tirelessly decipher the "numbers of the world," descend, through the avatars we have examined, from the ancient fraternity of Croton and Metapontum.

The foliage of these main branches that have grown in the secret of the night like separate trees has combined together again, and through the mists of the history that we have attempted to sound appears the somber majesty of the common trunk sprouted from the Pythagorean rootstock. Shining on the heads of the *nagas* is the old symbol of life and harmony, and to this black star of the Satanists (upside-down pentagram), blazing star of the Freemasons (pentagram with the central *G*), and five-petalled rose of the Rosicrucians and pansophists,[37] the Church opposes the star of Bethlehem that guided the three magi to the crèche of the divine child. Similar to the white phalanxes forming the white rose Beatrice showed Dante, the processions of initiates, behind their torchbearers, looming out of the night, are arrayed before us. The chains of thinkers and builders we have followed since the apparent destruction of the Pythagorean society are arrayed like the various arches, ribs, and radii of a gigantic stained-glass window, in which the detached fragments are rejoined, and where dogmas and symbols have been adjusted and reattached. In the heart of the common symbol of the rose window appears (as in the great rose window of Amiens) the pentagram, the seal of the Master, the exact image of proportion, of infinitely recurring analogy between

37. It is fairly curious that the Soviets also chose the "blazing star" with five points as a principal emblem. In their military order (the order of the "Red Flag"), the star is upside down, meaning two points are on top like the "Satanic" pentagram of black magic, the one that diagrams the head of the bisexual goat.

the whole and all its parts, of living harmony, of love, the preeminent "sign" of the peoples I call "Mediterranean."[38]

In this book I have endeavored to contribute to a somewhat clearer view of the logical and emotional invariants and concepts that have been the driving ideas of this Western world whose wreck or at least twilight was announced a few years ago. Among these idea-forces I have stressed those that distinguish us, that properly belong to us (and in this broadly stated Mediterranean "we," I include not only the European Celts and Nordics, all the European settlers and descendants of America and the young Anglo-Saxon nations of the South Pacific, but the Semites of the Eastern Mediterranean, both Hebrews and Arabs),[39] and characterize us in relation to the other peoples of the world. These ideas being first and foremost, to repeat this litany, those of number, rhythm, proportion, and harmony (what I have condensed as the law of number), and, grafted onto this geometric rigor, the fervent worship of the beauty of forms, extending, finally, through the subdivision of this fervor into fraternal tenderness for all living things, to the final illumination, knowledge and love, in communion with divine harmony.

And the two directing concepts, number and love, from which emerged, respectively, the three major flowerings: Western geometry (and mathematics), Greek and Gothic art, and the Christian religion, are already found in their pure state as a focal point of what I have called the light of Pythagoras.[40]

---

38. Dante also sees from afar that a star is at the center of the "eternal rose" of Empyrea: "*O trina luce, che in unica stella scintillando*" (*Paradiso,* canto 31). We know that each of the three books of the *Divine Comedy* ends with the word *stelle,* "the stars."

39. We have seen that Pharaonic, Ptolemaic, and Coptic Egypt forms one of the principal factors of "Mediterranean" culture, especially with regard to "magic." We should not forget in this regard the very large role the magic of the black races played in Egyptian occultism, an influence underscored in the old tale of the "Duel of the Magicians." The effect was obviously reciprocal; the magic rites of Voodoo (cf. William Seabrook, *The Magic Island*) reveal curious survivals: the androgynous goddess Nebo, the horizontal copper serpent, and so on.

40. Let me again state that the doctrines of fraternity and love and the "musical" conception of the cosmos aside, the rest of Pythagorean ideas and rites can be connected

This is what I intended to demonstrate. To achieve my design, I had to pick up, one by one, arid notions and definitions (for which I tendered my excuses at the beginning) and extricate the teachings of dust-covered or petrified symbols that had lost their meaning and radiance, so they could be juxtaposed, reconnected, overlaid, and proportioned in an autonomous, logical microcosm.

To do this, I have had to drag my reader through dark and suspect woodlands, riddled with marshes, haunted by will-o'-the-wisps, which serious historians avoid like cautious travelers avoided the Insulinde jungles where flowers the seductive and sinister pentagram of the *Rafflesia arnoldii* (corpse lily).

---

to Egyptian sources. Herodotus and Plutarch both cited many examples in this regard. I showed in my *Esthétiques des proportions* that the proportions of the Great Pyramid revealed a theme strictly derived from the golden section.

The pentagram can be found in a radial, condensed form (reminiscent of some starfish) as a hieroglyph. This diagrammatic five-pointed star is studded all over the body of the goddess Nut and the entire vault of the funerary chamber of the pharaoh Zoser in Saqqara. I mentioned earlier that this star was assigned to the goddess Hathor as Sothis or Set (Sirius).

The overall persistence of forms, phrases, and techniques from ancient Egypt into modern European life would provide the subject for an interesting study. The preamble of the treaty of Megiddo between Ramses II and Khattusil, the Hittite king, has passed directly into the bulk of diplomatic works up to the Treaty of Versailles (the latter intentionally omits the clause concerning eternal friendship). Other survivals include the flabellum and tiara of the popes, the cross and miter of the bishops, and so forth.

In our pharmacopeia we can see find complete prescriptions and formulations whose trails can be followed from the medical manuals of the Middle Ages, the Arabic and Greek manuscripts, the Ebers Papyrus, and so on back to the pharmacopeia of pharaonic Egypt. All the herbs in the book!

Regarding Freemasonry, I already mentioned several of its rites and symbols that, sometimes by way of Eleusis, sometimes by way of the ancient guilds, come from ancient Egypt: the acacia of Osiris, the initiatory holding of hands, apron, underground chamber, and so on; Isis and Maat were the two patron goddesses of initiation.

And we can see echoes as an Egyptian amulet, the emblem of Maat, goddess of measure and harmony (same root as the "word of power" that grants life to "golems," cf. chapter 6) in the London Museum of Magic and Medicine, the Masonic square (Pythagoras's angle of equity) as it is depicted on the tombs of the "master builders," the abbey tower of La-Charité-sur-Loire, and on the sarcophagi of Roman architects.

Some will be surprised that I have granted such importance to the mysteries of antiquity; others will shrug their shoulders or kindly nod their heads when seeing me mention magic, Kabbalah, and Gnosticism so often.

How to answer?

People who see symbols everywhere are quite exhausting. But those who deny their presence and role on principle fall into an extreme position that may be equally dangerous.

Between the attitude of the visionary and the stance of those who are blinded by conviction there is a place for a geodesic of thought.

In any case: seeking to comprehend the world of antiquity by ignoring the influence of Eleusis and Pythagoreanism is as futile as trying to describe the Middle Ages without taking Christian mysticism into account, or (for the historians of the year 3000) describing the history of our own era while overlooking the Marxist mystique.

Finally, such as I have been able to craft and rig it out, my ideological overview is ready to set sail. I am cutting its moorings, and—whether it turns out to be a monstrous fireship* destined for Erebus, or a smart caravel toward the virgin continent of the Absolute—I wish it good fortune on the ocean where the hypotheses cross.

[*The word for "fireship," *brûlot,* also means "polemical tract"—*Trans.*]

Conclusion

We could—to go back to the allegory of the "Duel of the Magicians" I used earlier to summarize the current rebirth of the antagonistic groups descended from the early Pythagorean "brotherhood," or heirs to various fragments of its ideal—ask ourselves if, in the aftermath of these attempted reorganizations encouraged by the trends of current science and philosophy, a complete resurrection of Pythagoreanism as a synthesis of philosophy *and* religion would be possible. I have said enough over the course of the various chapters of this book to show there is nothing absurd about this question, and that in the West (Europe and North America), a pantheistic, Bergsonian-Pythagorean, vitalist religious philosophy, with an initiatory ritual and an ethic based on harmony and love, would have at least as much chance at life as the bland positivism of August Comte once did. The proof that this possibility exists is that the possible danger of such a crystallization is coolly envisioned by Henri de Guilledert, the most erudite collaborator on the Catholic journal devoted specially to the study of the secret societies and "Gnostic" sects that are literally proliferating in France, Germany, and the English-speaking countries. Here is his warning, which comes, moreover, after an account of the discovery of the Pythagorean church by the Porta Maggiore, which was discussed in chapter 7:

> It would be perilous to wait for theosophy or contemporary spiritualism to take possession of these new data concerning the natural survival of the soul, and this overly humanistic painting of immortality. So long as theosophy remains ensnared in our land by the Hindu reveries of the Mahatmas, Blavatsky-Besants, Krishnamurtis, and other Leadbeaters; so long as spiritualism lingers among the phantasmagoric illusions of Allan Kardec and Léon Denis, their hold on the steady mind of the Occident can hardly be cause for alarm. But let a man of talent, if not genius take the stage one day who has taken hold of the sentiment of universal anguish that tortures the religious animal fallen prey to disbelief, a man who knows how to offer people a solution borrowed from the so-called good sense of antiquity and humanity's oldest traditions with the prestigious magic of a deft tongue and a clear mind, the unsuspecting guardians of the stockroom of faith will be horrified at the sudden and irreparable ravages this heresy will cause anew in our decadent society, which is also greedy to redo, even at the cost of numerous passages through the beyond, its failed life, and once again find hope!
>
> Beyond the curtain of popular idolatries, the ideal pursued by the greatest souls of antiquity is clearly capable of seducing another generation starving for the God that secularism has stolen from it, and ready to acknowledge him, if need be, in the neo-Pythagoreanism crafted slowly in the shadows by a long series of dark and wicked prophets for their own purposes.[1]

We can see in the circles of neo-Gnostic, pansophist, occult, Masonic "initiates" a shared Hermetic tradition (I am excluding the Theosophists whose discipline and metaphysics are inspired by esoteric Hinduism,

1. *Revue Internationale des Sociétés secrètes,* supplement, September 1, 1928.

as well as the Anthroposophists of Steiner's lineage and the spiritists like Arthur Conan Doyle, to only consider the Egyptian-Pythagorean-Platonic lineage, the one that keeps as its guide the *law of number*) that is the basis for their strivings toward serious collective activity and analytical thought. Their sister and implacable enemy, the Church of Rome (acting head and continuous soul of Christianity), similarly has a lucid determination to collect all knowledge concerning the past, all information concerning the current evolution of the Hermetic complex, and to carry the battle into the domain of its hereditary enemy by attacking its vulnerable point, by trying to pour floods of light on the secrecy and mystery that rules, by principle, over its rites and organization. I will not seek to determine if there truly exists above the known Masonic high grades, above the "invisible degrees,"[2] and directing all the world's lodges (they have all actually and publicly been federated since 1921 into one International Masonic Association), a "Council of Seven," a supreme occult committee answering to an "unknown patriarch." But it appears to me that as a summary, a viewpoint not devoid of interest, of the contemporary battle of ideas in the Western world, the evocation of the "Duel of the Magicians" is more than a simple allegory. It is a war without mercy, whose eddies can at least be followed by those interested.

The duel's spectators can realize that the Catholic Church, despite the persecutions and attacks it has suffered during the nineteenth century, has lost nothing of its actual vigor and spiritual power, and that from the "Bergsonian" perspective, which distinguishes efficient organisms from those facing ankylosis, the biological cul-de-sac, it appears to be more alive than ever.[3]

---

2. I would like to point out that the 27th degree of Scottish Masonry is that of the "Great Commander of the Temple," that the 30th degree (Knight Kadosh), whose "duty" is to prepare "the material realization of the Gnostic doctrines," is also the "degree of vengeance," and that finally the 33rd degree (the last "visible" grade) has for its password "De Molay."

3. It may be interesting to quickly scan the following statistics (from 1927). Humanity numbered:

304 million Catholics
212 million Protestants

I am again citing the Church of Rome as being the head and symbol of Christianity in general. It is also the Church that has drawn from the initiatory traditions of Greco-Egyptian culture, which was its cradle, and has kept alive the necessary minimum (is it sufficient?) of incantatory ritual and magic.

Perhaps this Christian form of Pythagoreanism is more adapted to the needs of the average man. It is also acceptable to think that, practically speaking, the Christian family still represents a type of social cell more interesting for humanity that the asexual state-based solution of the social insects and communists, or the hypersexual concept of the "communitarians."

The initiatory selection and the initiatory state of mind, which are so attractive to elitists, appear by that fact to be destined to lead these elites onto the road of pride, the cruel and voluptuous pride of the cursed races who vanished or sank in mysterious catastrophes.

We know that after providing a magnificent description at the beginning of the *Critias* of the glorious civilization of Atlantis, the perfect organization of its empire, the splendor of Poseidonis, and of its sanctuary and rites, Plato tells how the pride of this privileged race was the cause of its downfall. We also know that his story comes to an abrupt end and abandons us before the most nagging enigma of history.

I have reminded my readers over the course of this book that some of the riddles posed by Plato were only solved recently. We may perhaps never know if his vibrant description of the lost continent was only his most grandiose allegory. But, perhaps to the contrary, one day our descendants may see in the steel net of a trawler gleaming out of the

---

(cont. from p. 415) 157 million Orthodox Christians and Eastern dissidents
227 million Muslims
210 million Hindu-Brahmins
120 million Buddhists
279 million adherents to other Asian religions

the remainder (159 million) include fetishists, animists, and so forth. The Catholic world is administered by 1,158 bishops, 218 apostolic vicars, and 312,000 priests.

pelagic mud the immortal smile of one of the Sea's golden daughters whose round dance encircled the large statue of Poseidon, or an orichalcum stele with unknown hieroglyphs.[4]

Among other mysteries, this would shed light on the enigma of the Egyptian civilization at the cusp of history radiating a perfection so royally sure of its rite and rhythm that we are still advancing, like lost children, in the direction formerly indicated by its beacons, and also the tenacious legend of a golden age in the blessed islands of the West, with the strictly corresponding tradition (and the same name "Atlan") among the Yucatan Maya and the extinct race of the Caribs. This would also finally explain the harmonious affinity with which were wed as alternating and complementary waves during our Mediterranean evolution, the "Hyperborean" contribution, the Egyptian contribution, Celtic-Nordic idealism, and geometric rigor, all meeting and avidly blending like the accidentally separated branches of one same root. Swan and bull, purifying snow and generating fire, formerly combined on the sacred mountain, periodically rediscover each other after millennia of separation: the gold arrows of Abaris, the scepter of the pharaohs, would be the tips of Poseidon's broken trident.

It is likely that no tangible proof will ever come to confirm the saga of the *Critias,* and it will remain the eternal mystery, and also (the starting point of this last digression) a melancholy allegory illustrating, like its "Hyperborean" echo in the collapse of the Norse gods' Valhalla, the tragic destiny of great pride.

To revisit the question raised by the "Duel of the Magicians," we do not have the impression that a superman, daimon, or demigod, combining Plato and Leonardo's love of Knowledge and Beauty, Saint Francis's gentle love of creation and the Creator, the musical mysticism of a Steiner, the scientific erudition of an Einstein, a Poincaré, an Eddington with the royal genius and purity of a Pythagoras, is close to

4. I would like to take this opportunity to point out to those interested in this question the monthly magazine *Atlantis* in which Paul Le Cour and his colleagues tirelessly and impartially study everything relating to the problem of Atlantis.

being born in the initiatory groups. And the rite of the Great Pan will never leave the highest minds with anything but a melancholy aftertaste of the fleeting "wild strawberry."

Pythagoreanism was a religion of fraternity but not equality; the principle of initiatory religions can never truly harmonize with democracies. Christianity and Buddhism are essentially democratic.

The Christian church, again (like its Asian rival in practical success, Buddhism) probably holds for normal minds, for the great majority, a reservoir of love and kindness with a sufficient number of cathartic rites, precisely adapted to the needs of average humanity.[5] Its imperfections, its contradictions, its human puerilities do not prevent it from being the authentic, albeit weakened and distorted, voice of the Great Love, echo of the divine voice, just as our science is the projection through our earthly intelligence of a reality that is both simpler and less accessible.

It is always Plato's cave!

God and Reality are the Sun shining above our cave, whose brilliance we are unable to tolerate. But the projected shadows dancing on the screen visible to our senses, and their rhythms, are emanations of them all the same.

The duel of the magicians continues without end, necessary perhaps as one of those great antagonisms that create harmony through

---

5. It could be said that the images of gods or God (for example, the one the Church permits to filter out) are as real or subjective as our fumbling perceptions of the outside world, the one that modern science modestly reduces to the perception of certain relationships between certain symbols.

Hence the recent quip of an English philosopher (Hobhouse): "I would go so far to say that some of the things I've been asked to 'swallow' (by modern physics) as being absolutely necessary for my intellectual salvation appear harder to clearly understand than those dogmas of the Council of Nicaea from which I had thought I had escaped once and for all."

And Ramon Fernandez: "Among the causes of so many returns to Catholicism and Christian confessions we should perhaps include, beyond the insufficiency of a purely human religion, the fact that Christian mythology does not appear any more unlikely today than the humanitarian mythology, and in any case, we can find a greater 'psychological truth' in it." See "La Pensée et la Révolution," *Nouvelle Revue Française,* September 1, 1930.

the fruitful opposition of the same and the other, the "Yes and No" of Pythagoras. Alongside those who always find their harmony in the disciplines of established religions, there will always be passionate, proud seekers who wish to personally take a bite from the fruit of knowledge (it is the pomegranate, incidentally), and who prefer to the tranquil channels of accepted dogmas, the solitary road overlooking the "great sea of blessed delirium," the absolute hydra "drunk on its blue flesh," whose tumultuous rhythm and life-giving foam ends up wresting from the bitter skepticism of a Paul Valéry the virile acquiescence that tatters in a fanfare his musings among the tombs of the "Cemetery by the Sea," and commands the soul to not only "try to live" but to prepare itself to fearlessly confront—with the serenity of those who feel that they, too, are of the heavenly race—the "Leucadian leap" into the torrent of no return of the Great Adventure.

Index